CHEMISTRY

Phase 1

NoMoreClass™

NOTION PRESS

NOTION PRESS

India. Singapore. Malaysia.

ISBN xxx-x-xxxxx-xx-x

Dedicated to our Beloved Prime Minister Shree Narendra Modi ji, whose life gives the Inspiration to every Individual

*If a Man **Decides** to achieve something in his life nothing is Impossible*

CHAPTORS

1. STOICHIOMETRY AND REDOX REACTIONS

THE MOLE AND EQUIVALENT CONCEPT

1. INTRODUCTION

Some important terminologies required to master the concepts of chemistry are as follows:

(a) Chemical Equation: It is the symbolic representation of a true chemical reaction. The equation provides qualitative and quantitative information about a chemical change in a simple manner. For e.g. in the reaction,

$$3BaCl_2 + 2Na_3PO_4 \longrightarrow Ba_3(PO_4)_2 + 6NaCl$$

The '+' sign on left hand side means 'react with' and on the right of arrow as 'produces'. The substances which react ae called 'reactants' and those produced in the reaction are called 'products' and they are represented on the LHS and RHS of the arrow respectively.

(b) Thermo Chemical Equation: The equations which represent chemical as well as thermal changes are called thermo chemical equations. The energy change is shown by putting value on the RHS of the reaction. For eg.

$$\text{(Exothermic)}: C(s) + O_2(g) \longrightarrow CO_2(g); \quad \Delta H = -ve$$

$$\text{(Endothermic)}: C(s) + 2S(s) \longrightarrow CS_2(g); \quad \Delta H = +ve$$

(c) Molecule and Atom: A molecule is defined as the smallest possible particle of a substance which has all the properties of that substance and can exist freely in nature. On the other hand an atom is the smallest particle of an element (made up of still smaller particles like electrons, protons, neutrons, etc.) which can take part in a chemical reaction. It may or may not exist free in nature.

(d) Atomicity of an Element: The term refers for the number of atoms present in one molecule of an element, e.g., atomicity of H_2, O_3, S_8, P_4 is 2, 3, 8 and 4 respectively.

(e) Mole: A mole is defined as the amount of matter that contains as many objects (atoms, molecules, electron, proton or whatever, objects we are considering) as the same number of atoms in exactly 12 g of C^{12}. This number is also known as **Avogadro's number (N_A)**. Avogadro's number = 6.023×10^{23} molecule/mole. Thus, 1 mole of an entities contains N_A particles of that entity.

Number of objects (N) in n mole = number of mole (n) × number of objects per mole (N_A)

$N = n \times N_A = \frac{w}{M} \times N_A$, $\left[n = \frac{w}{M}\right]$ where N_A is Avogadro's number.

(f) Molar Mass: The mass in gram of 1 mole of a substance is known as molar mass or molecular weight of substance.

(g) Atomic Weight: The atomic mass of an element is defined as the average relative mass of its atoms as compared to the mass of a carbon atom taken as 12. Note that the atomic weight of an element is a relative weight of one atom and not the absolute weight.

(h) Gram Molecular Weight: Molecular weight or gram molecular weight of a substance is weight of N-molecules of that substance in grams.

Illustration 1: Calculate the mass of single atom of sulphur and a single molecule of carbon dioxide.

(JEE MAIN)

Sol: Know the gram atomic and the molecular mass of sulphur and CO_2 and then divide by N_A.

Mass of one sulphur atom $= \frac{\text{Gram atomic mass}}{6.02\times10^{23}} = \frac{32}{6.02\times10^{23}} = 5.33\times10^{-23}\,g$

Formula of carbon dioxide $= CO_2$

Molecular mass of CO_2 $= 12 + 2 \times 16 = 44$

Gram-molecular mass of CO_2 $= 44\ g$

Mass of one molecule of CO_2 $= \frac{\text{Gram molecular mass}}{6.02\times10^{23}} = \frac{44}{6.02\times10^{23}} = 7.308\times10^{-23}\,g$

Illustration 2: How many electrons are present in 1.6 g of methane? **(JEE ADVANCED)**

Sol: Know the molar mass of methane, the moles and the no. of molecules.Then calculate no. of electrons for the solved no. of molecules.

Gram molecular mass of methane, $(CH_4) = 12 + 4 = 16\ g$

Number of moles in 1.6 g of methane $= \frac{1.6}{16} = 0.1$

Number of molecule of methane in 0.1 mole $= 0.1\times6.023\times10^{23} = 6.02\times10^{22}$

One molecules of methane has $= 6 + 4 = 10$ electrons

So, 6.02×10^{22} molecules of methane have $= 10\times6.02\times10^{22}$ electrons $= 6.02\times10^{23}$ electrons

2. STOICHIOMETRY

The quantitative aspect, dealing with mass and volume relations among reactants and products is termed stoichiometry. Consider for example, the reaction represented by a balanced chemical equation:

Chemical Equation	$2H_2(g)$	+	$O_2(g)$	$\longrightarrow$	$2H_2O(g)$	
Mole ratio :	2 mol or		1 mol or		2 mol or	
Molecule ratio :	$2\times6.023\times10^{23}$ molecules or 2molecules		$1\times6.023\times10^{23}$ molecules or 1molecules		$2\times6.023\times10^{23}$ molecules or 2molecules	
Weight ratio :	4g		32g		36 g	
Volume ratio :	2 vol		1 vol		2 vol	(valid only for gaseous state at same P and T)

The given reaction suggests the combination ratio of reactants and formation ratio of products in terms of:

(a) Mole ratio: 2 mol H_2 reacts with 1 mol of O_2 to form 2 mol of H_2O vapors.

(b) Molecular ratio: 2 molecule of H_2 reacts with 1 molecule of O_2 to form 2 molecules of H_2O vapors.

(c) Weight ratio: 4 g H_2 reacts with 32 g O_2 to form 36 g of H_2O vapors.

(d) Volume ratio: In gaseous state 2 volume H_2 reacts with 1 volume O_2 to form 2 volume H_2O vapors at same conditions of P and T.

Therefore, coefficients in the balanced chemical reaction can be interpreted as the relative number of moles, molecules or volume (if reactants are gases) involved in the reaction. These coefficients are called stoichiometrically equivalent quantities and may be represented as:

$$2 \text{ mol } H_2 \equiv 1 \text{ mol } O_2 \equiv 2 \text{ mol } H_2O$$

Or $\quad$ Mole of H_2: Mole of O_2: Mole of H_2O = 2: 1: 2

Where the symbol ≡ is taken to mean 'stoichiometrically equivalent to'. The stoichiometric relation can be used to give conversion factors for relating quantities of reactants and products in a chemical reaction.

2.1 Some Important Laws

(a) Law of conservation of mass: "In all physical and chemical changes, the total mass of the reactants is equal to that of the products" or "matter can neither be created nor destroyed."

(b) Law of constant composition/definite proportion: "A chemical compound is always found to be made up of the same elements combined together in the same fixed ratio by weight".

(c) Law of multiple proportions: "When two elements combine together to form two or more chemical compounds, then the weight of one the elements which combine with a fixed weight of the other bear a simple ratio to one another".

(d) Law of reciprocal proportions: The ratio of the weights of two elements A and B which combine with a fixed weight of the third element C is either the same or a simple multiple of the ratio of the weights of A and B which directly combine with each other.

(e) Gay-Lussac's law of gaseous volumes: "When gases react together, they always do so in volumes which bear a simple ratio to one another and to the volumes of the products, if gaseous, all measurements are made under the same conditions of temperature and pressure".

Illustration 3: What mass of sodium chloride would be decomposed by 9.8 g of sulphuric acid, if 12 g of sodium bisulphate and 2.75 g of hydrogen chloride were produced in a reaction assuming that the law of conservation of mass is true? **(JEE MAIN)**

Sol: Apply the law of conservation of mass.

$$NaCl + H_2SO_4 \rightleftharpoons NaHSO_4 + HCl$$

According to law of conservation of mass, Total mass of reactant = Total mass of product

Let the mass of NaCl decomposed be x, so

$$x + 9.8 = 12 + 2.75$$
$$= 14.75$$
$$x = 4.95 \text{ g}$$

Illustration 4: How much volume of oxygen will be required for complete combustion of 40 mL of acetylene (C_2H_2) and how much volume of carbon dioxide will be formed? All volumes are measured at NTP. **(JEE ADVANCED)**

Solution: Write the balanced chemical reaction and from the given data determine the volume.

$$2C_2H_2 + 5CO_2 \rightleftharpoons 4CO_2 + 2H_2O$$

$2C_2H_2$	$5CO_2$	$4CO_2$
2 vol	5 vol	4 mol
40 ml	$\frac{5}{2} \times 40\text{ml}$	$\frac{4}{2} \times 40\text{ml}$
40ml	100ml	80ml

So, for complete combustion of 40 mL of acetylene, 100 mL of oxygen are required and 80 mL of carbon dioxide is formed.

2.2 Avogadro's Hypothesis

"Equal volumes of all gases/vapors under similar conditions of temperature and pressure contain equal number of molecules."

This statement leads to the following facts:

(a) One mole of all gases contain Avogadro's number of molecules, i.e., 6.023×10^{23} molecules.

(b) The volume of 1 mole of gas at NTP or STP is 22.4 litre.

(c) NTP or STP refers for $P = 1$ atm, $T = 0°C$ or 273 K.

(d) Molecular weight = 2 × vapour density (for gaseous phase only)

It provides a method to determine the atomic weights of gaseous elements.

2.3 Dulong and Petit's Law

This law is valid for metals only. According to this law, atomic weight × specific heat (in cal/g) ≈ 6.4. Also, heavier the element, lesser will be its specific heat. Therefore, $C_{Hg} < C_{Cu} < C_{Al}$.

2.4 Equivalent Weight

For comparing reacting weights of substances participating in a chemical reaction, chemists coined the term 'equivalent weight'. The substances react in their equivalent weight ratios.

2.4.1 Equivalent Weight of an Element or Compound in a Non-Redox Change

Equivalent weight of an element is its weight which reacts with 1 part by weight of hydrogen, 8 parts by weight of oxygen and 35.5 parts by weight of chlorine. This definition leads to following important generalisations.

(a) Equivalent weight of an element: $'E' = \frac{\text{Atomic weight of element}}{\text{Valence of element}}$

(b) Equivalent weight of an ionic compound: $(E) = \frac{\text{Formula weight of compound}}{\text{Total charge on cationic or anionic part}}$

Also, Eq. wt. of compound $E = \text{Eq. wt. of I part} + \text{Eq. wt. of II part}$

(c) Equivalent weight of an acid or base:

$$E_{Acid} = \frac{\text{Molecular weight}}{\text{Basicity}}$$

Basicity = Number of H-atoms replaced from one molecule of acid

$$E_{Base} = \frac{\text{Molecular weight}}{\text{Acidity}}$$

Acidity = Number of OH-groups replaced from one molecule of base.

(d) Equivalent weight of acid salt: $= \frac{\text{Molecular weight of acid salt}}{\text{Replaceable H-atom in it}}$

An acid salt is one which has replaceable H-atom, e.g., $NaHCO_3$, $NaHSO_4$, Na_2HPO_4, Na_2HPO_3 is not an acid salt, since it does not have replaceable H-atom.

Note: An acid salt possesses acidity as well as basicity both.

(i) Equivalent weight of basic salt: $= \dfrac{\text{Molecular weight of basic salt}}{\text{Replacable OH gps in basic salt}}$

A basic salt is one which has replaceable OH gps e.g., Ca(OH)Cl, $Al(OH)_2Cl$, $Al(OH)Cl_2$ etc.

2.4.2 Equivalent Weight of an Element or Compound in a Redox Change

For a redox change, the equivalent weight of a substance is given by,

$$\text{Equivalent weight of an oxidant or reductant} = \frac{\text{Molecular weight}}{\text{Number of electrons lost or gained by one molecule of oxidant or reductant}}$$

2.4.3 Gram Equivalent Weight

The equivalent weight of a substance expressed in grams is called gram eq. wt. or one gram equivalent. Now we can define gram-equivalent (g meq) in gms of a substance whose equivalent weight is as follows: No. of equivalents $= \dfrac{g}{E}$

Illustration 5: An unknown element forms an oxide. What will be the equivalent mass of the element if the oxygen content is 20% by mass? **(JEE MAIN)**

Sol: Use the equation of equivalent weight.

$$\text{Equivalent mass of element} = \frac{\text{Mass of element}}{\text{Mass of oxygen}} \times 8 = \frac{80}{20} \times 8 = 32$$

Illustration 6: The equivalent weight of a metal is double than that of oxygen. How many times is the weight of its oxide greater that the weight of metal? **(JEE ADVANCED)**

Sol: First calculate the equivalent weight of the metal and from the molecular formula, determine the ratio.

Equivalent mass of metal $= 16 = \dfrac{x}{n}$

Where x= atomic mass of metal

N = valency of metal

Molecular formula of metal oxide $= M_2O_n$

$$\frac{\text{Mass of metal oxide}}{\text{Mass of metal}} = \frac{2(16n) + 16(n)}{2(16n)} = 1.5$$

2.5 The Limiting Reagent

The reagent producing the least number of moles of products is the limiting reagent. For example, consider a chemical reaction given below, containing 10 mol of H_2 and 7 mol of O_2. Since, 2 mol H_2 reacts with 1 mol O_2, thus,

	$2H_2(g)$ +	$O_2(g)$ $\longrightarrow$	$2H_2O(V)$
Moles before reaction	10	7	0
Moles after reaction	0	2	10

It is thus, evident that the reaction stop only after consumption of 5 moles of O_2 since, no further amount of H_2 is left to react with unreacted O_2. The substance that is completely consumed in a reaction is called **limiting reagent** because it determines or limits, the amount of product. The other reactants present in excess are sometimes called as **excess reagents**.

Calculation of limiting reagent

(a) By calculating the required amount by the equation and comparing it with given amount. [Useful when only two reactants are there]

(b) By calculating amount of anyone product obtained taking each reactant one by one irrespective of other reactants. The one giving least product is limiting reagent.

(c) Divide given moles of each reactant by their stoichiometric coefficient, the one with least ratio is limiting reagent. [Useful when numbers of reactants are more than two].

2.6 Reaction Yield

The theoretical yield of a product is the maximum quantity that can be expected on the basis of stoichiometry of a chemical equation. The percentage yield is the percentage of a theoretical yield actually achieved. The lower yield of a chemical reaction is due to side reactions.

$$\text{Percentage yield} = \frac{\text{Actual yield}}{\text{Theoretical yield}} \times 100$$

Illustration 7: 10 mL N_2 and 25 mL H_2 at same P and T are allowed to react to give NH_3 quantitatively. Predict (i) the volume of NH_3 formed, (ii) limiting reagent. **(JEE MAIN)**

Sol: Frame the reaction and lay down the conditions, due to which volume of NH_3 can be found which leads to the limiting reagent.

$$\begin{array}{lccc} & N_2 & +3H_2 \longrightarrow & 2NH_3 \\ \text{V at t} = 0 & 10 & 25 & 0 \\ \text{V at final condition} & \left[10 - \frac{25}{3}\right] & 0 & \frac{50}{3} \end{array}$$

$\therefore$ Volume of $2NH_3$ formed $= \frac{50}{3}$ mL

Limiting reagent is H_2.

Illustration 8: A chloride of an element contains 49.5% chlorine. The specific heat of the element is 0.056. Calculate the equivalent mass, valency and atomic mass of the element. **(JEE ADVANCED)**

Sol: Calculate mass of the metal from the given percentage and the equivalent mass of the metal. Using the Dulong and Petit's law, specific heat gives the atomic mass of the metal and then the valency can be found.

Mass of chlorine in the metal chloride = 49.5%

Mass of metal $= (100 - 49.5) = 50.5$

Equivalent mass of the metal $= \frac{\text{Mass of metal}}{\text{Mass of chlorine}} \times 35.5 = \frac{50.5}{49.5} \times 35.5 = 36.21$

According to Dulong and Petit's law,

Approximate atomic mass of the metal $= \frac{6.4}{\text{Specific heat}} = \frac{6.4}{0.056} = 114.3$

$$\text{Valency} = \frac{\text{Approximate atomic mass}}{\text{Equivalent mass}} = \frac{114.3}{36.21} = 3.1 \approx 3\ \text{v}$$

Hence, exact atomic mass = 36.21 × 3 = 108.63

3. METHODS OF EXPRESSING CONCENTRATION OF SOLUTION

(a) Strength of Solution: Amount of solute present in one litre solution

$$S = \frac{\text{Weight of solute}}{\text{Volume of solution in litre}} = \frac{w}{V\ \text{in (l)}}$$

$$= \text{Normality} \times \text{Equivalent weight}$$

$$= \text{Molarity} \times \text{Molecular weight}$$

(b) Mass Percentage or Percent by Mass:

$$\%(w/w)\ \text{Mass percentage of solute} = \frac{\text{Mass of solute}}{\text{Mass of solution}} \times 100$$

(c) Percent Mass by Volume: $\%(w/v) = \dfrac{\text{Mass of solute}}{\text{Volume of solution}} \times 100$

(d) Parts Per Million (ppm): $= \dfrac{\text{Mass of solute}}{\text{Mass of solution}} \times 10^6$

(e) Molarity: It is expressed as moles of solute contained in one litre of solution or it is also taken as millimoles of solute in 1000 cc(mL) of solution. It is denoted by M.

$$\text{Molarity} = \frac{\text{Moles of solute}}{\text{Litres of solution}} = \frac{\text{Millimoles of solute}}{\text{Millilitres of solution}}; \qquad M = \frac{n_B}{V_{lt}} = \frac{g_B / m_B}{V_{lt}}$$

(f) Molality: It is the number of mole present in 1kg solvent.

$$\text{Molality(m)} = \frac{\text{No. of moles of solute}}{\text{Weight (in kg) of solvent}}$$

Let w_A grams of the solute of molecular mass m_A be present in w_B grams of the solvent, then

$$\text{Molality(m)} = \frac{w_A}{m_A \times w_B} \times 1000$$

(g) Normality: It is define as number of equivalent of a solute present in one litre of solution.

$$N = \frac{\text{Equivalent of solute}}{\text{Volume of solution in litre}} = \frac{\text{Weight of solute}}{\text{Equivalent weight of solute} \times \text{V in litre}}$$

$$N = \frac{w}{E \times V\ \text{in (l)}} = \frac{w \times 1000}{E \times V\ \text{in mL}}$$

Note: A striking fact regarding equivalent and milli equivalent is equivalent and milli equivalent of reactants react in equal number to give same number of equivalent or milli equivalent of products separately.

(h) Formality: Since molecular weight of ionic solids is not determined accurately due to their dissociative nature and therefore molecular weight of ionic solid is often referred as formula weight and molarity as formality. $\text{Formality} = \dfrac{\text{Wt. of solute}}{\text{Formula wt.} \times \text{V(in l)}}$ i.e., molarity

(i) **Specific Gravity of Solution:** $= \frac{\text{Weight of solution}}{\text{Volume of solution}}$ i.e., weight of 1 mL solution.

NOMORECLASS CONCEPTS

- Molality, % by weight, mole fractions are independent of temperature since these involve weights.
- Rest all, i.e., normality, molarity, % by volume, % by strength and strength are temperature dependent, normally decrease with increase in temperature since volume of solution increases with T.
- Molar solution having normality 1N and molarity 1M respectively.
- On diluting a solution, eq. meq. mole or m mole of solute do not change however N and M change.

(j) **Mole Fraction:** It is the fractional part of the moles that is contributed by each component to the total number of moles that comprises the solution. In containing n_A moles of solvent and n_B moles of solute. Mole

fraction of B $= X_B = \frac{n_B}{n_A + n_B}$

Mole fraction of A $= X_B = \frac{n_A}{n_A + n_B}$

(k) **Ionic Strength:** The ionic strength (μ) of the solution obtained by mixing two or more ionic compounds is given by: $\mu = \frac{1}{2}\Sigma cZ^2$. Where c is the concentration (molarity) of that ion and Z is its valence.

Illustration 9: 30 mL of 0.1 N $BaCl_2$ is mixed with 40 mL of 0.2 N $Al_2(SO_4)_3$. How many g of $BaSO_4$ are formed? **(JEE MAIN)**

Sol: Frame the reaction and place the given data to find the milliequivalents at the end of the reaction. Using the formula below, weight of $BaSO_4$ can be found.

	$BaCl_2$	$+ Al_2(SO_4)_3$	$\longrightarrow BaSO_4$	$+ AlCl_3$
Meq. before reaction	$30 \times 0.1 = 3$	$40 \times 0.2 = 8$	$0 = 0$	$0 = 0$
Meq. after reaction	0	5	3	3

3 Meq. of $BaCl_2$ reacts with 3 Meq. of $Al_2(SO_4)_3$ to produce 3 Meq. of $BaSO_4$ and 3 Meq. of $AlCl_3$

∴ Meq. of $BaSO_4$ formed $= \frac{w_{BaSO_4}}{E_{BaSO_4}} \times 1000 = 3$

w_{BaSO_4} formed $= \frac{3 \times 233}{2 \times 1000} = 0.3495$ g

Illustration 10: 500 mL of aM solution and 250 mL of bM solution of the same solute are mixed and diluted to 2 litre. The diluted solution shows the molarity 1.6 M. If a: b is 2: 5, then calculate a and b. **(JEE ADVANCED)**

Sol: Using the mixture molarity formula $\frac{M_1 \times V_1 + M_2 \times V_2}{V_1 + V_2}$, a and b is calculated.

$$\frac{500 \times a + 250 \times b}{2000} = 1.6;\ 500\ a + 250\ b = 3200$$

If, $\frac{a}{b} = \frac{2}{5}$ then $\frac{500 \times b \times 2}{5} + 250\ b = 3200;\ 450\ b = 3200;\ b = 7.11$

Similarly, $500a + \frac{250 \times 5a}{2} = 3200;\ \therefore\ a = \frac{3200}{1125} = 2.84$

4. SOME CHARACTERISTIC APPLICATION OF MOLE CONCEPT

4.1 Gravimetric Analysis

Gravimetric analysis is an analytical technique based on the measurement of mass of solid substances and or volume of gaseous species. Gravimetric analysis is divided into three parts.

(a) Mass–Mass Relationship: It relates the mass of a species (reactant or product) with the mass of another species (reactants or products)

Let us consider a chemical reaction,

$$2NaHCO_{3(s)} \xrightarrow{\Delta} Na_2CO_{3(s)} + H_2O + CO_{2(g)}$$

Suppose the mass of $NaHCO_3$ being heated is 'x' g and we want to calculate the weight of Na_2CO_3 being produced by heating of 'a' g $NaHCO_3$.

The moles of $NaHCO_3 = \frac{x}{84}$

According to the above balanced equation 2 moles of $NaHCO_3$ upon heating gives 1 mole of Na_2CO_3

(b) Mass–Volume Relationship: It relates the mass of a species (reactant or product) and the volume of a gaseous species (reactant or product) involved in a chemical reaction. Suppose we are provided with 'a' gms of $NaHCO_3$ in a vessel of capacity VL and the vessel is heated, so that decomposes as $2NaHCO_3 \xrightarrow{\Delta} Na_2CO_3 + H_2O + CO_2$

Now, we want to calculate the volume of CO_2 gas being reduced.

Moles of $NaHCO_3$ taken $= \frac{X}{84}$

Now, since 2 moles of $NaHCO_3$ gives 1 mole of CO_2 at STP. Thus

Moles of CO_2 produced $= \frac{1}{2} \times \frac{X}{84}$

As we know that 1 mole of any gas at STP occupies a volume of 22.4 L.

So, volume of CO_2 produced $= \left(\frac{1}{2} \times \frac{X}{84} \times 22.4\right) L$

(i) Volume–Volume Relationship: It relates the volume of gaseous species (reactants or products) with the volume of another gaseous species (reactant or product) involved in a chemical reaction.

Illustration 11: An ore containing Mn_2O_3 is analysed for the manganese content by quantitatively converting the manganese to Mn_3O_4 and weighing it. A 1.52 g sample of ore yields 0.126 g Mn_3O_4. Calculate the percent of Mn and Mn_2O_3 in the sample. **(JEE ADVANCED)**

Sol: From the given data, find out the amount of Mn_2O_3 and calculate the %.

Equate the no. of moles of Mn_2O_3 with the no. of moles of Mn and hence find % of Mn. $3Mn_2O_3 \longrightarrow 2Mn_3O_4$

Mole ratio $Mn_2O_3 : Mn_3O_4 :: 3 : 2$

$\therefore$ Moles of $Mn_2O_3 = \frac{3}{2} \times$ Moles of $Mn_3O_4 = \frac{3}{2} \times \frac{0.126}{229} = 5.253 \times 10^{-4}$

$\therefore$ Amount of $Mn_2O_3 = 8.253 \times 10^{-4} \times 158 = 0.13$ g

$\therefore$ % of $Mn_2O_3 = \frac{0.13}{1.52} \times 100 = 8.58$

Also, $Mn_2O_3 \longrightarrow 2Mn$

$\therefore$ Mole of Mn = 2 × Mole of $Mn_2O_3 = 2 \times 8.253 \times 10^{-4} = 16.51 \times 10^{-3}$

$\therefore$ Amount of Mn $= 16.51 \times 10^{-3} \times 55$

$\therefore$ % Mn $= \frac{0.09}{1.52} \times 100 = 5.29$

Illustration 12: A 1.0 g sample of pure organic compound containing chlorine is fused with Na_2O_2 convert chlorine to NaCl. The sample is then dissolved in water, and the chloride precipitated with $AgNO_3$, giving 1.96 g of AgCl. If the molecular weight of organic compound is 147, how many chlorine atoms does each molecule contain?

(JEE MAIN)

Sol: Calculate the moles of chloride ions in silver chloride and the organic compound; and compare the ratio.

Moles of AgCl = Moles of chloride= $\frac{1.96}{143.5} = 0.0136$

Moles of organic compound $= \frac{1}{147} = 6.8 \times 10^{-3}$

Chlorine atoms in each molecules of organic compound $= \frac{0.0136}{6.5 \times 10^{-3}} = 2$

4.2 Volumetric Analysis

It is the process of determination of conc. of a solution with the help of another solution of known conc. It may also be defined as experimental method of determination of volume of a solution of known strength needed for a definite volume of another solution of unknown strength.

Titration: It is an operation forming the basis of volumetric analysis. The addition of measured amount of a solution of one reagent (called the **titrant**) from a burette to a definite amount of another reagent (called **analyte**) until the reaction between them is complete, i.e., till the second reagent (analyte) is completely used up, i.e., upto end point.

Type of Titrations: There are four general classes of volumetric titrations.

(a) **Acid-Base Titration:** Acid or base solutions are titrated against a standard solution of a strong base or strong acid using suitable acid-base indicator.

(b) **Precipitation Titration:** In such titration, the titrant forms an insoluble product with analyte e.g., titration of chloride ions against $AgNO_3$ solution.

(c) **Complexometric Titrations:** In such titrations, the titrant is a complexing agent and forms a water-soluble complex with the analyte, usually containing a metal ion. The titrant is often a chelating agent, e.g., ethylenediaminetetraacetic acid (EDTA).

(d) **Redox Titrations:** These involves the titration of an oxidizing agent against a reducing agent or vice versa.

Standard Solution: It is the solution of known strength.

Primary Standard Solution: The solution for which conc. is known is called primary standard solution.

Note: For primary standard solution,

(a) Solute should not be reactive towards solvent or air.

(b) Solute should not be hygroscopic.

(c) Temperature should be constant.

In **acids**, oxalic acid ($H_2C_2O_2.2H_2O$), benzoic acid (C_6H_5COOH), sulphamic acid (HNH_2SO_3), etc. are taken as primary standard solution.

In **bases**, washing soda ($NaCO_3.10H_2O$), borax ($Na_2B_4O_7$), etc are taken as primary standard solution.

In **oxidizing agents**, only potassium dichromate ($K_2Cr_2O_7$) is taken as primary standard solution.

In **reducing agents**, hypo ($Na_2S_2O_3.5H_2O$), Mohr's salt ($FeSO_4.(NH_3)_2SO_4.6H_2O$), sodium oxalate ($Na_2C_2O_4$), etc are taken as primary standard solutions.

End point: End point of titration is normally detected by a sudden change in color of the solution.

Indicator: These compound mixed in the solution in very small amount, which responses the sudden change in color of the solution and show the end point of titration. In acid–base titration, the indicators used are either weak organic acid or weak organic bases. Some examples are

Acidic Indicator: Phenolphthalein, litmus paper etc.

Basic Indicator: Methyl orange, methyl red, etc.

Acid	Base	Indicator
Strong	Strong	Any
Strong	Weak	Methyl orange, methyl red, etc
Weak	Strong	Phenolphthalein etc

Principle of Titration: Titration means stoichiometry and hence its problems may be solved by mole as well as equivalent concept. But for simplicity equivalent concept is preferred, according to which the number of g-equivalents of all reactants reacted will be equal and the same number of g-equivalents of each products will form. The number of g-equivalents of substances may be determined by using the following formulae:

$$\text{Number of g-equivalents} = \frac{\text{Wt.(in gm)}}{\text{Gm. eq.wt.}} = \frac{VS}{1000} = \frac{\text{Vol. of gas}}{\text{Eq. vol.}} = \text{Mole} \times x - \text{factor}$$

Where, S = strength in normality

Illustration 13: 30 mL of a certain solution of Na_2CO_3 and $NaHCO_3$, required 12 mL of 0.1 N H_2SO_4 using phenolphthalein as indicator. In presence of methyl orange, 30 mL of same solution required 40 mL of 0.1 N H_2SO_4. Calculate the amount of Na_2CO_3 per litre in mixture. **(JEE MAIN)**

Sol: Use titration principles to understand the numerical. Find out the milliequivalents of H_2SO_4 and Na_2CO_3.

Phenolphthalein as indicator:

Meq. of H_2SO_4 used = 12 × 0.1 = 1.2 for 30 mL mixture

$$\therefore \quad \frac{1}{2} \text{ Meq. of } Na_2CO_3 \text{ in 30 mL mixture} = 1.2 \qquad \text{...(i)}$$

Methyl orange as indicator: This time fresh solution is titrated with H_2SO_4 using methyl orange as indicator. By equating the data of the bases with the required acid, solve the milliequivalents and then calculate the strength of the bases.

$$\text{Meq. of } Na_2CO_3 + \text{Meq. of } NaHCO_3 = \text{Meq. of } H_2SO_4 \text{ used}$$

$$= 40 \times 0.1 = 4 \qquad \text{...(ii)}$$

By Eq. (i)

Meq. of Na_2CO_3 = 2.4

$$\therefore \quad \frac{w}{53} \times 1000 = 2.4$$

or $w_{Na_2CO_3} = 0.1272$ g in 30 mL

$\therefore$ Strength of Na_2CO_3 = 4.24 g litre^{-1}

Also, Meq. of $NaHCO_3$ = 4 – 2.4 = 1.6; $\frac{w}{84} \times 1000 = 1.6$

$\therefore$ $w_{NaHCO_3} = 0.1344$ g in 30 mL

Strength of $NaHCO_3 = \frac{0.1344 \times 1000}{30} = 4.48$ g litre^{-1}

Illustration 14: 0.5 g mixture of $K_2Cr_2O_7$ and $KMnO_4$ was treated with excess of KI in acidic medium. Iodine liberated required 150 cm^3 of 0.10N solution of thiosulphate solution for titration.

Find the percentage of $K_2Cr_2O_7$ in the mixture. **(JEE MAIN)**

Solution: Determine the equivalent weight of chromate and permanganate solution and compare the mili. Eq of each components to determine the % of chromate.

Reactions of $K_2Cr_2O_7$ and $KMnO_4$ with KI may be given as :

$$K_2Cr_2O_7 + 7H_2SO_4 + 6KI \rightarrow 4K_2SO_4 + Cr_2(SO_4)_3 + 7H_2O + 3I_2$$

$$2KMnO_4 + 8H_2SO_4 + 10KI \rightarrow 6K_2SO_4 + 2MnSO_4 + 5I_2$$

Thus equivalent wt. of $K_2Cr_2O_7 = \frac{294}{6} = 49$

Equivalent weight of $KMnO_4 = \frac{158}{5} = 31.6$

m.eq. of $K_2Cr_2O_7$ + m.eq. of $KMnO_4$ = m.eq. of I_2 = m.eq of hypo.

Let the mass of $K_2Cr_2O_7$ = x g

Mass of $KMnO_4$ = (0.5-x)g

$$\frac{x}{49} + \frac{(0.5 - x)}{31.6} = 150 \times 0.1 \times 10^{-3}\ x = 0.0732$$

$$\% \text{ of } K_2Cr_2O_7 = \frac{0.0732}{0.5} \times 100 = 14.64$$

4.3 Double Indicators Titration

For the titration of alkali mixtures (e.g., NaOH + Na_2CO_3) or (Na_2CO_3 + $NaHCO_3$), two indicators phenolphthalein and methyl orange are used. This will be discussed in detail in Ionic Equilibrium.

4.4 Eudiometry

Eudiometry or gas analysis involves the calculation based on gaseous reactions in which the amounts of gases are represented by their volumes, measured at STP. Some basic assumptions for calculations

(a) Gay-Lussac's law of volume combination holds good.

(b) For non-reacting mixture. Amagat's law holds good. According to this, the total volume of a non-reacting gaseous mixture is equal to the sum of partial volumes of all the component gases. The volume of solids or liquids is considered to be negligible in comparison to the volumes of gases.

Thus, we can summarize the above points as – eudiometry involves volume measurement during the reaction. Since, Volume of gas, V is directly proportional to number of moles at constant P, T and thus, volume ratio of gases can be directly used in place of mole ratio for analysis.

Illustration 15: A mixture of ethane (C_2H_6) and ethene (C_2H_4) occupies 40 litre at 1.00 atm and at 400 K. The mixture reacts completely with 130 g of O_2 to produce CO_2 and H_2O. Assuming ideal gas behaviour, calculate the mole fraction of C_2H_4 and C_2H_6 in the mixture. **(JEE MAIN)**

Sol: Using the Ideal gas equation, find out no. of moles. Frame the balanced combustion reactions of the hydrocarbons and lay down the values.Calculate the mole fraction accordingly.

For a gaseous mixture of C_2H_6 and C_2H_4

$$PV = nRT$$

$\therefore$ $1 \times 40 = n \times 0.082 \times 400$

Total mole of (C_2H_6 + C_2H_4) = 1.2195

Let mole of C_2H_6 and C_2H_4 be a and b respectively.

$$a + b = 1.2195 \quad \text{...(i)}$$

$$C_2H_6 + \frac{7}{2}O_2 \longrightarrow 2CO_2 + 3H_2O$$

$$C_2H_4 + 3O_2 \longrightarrow 2CO_2 + 2H_2O$$

$\therefore$ Mole of O_2 needed for complete reaction of mixture = 7a/2 + 3b

$$\therefore \quad \frac{7a}{2} + 3b = \frac{130}{32} \quad \text{...(ii)}$$

By Eqs. (i) and (ii), a = 0.808

b = 0.4115

$\therefore$ Mole fraction of C_2H_6 = $\frac{0.808}{1.2195} = 0.66$ and Mole fraction of C_2H_4 = 0.34

4.5 To Represent Concentration of H_2O_2 Solution

(a) In percentage: The amount of H_2O_2 present in 100 mL H_2O_2 solution is H_2O_2 concentration in percentage of H_2O_2 solution.

(b) In volume: The volume of O_2 at STP given by 1 mL H_2O_2 solution on decomposition is H_2O_2 concentration of H_2O_2 in volume.

Note:

(i) Direct conversions can be made by using following relations

- % strength $= \frac{17}{56} \times$ volume strength
- Volume strength = 5.6 × Normality
- Volume strength = 11.2 × Molarity

(ii) The volume strength of H_2O_2 solution decreases on long standing due to decomposition of H_2O_2 and O_2.

Illustration 16: Report the concentration of 1.5 N solution of H_2O_2 in terms of volume. **(JEE MAIN)**

Sol: From the given equivalent of H_2O_2, calculate the weight and then the volume of O_2. This itself can solve the volume strength of H_2O_2.

$\because$ Equivalent of H_2O_2 in 1 litre solution = 1.5

$\therefore$ $w_{H_2O_2}$ in 1 litre solution $= 1.5 \times \frac{34}{2} = 25.5$ g

$\therefore$ Volume of O_2 obtained by 1000 mL H_2O_2 solution $= \frac{22400 \times 25.5}{68} = 8400$ mL

$\therefore$ Volume strength of $H_2O_2 = \frac{8400}{1000} = 8.4$

4.6 To Represent the Concentration of Oleum

(100 – X%) of oleum means 'X' g H_2O reacts with equivalent amount of free SO_3 to give H_2SO_4.

Illustration 17: 0.5 g of fuming H_2SO_4 (oleum) is diluted with water. The solution requires 26.7 mL of 0.4N NaOH for complete neutralization. Find the % of free SO_3 in the sample of oleum. Also report % of oleum solution.

(JEE ADVANCED)

Sol: Principle of titration is used. Equation of oleum and the base in terms of their milliequivalents is done wherein the amount of oleum is found. % of SO_3 is thus found. Reaction of SO_3 with H_2O gives H_2SO_4. Lay down the calculated and the given values and solve the % of oleum.

Fuming H_2SO_4 contains H_2SO_4 and SO_3. Both react with NaOH. Let a g and b g SO_3 be present.

For reaction, $\therefore$ Meq. of H_2SO_4 + Meq. of SO_3 = Meq. of NaOH; $\frac{a}{98/2} \times 1000 + \frac{b}{80/2} \times 1000 = 26.7 \times 0.4$

$\therefore$ $80a + 98b = 41.87$ ...(i)

Also, $a + b = 0.5$ ...(ii)

$\therefore$ % of $SO_3 = \frac{0.1039}{0.5} \times 100 = 20.78\%$

$SO_3 + H_2O \longrightarrow H_2SO_4$

80 g SO_3 = 18 g H_2O

$\therefore$ 20.78 g $SO_3 = \frac{18 \times 20.78}{80} = 4.68$

% of oleum = 100 + 4.68 = 104.68%

4.7 To Determine Hardness of Water

Water, which gives foams easily with soap is called soft water and if not then hard water. The hardness of water is due to the presence of bicarbonates, chlorides and sulphates of Ca and Mg. The extent of hardness is known as **degree of hardness** defined usually as the no. of parts by weight of $CaCO_3$ present per million parts by weight of water. Hardness is expressed in ppm i.e., 1 ppm = 1 part of $CaCO_3$ in 10^6 part of hard water.

Note: The reason for choosing $CaCO_3$ as the standard to express hardness, inspite of the fact that $CaCO_3$ is not soluble in water but its molecular weight is 100 which makes calculation easy.

4.8 Mass Balance Equations

The principle of mass balance is based on the law of conservation of mass, i.e., the number of atoms of an element remains constant in a chemical reaction.

4.9 Charge Balance Equations

The principle of charge balance equations is based on the principle of electroneutrality, i.e., all solution are electrically neutral since sum of positive charges equals the sum of negative charges.

4.10 Saponification Value

It is the amount of KOH in mg required to neutralize a fatty acid obtained by the hydrolysis of 1 g of oil.

REDOX REACTIONS

1. INTRODUCTION

Molecular Equations: $2FeCl_3 + SnCl_2 \rightarrow 2FeCl_2 + SnCl_4$

The reactants and products have been written in molecular forms; thus, the equation is termed as **molecular equation**.

Ionic Equations: The reactions in which the reactants and products are present in the form of ions are called **ionic reactions**.

For example: $2Fe^{3+} + 6Cl^- + Sn^{2+} + 2Cl^- \rightarrow 2Fe^{2+} + 4Cl^- + Sn^{4+} + 4Cl^-$

Or $2Fe^{3+} + Sn^{2+} \rightarrow 2Fe^{2+} + Sn^{4+}$

Illustration 18: Represent the following equation in ionic form. **(JEE MAIN)**

$K_2Cr_2O_7 + 7H_2SO_4 + 6FeSO_4 = 3Fe_2(SO_4)_3 + Cr_2(SO_4)_3 + 7H_2O + K_2SO_4$

Sol: Knowing the oxidation numbers of the elements present, balanced ionic form can be represented. In this equation except H_2O, all are ionic in nature. Representing these compounds in ionic forms,

$$2K^+ + Cr_2O_7^{2-} + 14H^+ + 7SO_4^{2-} + 6Fe^{2+} + 6SO_4^{2-} \longrightarrow$$

$$6Fe^{3+} + 9SO_4^{2-} + 2Cr^{3+} + 3SO_4^{2-} + 2K^+ + SO_4^{2-} + 7H_2O$$

$2K^+$ ions and $13SO_4^{2-}$ ions are common on both sides, so these are cancelled. The desired ionic equation reduces to, $Cr_2O_7^{2-} + 14H^+ + 6Fe^{2-} \rightarrow 6Fe^{3+} + 2Cr^{3+} + 7H_2O$

Phenomenon of Oxidation and Reduction:

Oxidation or de-electronation is a process which liberates electrons.

Reduction or electronation is a process which gains electrons.

Oxidation	Reduction
a. $M \longrightarrow M^{n+} + ne^-$	$M^{n+} + ne^- \longrightarrow M$
b. $M^{n_1+} \longrightarrow M^{n_2+} + (n_2 - n_1)e^-$ $(n_2 > n_1)$	$M^{n_2+} + (n_2 - n_1)e^- \longrightarrow M^{n_1+}$ $(n_2 > n_1)$
c. $A^{n-} \longrightarrow A + ne^-$	$A + ne^- \longrightarrow A^{n-}$
d. $A^{n_1-} \longrightarrow A^{n_2-} + (n_1 - n_2)e^-$	$A^{n_2} + (n_1 - n_2)e^- \longrightarrow A^{n_2}$

Note: M may be an atom or a group of atoms; A may be atom or a group of atoms.

Oxidizing and Reducing Agent:

(a) If an element is in its highest possible oxidation state in a compound, it can function as an oxidizing agent, e.g. $KMnO_4, K_2Cr_2O_7, HNO_3, H_2SO_4, HClO_4$ etc.

(b) If an element is in its lowest possible oxidation state in a compound, it can function as a reducing agent, e.g. $H_2S, FeSO_4, Na_2S_2O_3, SnCl_2$ etc.

(c) If an element is in its intermediate oxidation state in a compound, it can function both as an oxidizing agent as well as reducing agent, e.g. $H_2O_2, H_2SO_3, HNO_3, SO_2$ etc.

(d) If highly electronegative element is in its higher oxidation state in a compound, that compound can function as a powerful oxidizing agent, e.g. $KClO_4, KClO_3, KIO_3$ etc.

(e) If an electronegative element is in its lowest possible oxidation state in a compound or in free state, it can function as a powerful reducing agent, e.g. I^-, Br^-, N_3^- etc.

2. MODERN CONCEPT OF OXIDATION AND REDUCTION

According to the modern concept, loss of electrons is oxidation whereas gain of electrons is reduction. Oxidation and reduction can be represented in a general way as shown below:

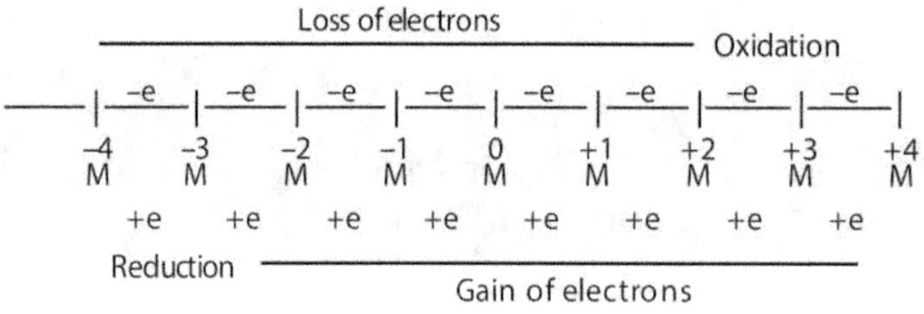

Figure 1.1: Oxidation and Reduction

NOMORECLASS CONCEPTS

- In a redox process the valency of the involved species changes. The valency of a reducing agent increases while the valency of an oxidising agent decreases in a redox reaction. The valency of a free element is taken as zero.
- Redox reaction involves two half reactions, one involving loss of electron or electrons (oxidation) and the other involving gain of electron or electrons (reduction).

3. ION ELECTRON METHOD FOR BALANCING REDOX REACTIONS

This method involves the following steps:

(a) Divide the complete equations into two half reactions

(i) One representing oxidation

(ii) The other representing reduction

(b) Balance the atoms in each half reaction seperately according to the following steps

(i) Balance all atoms other than oxygen and hydrogen

(ii) To balance oxygen and hydrogen

(c) **Acidic Medium**

(i) Add H_2O to the side which is oxygen deficient to balance oxygen atoms

(ii) Add H^+ to the side which is hydrogen deficient to balance H atoms

(d) Basic Medium

- **(i)** Add OH^- to the side which has less negative charge
- **(ii)** Add H_2O to the side which is oxygen deficient to balance oxygen atoms
- **(iii)** Add H^+ to the side which is hydrogen deficient

Illustration 19: $H_2C_2O_4 + KMnO_4 \longrightarrow CO_2 + K_2O + MnO + H_2O$ **(JEE MAIN)**

Sol:

Step 1: Select the oxidant, reductant atoms and write their half reactions, one representing oxidation and other reduction. i.e., $C_2^{+3} \longrightarrow 2C^{+4} + 2e^-$

$$5e^- + Mn^{+7} \longrightarrow Mn^{+2}$$

Step 2: Balance the no. of electrons and add the two equation.

$$5C_2^{+3} \longrightarrow 10C^{+4} + 10e^-$$

$$10e^- + 2Mn^{+7} \longrightarrow 2Mn^{+3}$$

$$5C_2^{+3} + 2Mn^{+7} \longrightarrow 10C^{+4} + 2Mn^{+2}$$

Step 3: Write complete molecule of the reductant and oxidant from which respective redox atoms were obtained.

$$5H_2C_2O_4 + 2KMnO_4 \longrightarrow 10CO_2 + 2MnO$$

Step 4: Balance other atoms if any (except H and O).

In above example K is unbalanced, therefore,

$$5H_2C_2O_4 + 2KMnO_4 \longrightarrow 10CO_2 + 2MnO + K_2O \text{ (Mentioned as product)}$$

Step 5: Balance O atom using H_2O on desired side.

$$5H_2C_2O_4 + 2KMnO_4 \longrightarrow 10CO_2 + 2MnO + H_2O + 5H_2O$$

4. OXIDATION STATE AND OXIDATION NUMBER

4.1 Oxidation State

It is defined as the charge (real or imaginary) which an atom appears to have when it is in combination. In the case of electrovalent compounds, the oxidation number of an element or radical is the same as the charge on the ion.

4.2 Oxidation Number

(a) Oxidation number of an element in a particular compound represents the number of electrons lost or gained by an element during its change from free state into that compound or Oxidation number of an element in a particular compound represent the extent of oxidation or reduction of an element during its change from free state into that compound.

(b) Oxidation number is given positive sign if electrons are lost. Oxidation number is given negative sign if electrons are gained.

(c) Oxidation number represent real change in case of ionic compounds. However, in covalent compounds it represents imaginary charge.

Rules for Calculation of Oxidation Number:

Following rules have been arbitrarily adopted to decide oxidation number of elements on the basis of their periodic properties.

(a) In uncombined state or free state, oxidation number of an element is zero.

(b) In combined state oxidation number of-

- **(i)** F is always –1.
- **(ii)** O is –2. In peroxide it is –1, in superoxides it is –1/2. However in F_2O it is +2.
- **(iii)** H is +1. In ionic hydrides it is –1. (i.e., IA, IIA and IIIA metals).
- **(iv)** Halogens as halide is always –1.
- **(v)** Sulphur as sulphide is always –2.
- **(vi)** Metal is always +ve.
- **(vii)** Alkali metals (i.e., IA group – Li, Na, K, Rb, Cs, Fr) is always +1.
- **(viii)** Alkaline earth metals (i.e., IIA group – Be, Mg, Ca, Sr, Ba, Ra) is always +2.

(c) The algebraic sum of the oxidation number of all the atoms in a compound is equal to zero. e.g. $KMnO_4$.

Ox. no. of K + Ox. no. of Mn + (Ox. no. of O) × 4 = 0

(+1) + (+7) + 4x (–2) = 0

(d) The algebraic sum of all the oxidation no. of elements in a radical is equal to the net charge on the radical. e.g. CO_3^{-2}.

Oxidation no. of C + 3 × (Oxidation no. of O) = –2(4) + 3x (–2) = –2

(e) Oxidation number can be zero, +ve, –ve (integer or fraction)

(f) Maximum oxidation no. of an element is = Group no. (Except O and F)

Minimum oxidation no. of an element is = Group no. –8 (Except metals)

Redox reactions involve oxidation and reduction both. Oxidation means loss of electrons and reduction means gain of electrons. Thus redox reactions involve electron transfer and the number of electrons lost are same as the number of electrons gained during the reaction. This aspect of redox reaction can serve as the basis of a pattern for balancing redox reactions.

Oxidation number of Mn in $KMnO_4$: Let the oxidation number of Mn be x. Now we know that the oxidation numbers of K is +1 and that of O is –2.

K	Mn	O_4		K	Mn	O_4
+1	+x	+4 × –2	or	+1	+x	–8

Now to the sum of oxidation numbers of all atoms in the formula of the compound must be zero, i.e. +1 +x – 8 = 0. Hence, the oxidation number of Mn in $KMnO_4$ is +7.

Illustration 20: What is the oxidation number of Cr in $K_2Cr_2O_7$? **(JEE MAIN)**

Sol: Let the Ox. no. of Cr in $K_2Cr_2O_7$ be x.

We know that, Ox. no. of K = +1

Ox. no. of O = –2

So, 2(Ox. no. K) + 2(Ox. no. Cr) + 7(Ox. no. O) = 0

2(+1) 2(x) 7(–2) = 0

or +2 + 2x – 14 = 0

or $2x = +14 - 2 = +12$

or $x = +\frac{12}{2} = +6$ Hence, oxidation number of Cr in is +6.

Illustration 21: H_2S act only as reductant, whereas SO_2 acts as oxidant and reductant both. **(JEE ADVANCED)**

Sol: Oxidation number of S is –2 in H_2S. It can increase only oxidation number up to +6.

Oxidation number of S is +4 in SO_2. It can increase or decrease as it lies between maximum (+6) and minimum (–2) oxidation number of S.

Illustration 22: Which compound amongst the following has the highest oxidation number of Mn?
$KMnO_4$, K_2MnO_4, MnO_2 and Mn_2O_3. **(JEE MAIN)**

Sol:

		Ox. no. of Mn
$KMnO_4$	+1+x-8=0 x=+7	+7
K_2MnO_4	+2+x-8=0 x=+6	+6
MnO_2	x-4=0 x=+4	+4
Mn_2O_3	2x-6=0 x=+3	+3

Thus, the highest oxidation number for Mn is in $KMnO_4$.

4.3 Balancing of Redox Reactions by Oxidation State Method

This method is based on the fact that the number of electrons gained during reduction must be equal to the number of electrons lost during oxidation. Following steps must be followed while balancing redox equations by this method.

(a) Write the skeleton equation (if not given, frame it) representing the chemical change.

(b) With the help of oxidation number of elements, find out which atom is undergoing oxidation/reduction, and white separate equations for the atom undergoing oxidation/reduction.

(c) Add the respective electrons on the right for oxidation and on the left for reduction equation. Note that the net charge on the left and right side should be equal.

(d) Multiply the oxidation and reduction reactions by suitable integers so that total electrons lost in one reaction is equal to the total electrons gained by other reaction.

(e) Transfer the coefficients of the oxidizing and reducing agents and their products as determined in the above step to the concerned molecule or ion.

(f) By inspection, supply the proper coefficient for the other formulae of substances not undergoing oxidation and reduction to balance the equation.

Illustration 23: $Cr_2O_7^{2-} + I^- + H^+ \longrightarrow Cr^{3+} + I_2$ **(JEE MAIN)**

Sol: (i) Find the oxidation state of atoms undergoing redox change

$$\overset{+6x2}{Cr_2}O_7^{2-} + \overset{-1}{I^-} \longrightarrow \overset{+3}{Cr^{3+}} + \overset{0}{I_2}$$

(ii) Balance the number of atoms undergoing redox change.

$$\overset{(+6)\times 2}{Cr_2}\ O_7^{2-} + \overset{2\times(-1)}{2I^-} \longrightarrow \overset{(+3)\times 2}{2Cr^{3+}} + \overset{0\times 2}{I_2}$$

(iii) Find the change in oxidation state and balance the change in oxidation states by multiplying the species with a suitable integer.

$$\overset{+12}{Cr_2}\ O_7^{2-} + \overset{-2}{2I^-} \longrightarrow \overset{+6}{2Cr^{3+}} + \overset{0}{I_2}$$

Change in ox. state = 6   Change in ox. state = 2×3

As the decrease in oxidation state if chromium is 6 and increase in oxidation state of iodine is 2, so we will have to multiply I^- / I_2 by 3 equalize the changes in oxidation state.

$$Cr_2O_7^{2-} + 6I^- \longrightarrow 2Cr^{3+} + 3I_2$$

(iv) Find the total charges on both the sides and also find the difference of charges.

Charge on LHS $= -2 + 6 \times (-1) = -8$

Charge on RHS $= 2 \times (+3) = +6$

Difference in charge $= +6 - (-8) = 14$

(v) Now, as the reaction is taking place in acidic medium, we will have to add the ions, to H^+ the side falling short in positive charges, so we will add $14H^+$ and LHs to equalize the charges on both sides.

$$Cr_2O_7^{2-} + 6I^- + 14H^+ \longrightarrow 2Cr^{3+} + 3I_2$$

(vi) To equalize the H and O atoms, add $7H_2O$ on RHS

$$Cr_2O_7^{2-} + 6I^- + 14H^+ \longrightarrow 2Cr^{3+} + 3I_2 + 7H_2O$$

Illustration 24: Balance the following equation by oxidation number method:

$$Cl_2 + IO_3^- + OH^- \longrightarrow IO_4^- + Cl^- + H_2O$$ **(JEE ADVANCED)**

Sol: Writing oxidation numbers of all atoms,

$$\overset{0}{Cl_2} + \overset{+5}{I}\ \overset{-2}{O_3^-} + \overset{-2}{O}\ \overset{+1}{H^-} \longrightarrow \overset{+7}{I}\ \overset{-2}{O_4^-} + \overset{-1}{Cl^-} + \overset{+1}{H_2}\ \overset{-2}{O}$$

Oxidation numbers of Cl and I have changed.

$$\overset{0}{Cl_2} \longrightarrow \overset{-1}{2Cl^-} \quad ...(i)$$

$$\overset{+5}{I}O_3^- \longrightarrow \overset{+7}{I}O_4^- \quad ...(ii)$$

Decrease in Ox. no. of Cl = 2 units per Cl_2 molecule

Increase in Ox. no. of I = 2 units per IO_3^- molecule

$$Cl_2 + IO_3^- \longrightarrow IO_4^- + 2Cl^-$$

To balance oxygen, $2OH^-$ ions be added on LHS and one H_2O molecule on RHS. Hence, the balanced equation is $Cl_2 + IO_3^- + 2OH^- \longrightarrow IO_4^- + 2Cl^- + H_2O$

5. TYPES OF REACTIONS

The redox reactions are of the following types:

(a) Combination reactions: A compound is formed by chemical combination of two or more elements. The combination of an element or compound with oxygen is called combustion. The combustion and several other combinations which involve change in oxidation state are called redox reactions.

e.g.,

$$\overset{-4}{C}\overset{+1}{H_4} + 2\overset{0}{O_2} \longrightarrow \overset{+4}{C}\overset{-2}{O_2} + 2\overset{+1}{H_2}\overset{-2}{O}$$

$$\overset{0}{C}(s) + \overset{0}{O_2}(g) \longrightarrow \overset{+4}{C}\overset{-2}{O_2}(g)$$

$$3\overset{0}{Mg} + \overset{0}{N_2} \longrightarrow \overset{+2}{Mg_3}\overset{-3}{N_2}$$

$$\overset{0}{H_2} + \overset{0}{Cl_2} \longrightarrow 2\overset{+1}{H}\overset{-1}{Cl}$$

(b) Decomposition reactions: Decomposition is the reverse process of combination, it involves the breakdown of the compound into two or more components. The product of decomposition must contain at least one component in elemental state.

e.g.,

$$2\overset{+1}{H_2}\overset{-2}{O}(g) \xrightarrow{\Delta} 2\overset{0}{H_2}(g) + \overset{0}{O_2}(g); \qquad 2NaH(s) \xrightarrow{\Delta} 2\overset{0}{Na}(s) + \overset{0}{H_2}(g)$$

$$2\overset{+1}{K}\overset{+5}{Cl}\overset{-2}{O_3}(s) \longrightarrow 2\overset{+1}{K}\overset{-1}{Cl} + 3\overset{0}{O_2}(g)$$

In above example, there is no change in oxidation state of potassium. Thus, it should be noted that the decomposition does not result into change in the oxidation number of each element.

(c) Displacement reactions: The reactions in which an atom or ion in a compound is displaced by another atom or ion are called displacement reactions. The displacement reactions are of 2 types:

(i) Metal displacement: In these reactions, a metal in a compound is replace by another metal in an uncombined state. It is found that a metal with stronger reducing character can displace the other metal having a weaker reducing character.

e.g.,

$$\overset{+3}{Cr_2}\overset{-2}{O_3} + 2\overset{0}{Al}(s) \longrightarrow \overset{+3}{Al_2}\overset{-2}{O_3}(s) + 2\overset{0}{Cr}(s)$$

$$\overset{+2}{Cu}\overset{+6}{S}\overset{-2}{O_4} + \overset{0}{Zn}(s) \longrightarrow \overset{+2}{Zn}\overset{+6}{S}\overset{-2}{O_4}(aq) + \overset{0}{Cu}(s)$$

(ii) Non-metal displacement: These displacement reactions generally involve redox reactions, where the hydrogen is displaced. Alkali and alkaline earth metals are highly electropositive, they displace hydrogen from cold water.

$$2\overset{0}{Na}(s) + 2\overset{+1}{H_2}\overset{-2}{O}(l) \longrightarrow 2\overset{+1}{Na}\overset{-2}{O}\overset{+1}{H}(aq) + \overset{0}{H_2}(g)$$

$$\overset{0}{Ca}(s) + 2\overset{+1}{H_2}\overset{-2}{O}(l) \longrightarrow \overset{+1}{Ca}(\overset{-2}{O}\overset{+1}{H})_2(aq) + \overset{0}{H_2}(g)$$

(d) Disproportionation and Oxidation–Reduction: One and the same substance may act simultaneously as an oxidizing agent with the result that a part of it gets oxidized to a higher state and rest of it is reduced to lower state of oxidation. Such a reaction, in which a substance undergoes simultaneous oxidation and reduction is called disproportionation and the substance is said to **disproportionate**.

The following are some of the examples of disproportionation:

(a)

Increase

$$\overset{-1}{H_2O_2} + \overset{-1}{H_2O_2} \longrightarrow 2H_2\overset{-2}{O} + \overset{0}{O_2}$$

Decrease

(b)

Decrease

$$3\overset{0}{Cl_2} + \underset{\text{(conc.)}}{6NaOH} \xrightarrow{\text{Hot}} 5Na\overset{-1}{Cl} + Na\overset{+5}{Cl}O_3 + 3H_2O$$

Increase

(e) Oxidation state of chlorine lies between –1 to +7; thus out of ClO^-, ClO_2^-, ClO_3^-, ClO_4^-; ClO_4^- does not undergo disproportionation because in this oxidation state of chlorine is highest, i.e., +7. Disproportionation of the other oxoanions are:

$$3\overset{+1}{Cl}O^- \longrightarrow 2\overset{-1}{Cl} + \overset{+5}{Cl}O_3^-$$

$$6\overset{+3}{Cl}O_2^- \longrightarrow 4\overset{+5}{Cl}O_3^- + 2\overset{-1}{Cl^-}; \qquad 4\overset{+5}{Cl}O_3^- \longrightarrow \overset{-1}{Cl^-} + 3\overset{+7}{Cl}O_4^-$$

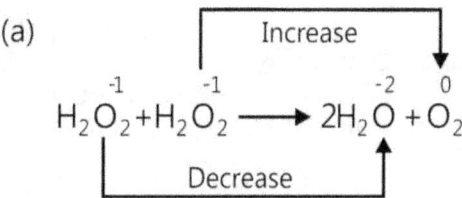

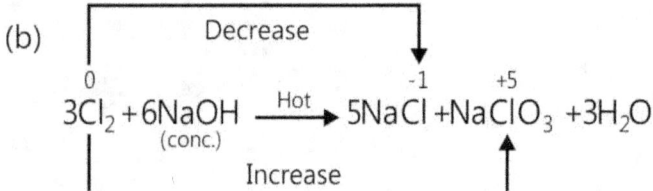

FORMULAE SHEET

Mass

Atomic mass

Atomic mass $= \dfrac{\text{Average mass of an atom}}{1/12 \times \text{Mass of an atom of } C^{12}}$

Molecular mass

Molecular mass = The sum of the atomic masses of each atom in the molecule

Equivalent weight

Equivalent weight $= \dfrac{\text{Molecular weight}}{\text{Valency factor or n factor}}$

Methods

Atomic mass

(i) Dulong and Petit's method
At mass x Sp.heat (cal/gram) = 6.4 (approx.)

(ii) Vapour density method
-suitable for elements whose chlorides are volatile.

(iii) Valency of the element

$= \dfrac{\text{Molecular mass of chloride}}{\text{Equivalent mass metal} + 35.5}$

$= \dfrac{2 \times \text{vapour density of chloride}}{\text{Equivalent mass metal} + 35.5}$

Atomic mass = Equivalent mass of metal x Valency

Molecular mass

(i) Dulong Method
(For Gases)- The ratio of rates of diffusion of two gases is inversely proportional to the square root of their molecular mass or densities.

$$\frac{r_1}{r_1} = \sqrt{\frac{M_2}{M_1}} = \sqrt{\frac{d_2}{d_1}}$$

(ii) Vapour density method (For gases only)- Mass of a fixed volume of the vapour is compared with the mass of the same volume of hydrogen under same condition; The ratio of these masses is called vapour density or Relative desity.

(iii) Molecular mass = 2 x Vapour density

Equivalent weight

(i) EM of an element

$= \dfrac{\text{Atomic mass}}{\text{Valency}}$

(ii) EM of an acid

$= \dfrac{\text{Molecular mass}}{\text{Basicity}}$

(Basicity of acid is the number of replaceable hydrogen atom in one molecule of the acid).

$= \dfrac{\text{Molecular mass}}{\text{Acidity}}$

(Acidity of a base is the number of replaceable OH group in one molecule of the base).

(iv) EM of an salt $= \dfrac{\text{Formula mass}}{\text{Total positive or negative charge}}$

(v) EM of an oxidizing agent= $\dfrac{\text{Formula mass}}{\text{Number of electrons gained per molecular or Total O.N.}}$

Equivalent mass of common oxidizing agent changes with the medium of the reaction.

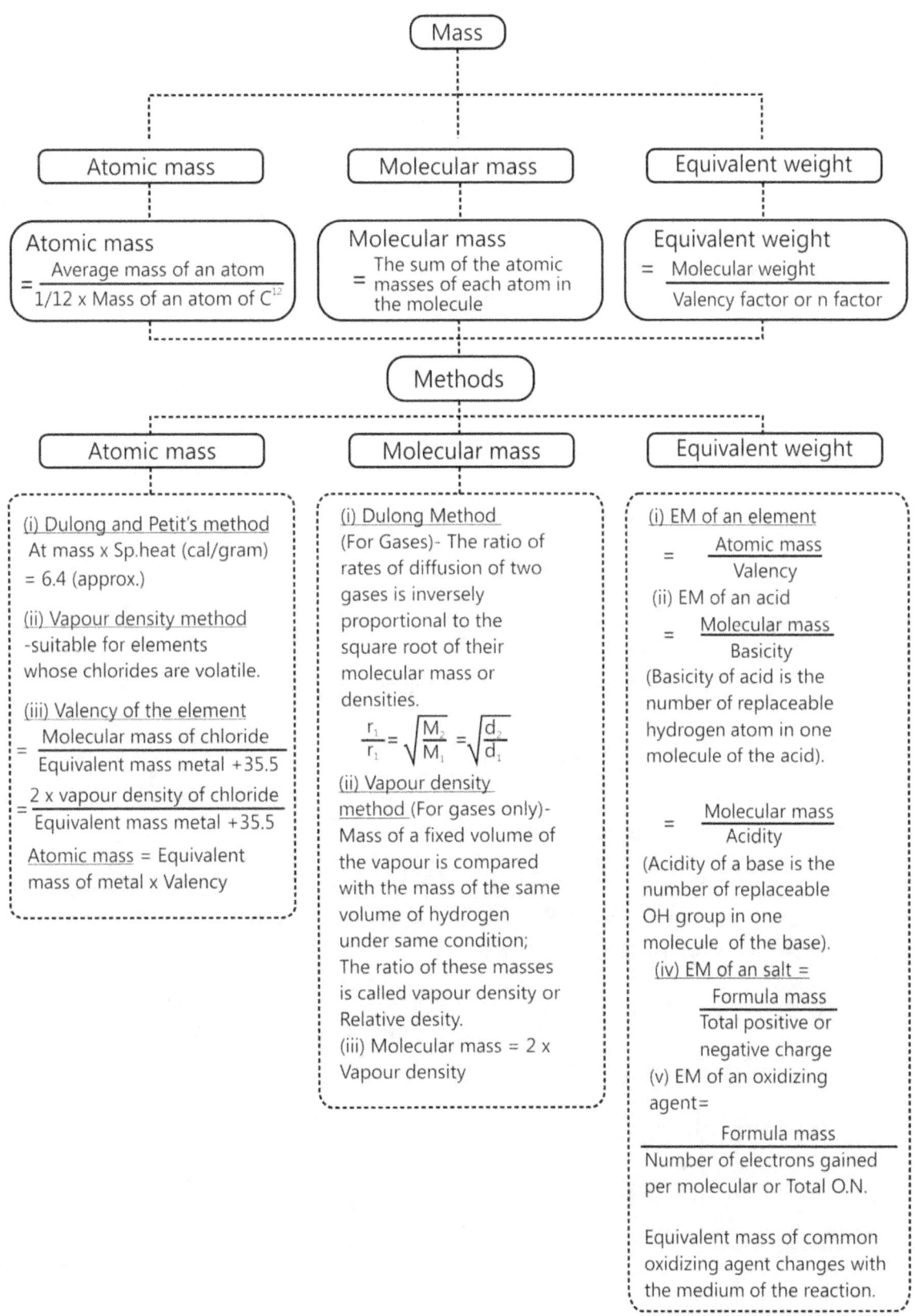

Mole and Concentration

Number of Moles = $\dfrac{\text{Weight (grams)}}{\text{Atomic or molecular weight}}$

= $\dfrac{\text{Volume in liters}}{22.4}$ at NTP

(only for gases)

Molarity = $\dfrac{\text{Number of moles of solute}}{\text{Volume of solution in litre}}$

Specific gravity = $\dfrac{\text{Wt. of solution}}{\text{Vol. of solution}}$

= Wt. of 1ml of solution

Formality = $\dfrac{\text{Wt. of solution}}{\text{Formula wt. of solute} \times V_{in}}$

Normality = X × No. of millimoles

= X× Molarity

= $\dfrac{\text{Strength of solution in gm/litre}}{\text{Eq. wt of solute}}$

Where, X = $\dfrac{\text{Mol.wt.}}{\text{Eq.wt.}}$

X=valency or change in oxi. Number.

Mole fraction (X_i)

It is the fraction of moles of a particular component in a mixture as

$X_i = \dfrac{n_i}{\sum\limits_{i=1}^{\cdot} n_i}$

Strength of solution (in gm/lit)

= Normality ×Eq. Wt. of the solute

= Molarity ×Mol. wt. solute

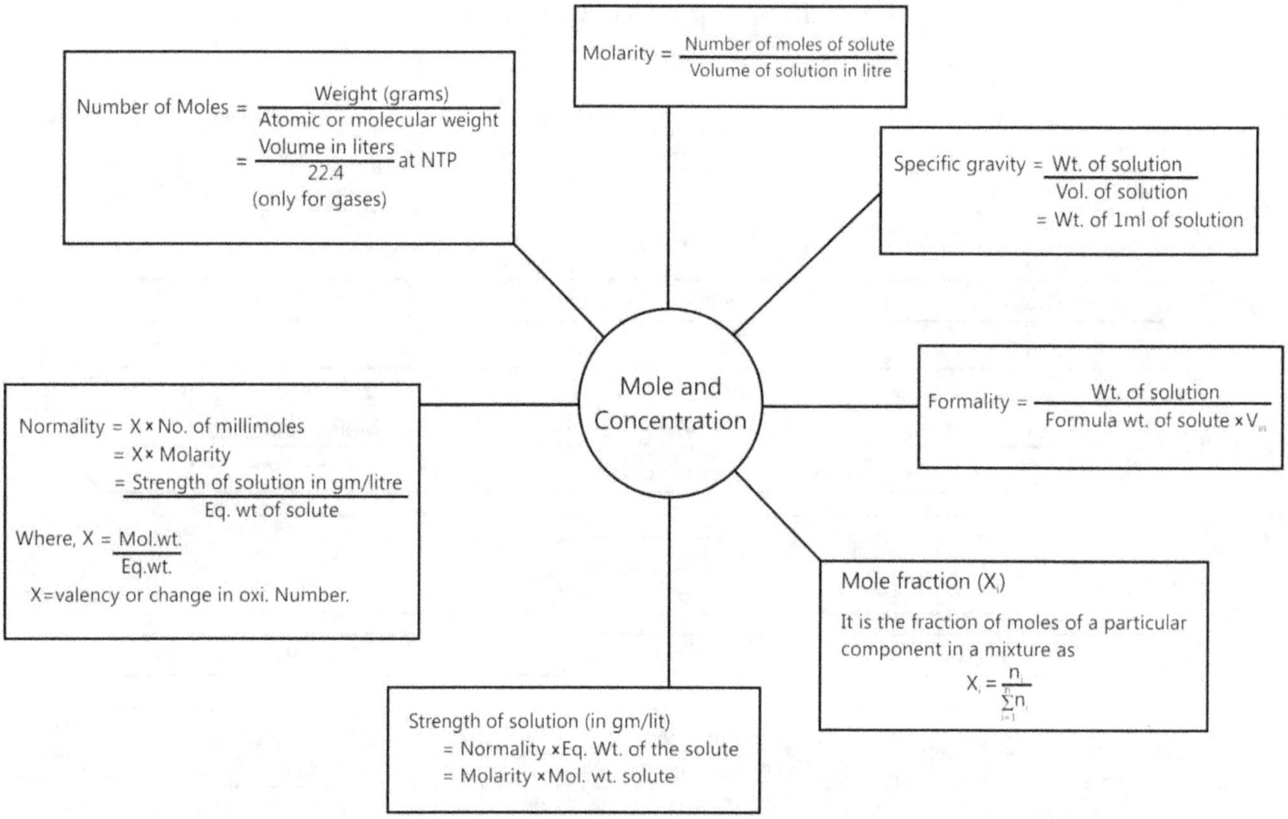

RULES IN BRIEF

The following are the definitions of 'mole' represented in the form of equations:

(a) Number of moles of molecules = $\dfrac{\text{Weight in g}}{\text{Molecular weight}}$

(b) Number of moles of atoms = $\dfrac{\text{Weight in g}}{\text{Atomic weight}}$

(c) Number of moles of gases = $\dfrac{\text{Volume at NTP}}{\text{Standard molar volume}}$

(Standard molar volume is the volume occupied by 1 mole of any gas at NTP, which is equal to 22.4 litres.)

(d) Number of moles of atoms / molecules / ions / electrons = $\dfrac{\text{No. of atoms / molecules / ions / electrons}}{\text{Avogadro constant}}$

(e) Number of moles of solute = Molarity × Volume of solution in litres

Or   No. of millimoles = Molarity × Volume in mL.

$\dfrac{\text{Millimoles}}{1000}$ =moles

(f) For a compound M_x , N_y , x moles of N = y moles of M

It is the reactant that is consumed completely during a chemical reaction. If the supplied mass ratio of reactant is consumed completely leaving parts of other uncreated, one that is consumed completely is known as "limiting reactant". It determines the amount of product in a given chemical reaction.

If two or more solution of different molarities M_1, M_2, M_3,..... are mixed together, molarity of the resulting solution can be worked out as

$$M = \frac{M_1V_1 + M_2V_2 + M_3V_3 ...}{V_1 + V_2 + V_3 ....}$$

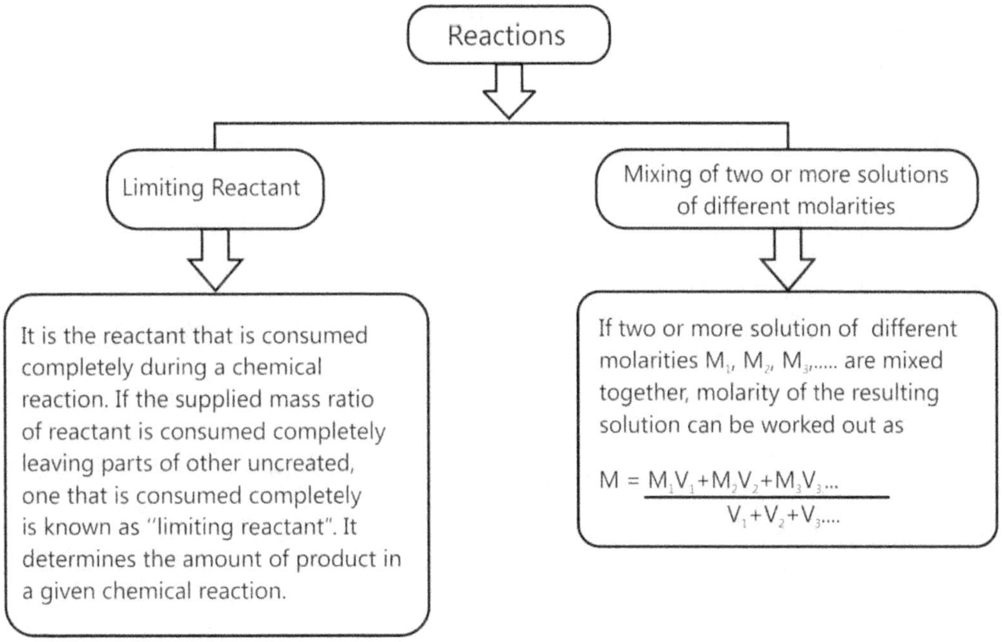

Solved Examples

JEE Main/Boards

Example 1: Calculate the composition of 109% oleum.

Sol: Let the mass of SO_3 in the sample be 'w' g, then the mass of H_2SO_4 would be (100 – w)g. On dilution,

$$\underset{80g}{SO_3} + \underset{18g}{H_2O} \longrightarrow H_2SO_4$$

Moles of SO_3 in oleum = $\frac{w}{80}$ = Moles of H_2SO_4 formed after dilution.

$\therefore$ Mass of H_2SO_4 formed on dilution = $\frac{98w}{80}$

Total mass of H_2SO_4 present in oleum after dilution

$= \frac{98w}{80} + (100 - w) = 109$; w = 40

Thus oleum sample contains 40% SO_3 and 60% H_2SO_4.

Example 2: 20g of a sample of $Ba(OH)_2$ is dissolved in 10 mL. of 0.5 N HCl sol. The excess of HCl was titrated with 0.2 N NaOH. The volume of NaOH used was 10 cc. Calculate the percentage of $Ba(OH)_2$ in the sample.

Sol: The titration principle is applied wherein milli-equivalents of the neutralization reactions is calculated.

Solving further, one gets the mass and % of the base.

Milli eq. of HCl initially = 10 × 0.5 = 5

Milli eq. of NaOH consumed

= Milli eq.of HCl in excess = 10 × 0.2 = 2

$\therefore$ Milli eq. of HCl consumed

= Milli eq. of $Ba(OH)_2$ = 5 – 2 = 3

$\therefore$ Eq. of $Ba(OH)_2$ = 3/1000 = 3 × 10^{-3}

Mass of $Ba(OH)_2$ = 3 × 10^{-3} (171/2) = 0.2565 g

% $Ba(OH)_2$ = (0.2565/20) × 100 = 1.28%

Example 3: One litre of mixture of CO and CO_2 is passed through red hot charcoal in tube. The new volume becomes 1.4 litre. Find out % composition of original mixture by volume. All measurements are made at same P and T.

Sol: Assuming the mixture contents as a and b, the reaction is framed and values are laid down.

Let the mixture contains

CO = a litre; CO_2 = b litre

$\therefore$ a + b = 1   ...(i)

On passing the mixture over charcoal only CO_2 reacts as:

$$CO_2 + C \longrightarrow 2CO$$

Vol. before reaction	b		0
Vol. after reaction	0		2b

$\therefore$ a + 2b = 1.4

By Eqs. (i) and (ii)

a = 0.6 litre or a = 60%

b = 0.4 litre or b = 40%

Example 4: 0.5 g sample containing is treated with HCl liberating Cl_2. The is passed into a Sol. of KI and 30.0 cm³ of 0.1 M are required to titrate the liberated iodine. Calculate the percentage of in sample. (At. wt. of Mn = 55).

Sol: Principle of titration is involved in which equating the neutralization reactions is done and milliequivalents of each species is calculated. Thus, weight is calculated and the purity is found.

$$MnO_2 \xrightarrow{HCl} Cl_2 \xrightarrow{KI} I_2$$

$$\xrightarrow{Na_2S_2O_3} NaI + Na_2S_4O_6$$

Redox change are: $2e^- + I_2^0 \longrightarrow 2I$

$$2S_2^{2+} \longrightarrow S_4^{(5/2)^+} + 2e^-$$

$$2e^- + Mn^{4+} \rightarrow Mn^{2+}$$

The reactions suggest that,

Meq. of MnO_2 = Meq. of Cl_2 formed

= Meq.of I_2 liberated = Meq. of $Na_2S_2O_3$ used

$$\therefore \quad \frac{w}{M/2} \times 1000 = 0.1 \times 1 \times 30$$

[$\because N_{Na_2S_2O_3} = M_{Na_2S_2O_3}$ since valency factor = 1, see redox changes for $Na_2S_2O_3$]

$$\text{Or } w_{MnO_2} = \frac{0.1 \times 1 \times 30 \times M}{2000} = \frac{0.1 \times 1 \times 30 \times 87}{2000}$$

($\because M_{MnO_2} = 87$); $w_{MnO_2} = 0.1305$

$$\therefore \text{Purity of } MnO_2 = \frac{0.1305}{0.5} \times 100 = 26.1\%$$

Example 5: 10 mL mixture of CH_4, C_2H_4 and C_3H_8 in the ratio 1: 1.5: 2.5 respectively is burnt in excess of air. Calculate the volume of air used and volume of CO_2 formed after combustion. All measurements are made at same P and T.

Sol: Using the given ratios, find the volumes of the hydrocarbons and frame the balanced combustion reactions.

The calculated O_2 level is 1/5th of the air.Hence volume of CO_2 is found.

$$\text{Volume of } CH_4 = \frac{1 \times 10}{5} = 2 \text{ mL}$$

$$\text{Volume of } C_2H_4 = \frac{1.5 \times 10}{5} = 3 \text{ mL}$$

$$\text{Volume of } C_3H_8 = \frac{2.5 \times 10}{5} = 5 \text{ mL}$$

$$CH_4 + 2O_2 \longrightarrow CO_2 + 2H_2O$$

$$C_2H_4 + 3O_2 \longrightarrow 2CO_2 + 2H_2O$$

$$C_3H_8 + 5O_2 \longrightarrow 3CO_2 + 4H_2O$$

$\therefore$ Volume of O_2 needed

= 2 × 2 + 2 × 3 + 3 × 5 = 38 mL

Since, O_2 is 1/5th part of air

$$\therefore \quad V_{air} = \frac{25 \times 100}{20} = 125 \text{ mL}$$

Volume of CO_2 formed

= 2 × 1 + 2 × 3 + 3 × 5 = 23 mL

Example 6: Select the species acting as reductant and oxidant in the reaction given below:

(i) $PCl_3 + Cl_2 \longrightarrow PCl_5$

(ii) $AlCl_3 + 3K \longrightarrow Al + 3KCl$

(iii) $SO_2 + 2H_2S \longrightarrow 3S + H_2O$

(iv) $BaCl_2 + Na_2SO_4 \longrightarrow BaSO_4 + 2NaCl$

(v) $3I_2 + 6NaOH \longrightarrow NaIO_3 + 5NaI + 3H_2O$

Sol: Calculate the oxidation numbers, find the loss/gain of electrons and thus identify the respective oxidants and reductants.

In a conjugate pair oxidant has higher oxidation number.

(i) $P^{+3} \longrightarrow P^{+5} + 2e^-$

$$2e^- + Cl_2^0 \longrightarrow 2Cl^{-1}$$

$\therefore$ PCl_3 is reductant and Cl_2 is oxidant.

$\because$ In a conjugate pair of redox, the one having higher ox. no. is oxidant.

(ii) For $AlCl_3 : Al^{+3} + 3e^- \longrightarrow Al^0$;

For $K : K^0 \longrightarrow K^{+1} + e^-$

Oxidant is $AlCl_3$ and reductant is K.

(iii) For $SO_3 : S^{+4} + 4e^- \longrightarrow S^0$;

For $H_2S : S^{-2} \longrightarrow 2e^-$

$\therefore$ SO_2 is oxidant and H_2S is reductant.

(iv) No change in ox. no. of either of the conjugate pair.

$\therefore$ None is oxidant or reductant.

(v) For $I_2 : I_2^0 \longrightarrow 2I^{+3}$ and $I_2^0 + 2e^- \longrightarrow 2I^{-1}$ I_2 acts as oxidant and reductant both.

Example 7: Balance the following reaction

$NO_3^- + Al \longrightarrow Al^{3+} + NH_4^+$ in basic medium.

Sol: Here NO_3^- is undergoing reduction and Al is undergoing oxidation.

(i) $NO_3^- \longrightarrow NH_4^+$ (ii) $Al \longrightarrow Al^{3+}$

by balancing each half reaction, we get

(iii) $NO_3^- + 7H_2O + 8e^- \longrightarrow NH_4^+ + 10\ OH^-$

(iv) $Al \longrightarrow Al^{3+} + 3e^-$

by multiplying equation (iii) by 3 and equation (iv) by 8, we get

(v) $3NO_3^- + 7H_2O + 24e^- \longrightarrow 3NH_4^+ + 30\ OH^-$

(vi) $8Al \longrightarrow 8Al^{3+} + 24e^-$

by combining these equations, we get

$8Al + 3NO_3^- + 21H_2O \longrightarrow 8Al^{3+} + 3NH_4^+ + 30\ OH^-$

Example 8: The composition of a sample of wurtzite is $Fe_{0.93}O_{1.00}$. What percentage of iron is present in the form of Fe III?

Sol: Oxidation no. of Fe in wustite is $\frac{200}{93} = 2.15$

It is an intermediate value between two oxidation state of Fe as, Fe (II) and (III).

Let percentage of Fe (III) be a, then

$2 \times (100 - 0) + 3 \times a = 2.15 \times 100$ Or a = 15

$\therefore$ Percentage of Fe(III) = 15%

Example 9: A 5.0 cm³ solution of H_2O_2 liberates 0.508g of iodine from an acidified KI solution. Calculate the strength of H_2O_2 solution in term of volume strength at STP.

Sol: Volume strength is the volume of oxygen released from 1 mL of hydrogen peroxide solution.

Meq. of H_2O_2 = Meq. of I_2

$$\frac{w}{17} \times 1000 = \left[\frac{0.508}{\frac{254}{2}}\right] \times 1000$$

$\therefore$ w = 0.068 g

$$H_2O_2 \longrightarrow H_2O + \frac{1}{2}O_2$$

$\because$ 34 g H_2O_2 gives 11.2 litre O_2,

$\therefore$ 0.068 g gives $\frac{11.2 \times 0.068}{34} = 0.0224$ litre = 22.4 ml O_2

$\therefore$ Volume strength of $H_2O_2 = \frac{22.4}{5} = 4.48\%$

Example 10: A 1.100 g sample of copper ore is dissolved and the Cu^{2+} is treated with excess KI. The liberated I_2 requires 12.12 mL of 0.10 M $Na_2S_2O_3$ solution for titration. What is % copper by mass in the ore?

Sol: The titration reaction is framed to identify the loss/gain of electrons.The milliequivalents of the respective ions are equated and the amount is calculated. % can be found by dividing the whole weight.

$Cu^{2+} + e^- \rightarrow Cu^+$; $2I^- \longrightarrow I_2 + 2e^-$

$2S_2O_3^{2-} \longrightarrow S_4O_6^- + 2e^-$

Meq. of Cu^{2+} = Meq. of liberated I_2 = Meq. of $Na_2S_2O_3$ = 12.12 × 0.1 × 1 = 1.212

$$\therefore \frac{w_{Cu^{2+}}}{63.6/1} \times 1000 = 1.212$$

$$\therefore w_{Cu^{2+}} = 0.077\ g = w_{Cu}\ (Cu \xrightarrow{H_2SO_4} CuSO_4)$$

$$\therefore \%\ Cu = \frac{0.077}{1.10} \times 100 = 7\%$$

JEE Advanced/Boards

Example 1: Chile salt petre, a source of $NaNO_3$ also contains $NaIO_3$. The $NaIO_3$ can be used as source of iodine, produced in the following reactions.

$IO_3^- + 3HSO_3^- \longrightarrow I^- + 3H^+ + 3SO_4^{2-}$ ...(i)

$5I^- + IO_3^- + 6H^+ \longrightarrow 3I_{2(g)} + 3H_2O$ ...(ii)

One litre of chile salt petre solution containing 5.80g $NaIO_3$ is treated with stoichiometric quantity of $NaHSO_3$. Now an additional amount of same solution is added to reaction mixture to bring about the second reaction.

How many grams of $NaHSO_3$ are required in step I and what additional volume of chile salt petre must be added in step II to bring in complete conversion of I^- to I_2?

Sol: The titration reaction is used to identify the loss/gain of electrons.The milliequivalents of the respective species are equated and the amount is calculated. Stepwise calculation gives the volume of $NaIO_3$.

Meq. of $NaHSO_3$ = Meq. of $NaIO_3$

$$= N \times V = \frac{5.8}{198/6} \times 1000$$

[Et. wt. of NaI = M/6 because $I^{3+} + 6e \longrightarrow I^-$]

Meq. of $NaHSO_3$ = 175.76

$$\therefore w_{NaHSO_3} = \frac{175.76 \times 104}{2000} = 9.14 \text{ g}$$

Also Meq. of formed in I step using valence factor 6 = 175.76

In II step valence factor of IO_3^- is 1 and valence factor of is 5.

Thus, Meq. of formed using valence factor $1 = \frac{175.76}{6}$

Also Meq. of $NaIO_3$ used in step II $= \frac{175.76}{6}$

$$\therefore N \times V = \frac{175.76}{6}; \Rightarrow \frac{5.8}{198/5} \times V = \frac{175.76}{6}$$

$$\therefore V_{NaIO_3} = 200 \text{ mL}$$

Example 2: What amount of substance containing 60% NaCl, 37% KCl should be weighed out for analysis so that after the action of 25 mL of 0.1 N $AgNO_3$ solution, excess of Ag^+ is back titrated with 5 mL of NH_4SCN. Given that 1 mL of NH_4SCN = 1.1 mL of $AgNO_3$.

Sol: Let a g of the given sample be weighed out for the purpose. The reaction of the chlorides with $AgNO_3$ will give AgCl. The unreacted Ag^+ ions will get consumed by NH_4SCN to give AgSCN. Proceeding accordingly, equate the milliequivalents and calculate a.

$$\therefore \text{Wt. of NaCl} = \frac{60}{100} \times a = 0.6 \text{ a g}$$

$$\therefore \text{Wt. of KCl} = \frac{37}{100} \times a = 0.37 \text{ a g}$$

Now this mixture reacts with $AgNO_3$, the excess of $AgNO_3$ is back titrated with NH_4SCN. Meq. of $AgNO_3$ added to mixture

= 25 × 0.1 = 2.5

Normality of NH_4SCN can be derived as

Meq. of NH_4SCN = Meq. of $AgNO_3$

$N \times 1 = 0.1 \times 1.1$

N = 0.11

Meq. of $AgNO_3$ left = Meq. of NH_4SCN

= 5 × N

∴ Meq. of $AgNO_3$ left = 5 × 0.11 = 0.55

∴ Meq. of $AgNO_3$ used for mixture = 2.5 – 0.55 = 1.95

Meq. of KCl + Meq. of NaCl is mixture

$$= 1.95; \frac{0.73a}{74.5} \times 1000 + \frac{0.6a}{58.5} \times 1000 = 1.95$$

∴ a = 0.128 g

Example 3: NaOH and Na_2CO_3 are dissolved in 200 mL aqueous solution In the presence of phenolphthalein indicator, 17.5 mL of 0.1 N HCl are used to titrate this solution. Now methyl orange is added in the same sol. titrated and it requires 2.5 mL of the same HCl. Calculate the normality of NaOH and Na_2CO_3 and their mass present in the solution.

Sol: The titration of a simple acid and a base using an indicator is seen over here. The milliequivalents of the acid is calculated and equated with that of the base. The volume and the mass is thus calculated.

Milli equivalent (a) of HCl used in the presence of phenolphthalein indicator.

= N × V (mL) = 0.1 × 17.5 = 1.75

1.75 (a) = milli. eq. of NaOH + 1/2 milli eq. of Na_2CO_3 ... (i)

Milli eq. (b) of HCl used in the presence of methyl orange indicator

= N × V (mL) = 0.1 × 2.5 = 0.25

0.25 (b) = 1/2 milli eq. of Na_2CO_3 ... (ii)

For Na_2CO_3 solution.; from equation (ii)

Milli eq. of acid used by Na_2CO_3 = 2b = 2 × 0.25 = 0.5

Volume of Na_2CO_3 solution = 200 mL

Suppose, Normality of Na_2CO_3 = N

Milli equivalents of = N × V (mL) =200 N

Putting equivalents of acid and Na_2CO_3 equal 200 N = 0.5

Or (Normality of Na_2CO_3 solution) $N = \frac{1}{400}$

Mass of Na_2CO_3 = N × E × V (litre)

(E for Na_2CO_3 = 53) $= \frac{1}{400} \times 5 \times 0.2 = 0.0265$ gram

For NaOH Sol.; from equation (i) and (ii)

Milli eq. acid used by NaOH = a – b = 1.75 – 0.25 = 1.50

Volume of NaOH solution = 200 mL

Suppose, Normality of NaOH solution = N

Milli eq. of NaOH = N × V (mL) = 200 N

Putting the milli eq. of NaOH and acid used equal 200 N = 1.5

(Normality of NaOH Sol.) $N = \frac{1.5}{200}$

Mass of NaOH = N × E × (V litres)

$= \frac{1.5}{200} \times 40 \times 0.2$ (E for NaOH = 40) = 0.06 g

Example 4: The molarity and molality of a solution are M and m respectively. If the molecular weight of the solute is M'. Calculate the density of the solution in terms of M, m and M'.

Sol: Let weight of solute be w g and weight of solvent be W g, volume of solution be V mL and density be D. Substitute as follows.

$$\therefore \quad M = \frac{w \times 1000}{M' \times V} \quad ...(i)$$

$$m = \frac{w \times 1000}{M' \times W} \quad ...(ii)$$

$$D = \frac{w + W}{V} \quad ...(iii)$$

By Eq. (i) $w = \frac{MM'V}{1000}$ ...(iv)

By Eq. (ii) $W - \frac{w \times 1000}{M' \times m}$

By Eq. (iv) $W = \frac{MM'V \times 1000}{1000 \times M' \times m} = \frac{MV}{m}$ ...(v)

$\therefore$ By Eq. (iii) $D = \frac{\frac{MM'V}{1000} + \frac{MV}{m}}{V}$; $D - M\left[\frac{1}{m} + \frac{M'}{1000}\right]$

Example 5: 1.249 g of a sample of pure $BaCO_3$ and impure $CaCO_3$ containing some CaO was treated with dil. HCl and it evolved 168 mL of CO_2 at NTP. From this solution $BaCrO_3$ was precipitated, filtered and washed. The precipitate was dissolved in dilute sulphuric acid and diluted to 100 mL. 10 mL of this solution when treated with KI solution, liberated iodine which required exactly 20 mL of 0.05 N $Na_2S_2O_3$. Calculate the percentage of CaO in the sample.

Sol: An acid-base titration accompanied with iodine titration gives the following equation.

$$n_{CaCO_3} + n_{BaCO_3} = n_{CO_2}$$

Calculating the equivalents of the involved species gives their amount and the %.

$$= \frac{168}{22400} = 7.5 \times 10^{-3} \quad ...(i)$$

$$2BaCO_3 \longrightarrow 2BaCrO_4 \longrightarrow BaCr_2O_7 \xrightarrow{KI} I_2 + Na_2S_2O_3$$

Eq. of $Na_2S_2O_3$ = Eq. of I_2 = Eq. of $BaCr_2O_7$

$$= \frac{20 \times 10^{-3} \times 0.05 \times 100}{10} = 1 \times 10^{-2}$$

Moles of $BaCr_2O_7 = \frac{1}{6} \times 10^{-2}$,

Moles of $BaCrO_4 = \frac{2}{6}(1 \times 10^{-2})$

Moles of $BaCO_3 = \frac{1}{3} \times 10^{-2} = 3.33 \times 10^{-3}$ ...(ii)

Weight of $BaCO_3$ = 0.650 gm

From equation (i) and (ii) we get $\Rightarrow n_{CaCO_3} = 4.17 \times 10^{-3}$

Weight of $CaCO_3 = 100 \times 4.17 \times 10^{-3} = 0.417$ g

Weight of $CaO = 1.249 - 0.656 - 0.417 = 0.176$

$$\%\text{ of CaO} = \frac{0.176}{1.249} \times 100 = 14.09\ \%$$

Example 6: Find out the percentage of oxalate ion in a given sample of oxalate salt of which 0.3 g dissolved in 100 mL of water required 90 mL of N/20 $KMNO_4$ for complete oxidation.

Sol: Redox changes are

$$5e^- + Mn^{+7} \longrightarrow Mn^{+2}$$

$$C_2^{+3} \longrightarrow 2C^{+4} + 2e^-$$

$\therefore$ Meq. of oxalate ion = Meq. of $KMNO_4$

$$\frac{w}{E} \times 1000 = 90 \times \frac{1}{20};\ E_{C_2O_4^{-2}} = \frac{\text{Ionic wt.}}{2}\ \frac{w}{\frac{88}{2}} \times 1000 = \frac{9}{2}$$

$\therefore w_{C_2O_4^{-2}} = 0.198$ g

$\therefore$ 0.3 g $C_2O_4^{-2}$ sample has oxalate ion = 0.198 g

$\therefore$ Percentage of $C_2O_4^{-2}$ in sample $= \frac{0.198 \times 100}{0.3} = 66\%$

Example 7: Balance the following redox equation, $AsO_3^{-3} + MnO_4^- \longrightarrow AsO_4^{-3} + MnO_2$ using ion-electron method (alkaline medium)

Sol: (i) Identify the oxidation and reduction halves.

Reduction half reaction: $MnO_4^- \longrightarrow MnO_2$

Oxidation half reaction: $AsO_3^{-3} \longrightarrow AsO^{-3}$

(ii) Atoms of the element undergoing oxidation and reduction are already balanced.

(iii) Balancing O atoms,

Reduction half reactions:

$$2H_2O + MnO_4^- \longrightarrow MnO_2 + 4OH^-$$

Oxidation half reactions:

$$2OH^- + AsO_3^{-3} \longrightarrow AsO_4^{-3} + H_2O$$

(iv) Balancing H atoms, H atoms are already balanced in both the half reactions.

(v) Balancing charge,

Reduction half reaction:

$$3e^- + 2H_2O + MnO_4^- \longrightarrow MnO_2 + 4OH^- \quad ...(ii)$$

Oxidation half reaction:

$$2OH^- + AsO_3^{-3} \longrightarrow AsO_4^{-3} + H_2O + 2e^- \quad ...(i)$$

(vi) Multiply equation (i) by 3 and equation (ii) by 2 and then add (i) and (ii).

$$3e^- + 2H_2O + MnO_4^- \longrightarrow MnO_3 + 4OH^-] \times 2$$

$$2OH^- + AsO_3^{-3} \longrightarrow AsO_4^{-3} + H_2O + 2e^-] \times 3$$

$$AsO_3^{-3} + 2MnO_4^- + H_2O \longrightarrow 3AsO_4^{-3} + 2MnO_2 + 2OH^-$$

Example 8: 1 g sample of $AgNO_3$ is dissolved in 50 mL of water. It is titrated with 50 mL of KI solution. The AgI precipitated is filtered off. Excess of KI in filtrate is titrated with M/10 KIO_3 in presence of 6M HCl till all I^- converted into ICl. It requires 50 mL of M/10 KIO_3 solution. 20 mL of the same stock solution of KI requires 30 mL of M/10 KIO_3 under similar conditions. Calculate % of $AgNO_3$ in sample. The reaction is:

$$KIO_3 + 2KI + 6HCl \longrightarrow 3KCl + 3H_2O$$

Sol: Follow the reaction $AgNO_3 + KI \longrightarrow AgI + KNO_3$

1. Ag present in $AgNO_3$ is removed as AgI by adding 50 mL KI of which 20 mL requires 30 mL of M/10 KIO_3.

2. The solution contains KI unused. The unused KI is converted into ICl by KIO_3.

$\therefore$ Meq. of KI in 20 mL = Meq. of KIO_3

$$4e^- + I^{+5} \longrightarrow I^{+1}$$

$$= 30 \times \frac{1}{10} \times 4 \quad \Big| \quad I^- \longrightarrow I^{+1} + 2e^-$$

$\therefore$ Meq. of KI in 50 mL added to $AgNO_3$

$$\therefore \text{Eq. wt. of KI} = \frac{M}{2} = \frac{30 \times 4 \times 50}{10 \times 20} = 30$$

Now, Meq. of KI left unused by $AgNO_3$ = 30 − 20

$\because$ Mole ratio of $AgNO_3$ and KI

$\therefore$ Meq. of $AgNO_3$ = 10

Reaction is 1: 1 and thus if Eq.

$$\therefore \frac{w}{170/2} \times 1000 = 10 \quad \Big| \quad \text{Wt. of KI is } M/2,$$

w = 0.85 g then Eq. wt. of $AgNO_3$ = M/2

$\therefore$ Percentage of purity of $AgNO_3$ in sample

$$= \frac{0.85 \times 100}{1} = 85\%$$

Example 9: Selenium in a 10.0 gm soil sample is distilled as the tetrabromide, which is collected in an aqueous solution, where it is hydrolysed to SeO_3^{-2}. The SeO_3^{-2} is estimated iodometrically, requiring 4.5 mL of standard $Na_2S_2O_3$ solution for the titration. If 1 mL of $Na_2S_2O_3$ = 0.049 mg of $K_2Cr_2O_7$, what is the concentration of Se in the soil in ppm?

Sol: Follow the reaction

$$Se \longrightarrow SeBr_4 \longrightarrow SeO_3^{-2}$$

$$SeO_3^{-2} + 4I^- + 6H^+ \longrightarrow Se + 2I_2 + 3H_2O$$

$$I_2 + 2Na_2S_2O_3 \longrightarrow Na_2S_4O_6 + 2NaI$$

$$1mL\ Na_2S_2O_3 \equiv \frac{0.049 \times 10^{-3} \times 6}{294} \text{ eq. of } K_2Cr_2O_7$$

$$\equiv \frac{0.049 \times 10^{-3} \times 6 \times 10^3}{294} \text{ Meq. of } K_2Cr_2O_7$$

$\therefore$ 4.5 mL $Na_2S_2O_3$

$$= \frac{0.049 \times 10^{-3} \times 6 \times 10^3 \times 4.5}{294} \text{ Meq. of } K_2Cr_2O_7 = 4.5 \times 10^{-3}$$

Meq. of $K_2Cr_2O_7$ or Meq. of $Na_2S_2O_3$

Meq. of Se = Meq. of SeO_3^{-2} = Meq. of KI = Meq. of I_2 = Meq. of $Na_2S_2O_3$

$$\frac{w_{Se}}{79} \times 1000 \times 4 = 4.5 \times 10^{-3}$$

$\therefore w_{Se} = 8.8875 \times 10^{-5}$ g

$$\therefore \text{ppm} = \frac{8.8875 \times 10^{-5} \times 10^6}{10} = 8.8875$$

Exercise 1

Mole Concept

Q.1 Express the following in S.I. units:

(i) 125 pounds, the average weight of an Indian boy (1 ℓb = 545 g)

(ii) 14 ℓb/m^2 (atmospheric pressure)

(iii) 5'8", the average height of ramp models.

Q.2 The isotropic distribution of potassium is 93.2% ^{39}K and 6.8% ^{41}K. How many ^{41}K atoms are there in 2g-atoms?

Q.3 How many oxygen atoms are present in 6.025 g of Barium phosphate (at. mass of Ba=137.5, P= 31, O = 16 amu)

Q.4 The vapour density of a mixture containing NO_2 and N_2O_4 is 3.83 at 27°C. Calculate the moles of NO_2 in 100 g mixture.

Q.5 Assume that the nucleus of the F atom is a sphere of radius 5×10^{-3} cm. Calculate the density of matter in F nucleus. (At. mass F = 19)

Q.6 20.0 mL of dil. HNO_3 is neutralised completely with 25 mL of 0.08 M NaOH. What is molarity of HNO_3?

Q.7 Gastric juice containing 3.0 g of HCl per litre. If a person produces about 2.5 litres of gastric juice a per day, how many antacid tablets each containing 400 mg of $Al(OH)_3$ are needed to neutralise all the HCl produced in one day.

Q.8 10 mL of HCl solution produced 0.1435 g of AgCl when treated with excess of Silver nitrate solution. What is the Molarity of acid solution [At. mass Ag = 100].

Q.9 A certain compound containing only carbon and oxygen. Analysis show it has 36% carbon and 64% oxygen. If its molecular mass is 400 then what is the molecular formula of the compound.

Q.10 0.44 g of a hydrocarbon on complete combustion with oxygen gave 1.8 g water and 0.88 g carbon dixoide. Show that these results are in accordance with the law of conservation of mass.

Q.11 A chloride of phosphate contains 22.57% P. Phosphine contains 8.82% hydrogen and hydrogen chloride gas contain 97.26% chlorine. Show that the data illustrate law of reciprocal proportions.

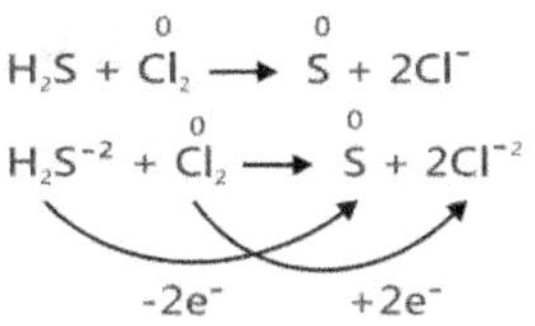
$H_2S + \overset{0}{Cl_2} \rightarrow \overset{0}{S} + 2Cl^-$
$H_2S^{-2} + \overset{0}{Cl_2} \rightarrow \overset{0}{S} + 2Cl^{-2}$
-2e$^-$
+2e$^-$

Q.12 1.375 g of cupric oxide was reduced by heating in a current of hydrogen and the mass of copper that remained was 1.098 g. In another experiment, 1.179 g of copper was dissolved in the nitric acid and the resulting copper nitrate converted into cupric oxide by ignition. The mass of cupric oxide formed was 1.476 g. Show that these results illustrate the law of constant composition.

Q.13 1.020 g of metallic oxide contains 0.540 g of the metal. Calculate the equivalent mass of the metal and hence its atomic mass with the help of Dulong and Petit's law. Taking the symbol for the metal as M, find the molecular formula of the oxide. The specific heat of the metal is 0.216 cal deg^{-1} g^{-1}.

Q.14 Potassium per magnate is a dark green crystalline substance whose composition is 39.7% K, 29.9% Mn and rest O. Find the empirical formula?

Q.15 Calculate the molarity of pure water at 4°C.

Q.16 (i) What is the mass in grams of one molecule of caffeine ($C_8H_{20}N_4O_2$)?

(ii) Determine the total number of electrons in 0.142 g Cl_2.

Q.17 Calculate the molarity of distilled water if its density is 10^3 kg/m^3.

Q.18 A plant virus if found to consist of uniform cylindrical particles of 150 Å in diameter and 5000 Å long. The specific volume of virus is 0.75 cm^3/g. If the virus is considered to be a simple particle, find the its molecular weight.

Q.19 Calculate the mass of two litre sample of water containing 25% heavy water D_2O in it by volume. Density of H_2O is 1.0 g cm^{-3} whereas that of D_2O is 1.06 g cm^{-3}.

Q.20 2.5 moles of sulphuryl chloride were dissolved in water to produce sulphuric acid and hydrochloric acid. How many moles of KOH will be required to completely neutralise the solution?

Q.21 100 g of a sample of common salt containing contamination of NH_4Cl and $MgCl_2$ to the extent of 2% each by mass is dissolved in water. How much volume of 5% by mass of $AgNO_3$ solution (d = 1.04 g cm^{-3}) is required to precipitate all chloride ions?

Q.22 A mixture of formic acid and oxalic acid is heated with concentrated H_2SO_4. The gases produced are collected and on treatment with KOH solution, the volume of the gases decreased by 1.6th. Calculate the molar ratio of the two acids in the original mixture.

Q.23 The mean molecular mass of a mixture of methane (CH_4) and ethene (C_2H_4) in the molar ratio of x: y is found to be 20. What will be the mean molecular mass if the molar ratio of the gases is reversed?

Q.24 1 g sample of $KClO_3$ was heated under such conditions that a part of it decomposes a $2KClO_3 \longrightarrow 2KCl + 3O_2$ while the remaining part decomposes as

$4KClO_3 \longrightarrow 3KClO_4 + KCl$.

If net oxygen obtained is 146.8 mL at STP.

Calculate the mass of $KClO_4$ in the residue.

Q.25 A mixture of FeO and Fe_3O_4 was heated in air to constant mass and it was found to gain 5% in its mass. Find the composition of the initial mixture.

Q.26 Equal masses of zinc (at. mass 65) and iodine (at. mass 127) were allowed to react till completion of reaction to form ZnI_2. Which substance is left unreacted and to what fraction of its original mass?

Q.27 Two gram each of P_4 and O_2 are allowed to react till none of the reactant is left. If the products are P_4O_6 and P_4O_{10}. Calculate the mass of each of the product.

Q.28 A piece of aluminium weighing 2.7 g was heated with 100 mL of H_2SO_4 (25% by mass, d = 1.18 g cm^{-3}). After complete dissolution of metal, the solution is diluted by adding water to 500 mL.

What is the molarity of free H_2SO_4 in resulting solution?

Q.29 Chemical reaction between ferrous oxalate and $KMnO_4$ has been given in the form of three partial equations. Write the complete balanced equation and thus find out the volume of 0.5 M $KMnO_4$ required to completely react with 1.5 mol of FeC_2O_4.

$KMnO_4 + H_2SO_4 \longrightarrow$

$K_2SO_4 + MnSO_4 + H_2O + (O)$

$FeC_2O_4 + H_2SO_4 \longrightarrow FeSO_4 + H_2C_2O_4$

$FeSI_4 + H_2C_2O_4 + H_2SO_4 + O \longrightarrow Fe_2(SO_4)_3 + CO_2 + H_2$

Redox Reactions

Q.1 Indicate the oxidation number of underlined in each case:

(i) $(\underline{N}_2H_5)_2SO_4$	(ii) $\underline{Mg_3}N_2$
(iii) $[\underline{Co}(NH_3)_5Cl]Cl_2$	(iv) $K_2\underline{Fe}O_4$
(v) $Ba(H_2\underline{P}O_2)_2$	(vi) $H_2\underline{S}O_4$
(vii) $C\underline{S}_2$	(viii) $\underline{S}^{-2}$
(ix) $Na_2\underline{S}_4O_6$	(x) $\underline{S}_2Cl_6$
(xi) $R\underline{N}O_2$	(xii) $\underline{Pb}_3O_4$
(xiii) $\underline{S}_2O_8^{-2}$	(xiv) $\underline{C}_6H_{12}O_6$
(xv) $Mg_2\underline{P}_2O_7$	(xvi) $K\underline{Cl}O_3$

Q.2 Write complete balanced equation for the following in acidic medium by ion-electron method:

(i) $Br^- + BrO_3^- + H^+ \longrightarrow Br_2 + H_2O$

(ii) $H_2S + Cr_2O_7^{-2} + H^+ \longrightarrow Cr_2O_3 + S_8 + H_2O$

(iii) $Au + NO_3^- + Cl^- + H^+ \rightarrow AuCl_4^- + NO_3 + H_2O$

(iv) $Cu_2O + H^+ + NO_3^- \longrightarrow Cu^{+2} + NO + H_2O$

(v) $MnO_4^{-2} \longrightarrow MnO_4^{-1} + MnO_2$

(vi) $Cu^{2+} + SO_2 \longrightarrow Cu^+ + SO_4^{-2}$

(vii) $Cl_2 + I_2 \longrightarrow IO_3^- + Cl^-$

(viii) $Fe(CN)_6^{-4} + MnO_4^- \rightarrow Fe^{+3} + CO_2 + NO_3^- + Mn^{+2}$

(ix) $Cu_3P + Cr_2O_7^{-2} \longrightarrow Cu^{+2} + H_3PO_4 + Cr^{+3}$

Q.3 Write complete balanced equation for the following in basic medium by ion-electron method:

(i) $Cu^{+2} + I^- \longrightarrow Cu^+ + I_2$

(ii) $Fe_3O_4 + MnO_4^- \longrightarrow Fe_2O_3 + MnO_2$

(iii) $C_2H_5OH + MnO_4^- \longrightarrow C_2H_3O^- + MnO_2(s) + H_2O$

(iv) $CrI_3 + H_2O_2 + OH^- \longrightarrow CrO_4^{-2} + IO_4^- + H_2O$

(v) $KOH + K_4Fe(CN)_6 + Ce(NO_3)_4 \longrightarrow$

$Fe(OH)_3 + Ce(OH)_3 + K_2CO_3 + KNO_3 + H_2O$

Q.4 Balance the following equations by oxidation method:

(i) $I^- + H_2O_2 \longrightarrow H_2O + I_2$ (Acid medium)

(ii) $Cu^{+2} + I^- \longrightarrow Cu^+ + I_2$

(iii) $CuO + NH_3 \longrightarrow Cu + N_2 + H_2O$

(iv) $H_2SO_3 + Cr_2O_7^{-2} \longrightarrow H_2SO_4 + Cr^{+3} + H_2O$ (Acid medium)

(v) $Cr_2O_7^{-2} + C_2H_4O + H^+ \longrightarrow C_2H_4O_2 + Cr^{+3}$ (Acid medium)

(vi) $SbCl_3 + KIO_3 + HCl \longrightarrow SbCl_5 + ICl + H_2O + KCl$ (Acid medium)

Q.5 Define disproportionation? Give one example.

Q.6 Define difference between ion electron method and oxidation method?

Q.7 What is the most essential conditions that must be satisfied in a redox reaction?

Q.8 Does the oxidation number of an element in any molecule or any poly atomic ion represents the actual charge on it?

Q.9 What is redox couple?

Q.10 Calculate the standard e.m.f. of the cells formed by different combinations of the following half cells.

Zn(g) / Zn^{2+}(aq)

Cu(s) / Cu^{2+}(aq)

Ni(s) / Ni^{2+}(aq)

Ag(s) / Ag^{2+}(aq)

Q.11 Balance the following equations in acidic medium by both oxidation number and ion electron methods & identify the oxidants and the reductants.

(i) $MnO_4^-(aq) + C_2H_2O_4(aq) \longrightarrow Mn^{2+}(aq) + CO_2(g) + H_2O(l)$

(ii) $H_2S(aq) + Cl_2(g) \longrightarrow S(s) + Cl(aq)$

Q.12 Write the half reactions for the following redox reactions:

(i) $2Fe^{3+}(aq) + 2I^-(aq) \rightarrow 2Fe^{2+} + (aq) + I_2(aq)$

(ii) $Zn(s) + 2H^+(aq) \rightarrow Zn^{2+}(aq) + H_2(g)$

(iii) $Al(s) + 3Ag^+(aq) \rightarrow Al^{3+}(aq) + 3Ag(s)$

Q.13 Define oxidation & reduction in term of oxidation number.

Q.14 Discuss the following redox reactions?

(i) Combination reactions

(ii) Decomposition reactions

(iii) Displacement reactions

(iv) Disproportionation reaction

Q.15 What is the difference between valence and oxidation number?

Q.16 H_2S acts only as reducing agent while SO_2 can act both as a reducing agent and oxidising agent. Explain.

Q.17 What are half reactions? Explain with examples?

Q.18 Explain the term:

(i) Oxidation (ii) Reduction

(iii) Oxidizing agent (iv) Reducing agent

Exercise 2

Mole Concept

Single Correct Choice Type

Q.1 If 'x' gms of an element A reacts with 16 gms of oxygen then the equivalent weight of element A is

(A) $\frac{x}{4}$ (B) $\frac{x}{2}$ (C) x (D) 2x

Q.2 The mass of CO containing the same amount of oxygen as in 88 gms of CO_2 is

(A) 56 gms (B) 28 gms (C) 112 gms (D) 14 gms

Q.3 When 8 gms of oxygen reacts with magnesium then the amount of MgO formed is

(A) 18 gm (B) 20 gm (C) 24 gm (D) 32 gm

Q.4 One gram of the silver salt of an organic dibasic acid yields, on strong heating 0.5934 g of silver. If the weight percentage of carbon in it 8 times the weight percentage of hydrogen and one half the weight percentage of oxygen, determine the molecular formula of the acid. [Atomic weight of Ag = 108]

(A) $C_4H_6O_4$ (B) $C_4H_6O_6$ (C) $C_2H_6O_2$ (D) $C_5H_{10}O_5$

Q.5 Mass of sucrose $C_{12}H_{22}O_{11}$ produced by mixing 84 gm of carbon, 12 gm of hydrogen and 56 liter O_2 at 1 atm and 273 K according to given reaction, is

$$C(s) + H_2(g) + O_2(g) \longrightarrow C_{12}H_{22}O_{11}(s)$$

(A) 138.5 (B) 155.5 (C) 172.5 (D) 199.5

Q.6 40 gm of carbonate of an alkali metal or alkaline earth metal containing some inert impurities was made to react with excess HCl solution. The liberated CO_2 occupied 12.315 lit. at 1 atm and 300 K. The correct option is

(A) Mass of impurity is 1 gm and metal is Be

(B) Mass of impurity is 3 gm and metal is Li

(C) Mass of impurity is 5 gm and metal is Be

(D) Mass of impurity is 2 gm and metal is Mg

Q.7 An hydride of nitrogen decomposes to give nitrogen and hydrogen. It was formed that one volume of the hydride gave one volume of N_2 and 2 volume of H_2 at STP. The hydride of nitrogen is

(A) NH_3 (B) N_2H_6 (C) NH_2 (D) N_2H_4

Q.8 5 volumes of a hydrocarbon on complete consumed 10 volumes of oxygen giving 5 volumes of CO_2 at STP. The hydrocarbon is

(A) C_2H_6 (B) C_2H_4 (C) CH_4 (D) C_2H_4

Q.9 The percentage by mole of NO_2 in a mixture of NO_2(g) and NO(g) having average molecular mass 34 is

(A) 25% (B) 20% (C) 40% (D) 75%

Q.10 The minimum mass of mixture of A_2 and B_4 required to produce at least 1 kg of each product is (Given At. mass of 'A' = 10; At. mass of 'B' = 120)

$$5A_2 + 2B_4 \longrightarrow 2AB_2 + 4A_2B$$

(A) 2120 gm (B) 1060 gm (C) 560 gm (D) 1660 gm

Q.11 74 gm of a sample on complete combustion given 132 gm CO_2 and 54 gm of H_2O. The molecular formula of the compound may be

(A) C_5H_{12} (B) $C_4H_{10}O$ (C) $C_3H_{10}O_2$ (D) $C_3H_7O_2$

Q.12 The volume of oxygen used when x gms of Zn is converted to ZnO is

(A) $\frac{2x}{65} \times 5.6$ litres (B) $\frac{x}{65} \times 5.6$ litres

(C) $\frac{4x}{65} \times 5.6$ litres (D) None of these

Q.13 A sample of clay was partially dried and then contained 50% silica and 7% water. The original clay contained 12% water. The silica is original sample is

(A) 51.69 (B) 47.31

(C) 63.31 (D) None of these

Q.14 The mass of CO_2 produced from 620 mixture of $C_2H_4O_2$ and O_2, prepared produce maximum energy is (combustion reaction is exothermic)

(A) 413.33 gm (B) 593.04 gm

(C) 440 gm (D) 320 gm

Q.15 In the quantitative determination of nitrogen, N_2 gas liberated from 0.42 gm of a sample of organic compound was collected over water. If the volume of N_2 gas collected was 100/11 mL at total pressure 860 mm Hg at 250 K, % by mass of nitrogen in the organic compound is

[Aq. tension at 250 K is 24 mm Hg and R = 0.08 L atm mol^{-1} K^{-1}]

(A) $\frac{10}{3}$% (B) $\frac{5}{3}$% (C) $\frac{20}{3}$% (D) $\frac{100}{3}$%

Q.16 300 mL of 0.1 M HCl and 200 mL of 0.3 M H_2SO_4 are mixed. The normality of the resulting mixture is

(A) 0.4 N (B) 0.1 N (C) 0.3 N (D) 0.2 N

Q.17 The volume of water which should be added to 300 mL of 0.5 M NaOH solution so as to get a solution of 0.2 M is

(A) 550 mL (B) 350 mL (C) 750 mL (D) 450 mL

Q.18 The mole fraction of a solution containing 3.0 gms of urea per 250 gms of water would be

(A) 0.00357 (B) 0.99643

(C) 0.00643 (D) None of these

Q.19 The mass of P_4O_{10} produced if 440 gm of P_4S_3 is mixed with 384 gm of O_2 is $P_4S_3 + O_2 \longrightarrow P_4O_{10} + SO_2$

(A) 568 gm (B) 426 gm

(C) 284 gm (D) 396 gm

Q.20 Calculate percentage change in M_{avg} of the mixture, if PCl_5 undergo 50% decomposition. $PCl_5 \longrightarrow PCl_3 + Cl_2$

(A) 50% (B) 66.66%

(C) 33.33% (D) Zero

Q.21 The mass of Mg_3N_2 produced if 48 gm of Mg metal is reacted with 34 gm NH_3 gas is $Mg + NH_3 \longrightarrow Mg_3N_2 + H_2$

(A) $\frac{200}{3}$ (B) $\frac{100}{3}$ (C) $\frac{400}{3}$ (D) $\frac{150}{3}$

Q.22 The molarity of a solution of conc. HCl containing 36.5% by weight of HCl would be

(A) 16.75 (B) 17.75 (C) 15.75 (D) 14.75

Q.23 0.35 gms of a sample of $Na_2CO_3.xH_2O$ were dissolved in water and the volume was made to 50 mL of this solution required 9.9 mL of $\frac{N}{10}$ HCl for complete neutralization. Calculate the value of x.

(A) 1 (B) 2 (C) 3 (D) None of these

Q.24 1.2 gms of a sample of washing soda was dissolved in water and volume was made upto 250 cc. 25 cc of this solution when titrated against N/10 HCl for required 17 mL. The percentage of carbonate is given sample is

(A) Approximately 70% (B) Approximately 66%

(C) Approximately 76% (D) None of these

Q.25 The number of carbon atoms present in a signature, if a signature written by carbon pencil weights 1.2×10^{-3} g is

(A) 12.40×10^{20} (B) 6.02×10^{19}

(C) 3.01×10^{19} (D) 6.02×10^{20}

Q.26 The average atomic mass of a mixture containing 79 mole % of ^{24}Mg is 24.31. % mole of ^{26}Mg is

(A) 5 (B) 20 (C) 10 (D) 15

Q.27 25 cc of solution containing NaOH and Na_2CO_3 when titrated against N/10 HCl. Using phenolphthalein as indicator required 40 cc. of HCl. The same volume of mixture when titrated against N/10 HCl using methyl orange required 45cc of this HCl. The amount of NaOH and Na_2CO_3 in one mixture is

(A) NaOH = 28 gm/L ; Na_2CO_3 = 10.6 gm/L

(B) NaOH = 10.6 gm/L ; Na_2CO_3 = 28 gm/L

(C) NaOH = 14 gm/L ; Na_2CO_3 = 5.3 gm/L

(D) None of these

Q.28 0.5 gms of a mixture of K_2CO_3 and Li_2CO_3 requires 30 mL of 0.25 NHCl solution for neutralization. The percentage composition of mixture would be

(A) K_2CO_3 = 96%; Li_2CO_3 = 4%

(B) K_2CO_3 = 4%; Li_2CO_3 = 96%

(C) K_2CO_3 = 50%; Li_2CO_3 = 25%

(D) K_2CO_3 = 50%; Li_2CO_3 = 74%

Q.29 How many mL of a 0.05 M $KMnO_4$ solution are required to oxidise 2.0 g of $FeSO_4$ in a dilute acid solution?

(A) 5.263 (B) 0.5263

(C) 52.63 (D) None of these

Redox Reaction

Single Correct Choice Type

Q.1 The equivalent weight of $FeSO_4$ when it is oxidised by acidified $KMnO_4$ will be equal to

(A) M_0 of $FeSO_4$ (B) $\frac{M_0 FeSO_4}{2}$

(C) $2M_0$ $FeSO_4$ (D) $\frac{M_0 FeSO_4}{4}$

Q.2 The equivalent weight of $K_2Cr_2O_7$ when it is converted Cr^{3+} will be equal to

(A) $M_{K_2Cr_2O_7}$ (B) $\frac{M_{K_2Cr_2O_7}}{3}$

(C) $\frac{M_{K_2Cr_2O_7}}{4}$ (D) $\frac{M_{K_2Cr_2O_7}}{6}$

Q.3 The amount of H_2S that can be oxidised to sulfur on oxidation using 1.58 gm of $KMnO_4$ as oxidising agent in acidic medium will be

(A) 0.85 gms (B) 1.7 gms

(C) 0.425 gms (D) None of these

Q.4 The amount of nitric acid required to oxidise 127 gms of I_2 to I_2O_5 will be _____. Assume that during the reaction HNO_3 gets converted to NO_2.

(A) 12.7 (B) 3.15 (C) 315 (D) 31.5

Q.5 10 mL of oxalic acid was completely oxidised by 20 mL of 0.02 M $KMnO_4$. The normality of oxalic acid solution is

(A) 0.05 N (B) 0.1 N (C) 0.2 N (D) 0.025 N

Q.6 0.2 g of a sample of H_2O_2 required 10 mL of 1N $KMnO_4$ in a titration in the presence of H_2SO_4. Purity of H_2O_2 is

(A) 25% (B) 65% (C) 85% (D) None of these

Q.7 The number of moles of $KMnO_4$ that will be needed to react completely with one mole of ferrous oxalate in acidic solution is

(A) $\frac{2}{5}$ (B) $\frac{3}{5}$ (C) $\frac{4}{5}$ (D) 1

Q.8 A metal oxide is reduced by heating it in a stream of hydrogen. It is found that after complete reduction, 3.15 g of the oxide has yielded 1.05 g of the metal. We may deduce that

(A) The atomic weight of the metal is 8

(B) The atomic weight of the metal is 4

(C) The equivalent weight of the metal is 4

(D) The equivalent weight of the metal is 8

Q.9 Oxidation involves

(A) Gain of electrons

(B) Loss of electrons

(C) Increase in the valency of negative part

(D) Decrease in the valency of positive part

Q.10 The oxidation number of Cr in $K_2Cr_2O_7$

(A) +2 (B) −2 (C) +6 (D) −6

Q.11 When $K_2Cr_2O_7$ is converted into $K_2Cr_2O_4$ the change in oxidation number of Cr is

(A) 0 (B) 6 (C) 4 (D) 3

Q.12 White P reacts with caustic soda. The products are PH_3 and NaH_2PO_2. This reaction is an example of

(A) Oxidation (B) Reduction

(C) Oxidation and reduction (D) Neutralization

Q.13 The oxidation number of carbon in CH_2O is

(A) −2 (B) +2 (C) 0 (D) +4

Q.14 The oxidation number of C in CH_4, CH_3Cl, CH_2Cl_2, $CHCl_3$, and CCl_4 are respectively

(A) 0,2,−2,4,−4 (B) −4,−2,0,+2,+4

(C) 2,4,0,−2,−4 (D) 4,2,0,−2,−4

Q.15 Which of the following reactions is not redox type

(A) $2BaO + O_2 \longrightarrow 2BaO_2$

(B) $4KClO_3 \longrightarrow 2KClO_4 + KCl$

(C) $BaO_2 + H_2SO_4 \longrightarrow BaSO_4 + H_2O_2$

(D) $SO_2 + 2H_2S \longrightarrow 2H_2O + 3S$

Q.16 In which of the following compounds iron has lowest oxidation state

(A) $K_4Fe(CN)_6$ (B) K_2FeO_4

(C) Fe_2O (D) $Fe(CO)_5$

Q.17 Select the compound in which chlorine is assigned the oxidation number +5

(A) HClO (B) $HClO_2$ (C) $HClO_3$ (D) $HClO_4$

Q.18 If three electrons are lost by a metal iron M^{3+} its final oxidation number would be

(A) 0 (B) +2 (C) +5 (D) +6

Q.19 The oxidation number of Mn in MnO_4^- is

(A) +7 (B) −5 (C) −7 (D) +5

Q.20 The oxidation number of carbon in $CHCl_3$ is

(A) +2 (B) +4 (C) +4 (D) −3

Q.21 Pb^{2+} loses two electrons in a reaction. What will be the oxidation number of lead after the reaction?

(A) +2 (B) 0 (C) +4 (D) −2

Q.22 The oxidation number of carbon in $C_{12}H_{22}O_{11}$ is

(A) 0 (B) −6 (C) +2 (D) +6

Q.23 The oxidation state of sulphur in SO_4^{2-} is

(A) +2 (B) +4 (C) +5 (D) +6

Q.24 If the following reaction 'X' is

$MnO_2 + 4H^+ + X \longrightarrow Mn^{2+} + H_2O$

(A) $1e^-$ (B) $2e^-$ (C) $3e^-$ (D) $4e^-$

Q.25 In the following reaction the value of 'X' is

$H_2O + SO_3^{-2} \longrightarrow SO_4^{-2} + 2H^+ + X$

(A) $4e^-$ (B) $3e^-$ (C) $2e^-$ (D) $1e^-$

Q.26 The oxidation state of sulphur is $S_2O_7^{2-}$ is

(A) +6 (B) −6 (C) −2 (D) +2

Q.27 The oxidation number and covalency of sulphur in S_8 are respectively

(A) 0 & 2 (B) 0 & 8 (C) 6 & 8 (D) 6 & 2

Q.28 The oxidation state of nitrogen in N_3H is

(A) 1/3 (B) +3 (C) −1 (D) −1/3

Q.29 The oxidation number of iron in potassium ferricyanide is

(A) +1 (B) +2 (C) +3 (D) +4

Q.30 Oxidation number of hydrogen in MH_2 is

(A) +1 (B) −1 (C) +2 (D) −2

Q.31 The oxidation state of phosphorus varies from

(A) −1 to +1 (B) −3 to +3 (C) −3 to +5 (D) −5 to +1

Q.32 Select the compound in which chlorine is assigned the oxidation number +5

(A) $HClO_4$ (B) $HClO_2$ (C) $HClO_3$ (D) HCl

Previous Years' Questions

Mole Concept

Q.1 If we consider that 1/6, in place of 1/12, mass of carbon atom is taken to be the relative atomic mass unit, the mass of one mole of a substance will ***(2002)***

(A) Decrease twice

(B) Increase two fold

(C) Remain unchanged

(D) Be a function of the molecular mass of the substance

Q.2 A molar solution is one that contains one mole of a solute in ***(1986)***

(A) 1000 g of the solvent (B) One litre of the solvent

(C) One litre of the solution (D) 22.4 litres of the solution

Q.3 In the reaction,

$2Al(s) + 6HCl(S) \longrightarrow 2Al^{3+}(aq) + 6Cl^-(aq) + 3H_2(g)$ ***(2007)***

(A) 6 l HCl (aq) is consumed for every 3L $H_2(g)$ produced

(B) 33.6 l $H_2(g)$ is produced regardless of temperature and pressure for every mole Al that reacts

(C) 67.2 l $H_2(g)$ at STP is produced for every mole Al that racts

(D) 11.2 $H_2(g)$ at STP is produced for every mole HCl (aq) consumed

Q.4 How many moles of magnesium phosphate, $Mg_3(PO_4)_2$ will contain 0.25 mole of oxygen atoms ***(2006)***

(A) 0.02 (B) 3.125×10^{-2}

(C) 1.25×10^{-2} (D) 2.5×10^{-2}

Q.5 If 10^{21} molecules are removed from 200 mg of CO_2, then the number of moles of CO_2 left are ***(1983)***

(A) 2.85×10^{-3} (B) 28.8×10^{-3}

(C) 0.288×10^{-3} (D) 1.68×10^{-2}

Q.6 In standardization of $NA_2S_2O_3$ using $K_2Cr_2O_7$ by iodometry, the equivalent weight of $K_2Cr_2O_7$ is ***(2000)***

(A) $\frac{MW}{2}$ (B) $\frac{MW}{3}$ (C) $\frac{MW}{6}$ (D) $\frac{MW}{1}$

Q.7 The molarity of a solution obtained by mixing 750 mL of 0.5(M) HCl with 250 mL of 2(M) HCl will be: ***(2013)***

(A) 0.875 M (B) 1.00 M (C) 1.75 M (D) 0.975 M

Q.8 A gaseous hydrocarbon gives upon combustion 0.72 g of water and 3.08 g of CO_2. The empirical formula of the hydrocarbon is: ***(2013)***

(A) C_2H_4 (B) C_3H_4 (C) C_6H_5 (D) C_7H_8

Q.9 Experimentally it was found that a metal oxide has formula $M_{0.98}$ O. Metal M, present as M^{2+} and M^{3+} in its oxide. Fraction of the metal which exists as M^{3+} would be: ***(2013)***

(A) 7.01 % (B) 4.08 % (C) 6.05 % (D) 5.08 %

Q.10 The ratio of masses of oxygen and nitrogen in a particular gaseous mixture is 1 : 4. The ratio of number of their molecule is 3 ***(2014)***

(A) 1 : 4 (B) 7 : 32 (C) 1 : 8 (D) 3 : 16

Q.11 The molecular formula of a commercial resin used for exchanging ions in water softening is $C_6H_7SO_3Na$ (Mol. Wt. 206). What would be the maximum uptake of Ca^{2+} ions by the resin when expressed in mole per gram resin? ***(2015)***

(A) $\frac{1}{103}$ (B) $\frac{1}{206}$ (C) $\frac{2}{309}$ (D) $\frac{1}{412}$

Q.12 At 300 K and 1 atm, 15 mL of a gaseous hydrocarbon requires 375 mL air containing 20% O_2 by volume for complete combustion. After combustion the gases occupy 330 mL. Assuming that the water formed is in liquid form and the volumes were measured at the same temperature and pressure, the formula of the hydrocarbon is: ***(2016)***

(A) C_2H_{12} (B) C_4H_8 (C) C_4H_{10} (D) C_3H_6

Redox Reactions

Q.13 Several blocks of magnesium are fixed to the bottom of a ship to ***(2003)***

(A) Keep away the sharks

(B) Make the ship lighter

(C) Prevent action of water and salt

(D) Prevent puncturing by under-sea rocks

Q.14 Which of the following chemical reactions depicts the oxidizing behaviour of H_2SO_4? ***(2006)***

(A) $2HI + H_2SO_4 \rightarrow I_2 + SO_2 + 2H_2O$

(B) $Ca(OH)_2 + H_2SO_4 \rightarrow CaSO_4 + 2H_2O$

(C) $NaCl + H_2SO_4 \rightarrow NaHSO_4 + HCl$

(D) $2PCl_5 + H_2SO_4 \rightarrow 2POCl_3 + 2HCl + SO_2Cl_2$

Q.15 The oxidation number of carbon in CH_2O is ***(1982)***

(A) −2 (B) +2 (C) 0 (D) 4

Q.16 The oxidation state of chromium in the final product formed by the reaction between KI and acidified potassium dichromate solution is ***(2005)***

(A) +4 (B) +6 (C) +2 (D) +3

Q.17 When $KMnO_4$ acts as an oxidising agent and ultimately forms $[MnO_4]^{-2}$, MnO_2, Mn_2O_3, Mn^{+2} then the number of electrons transferred in each case respectively is ***(2002)***

(A) 4, 3, 1, 5 (B) 1, 5, 3, 7

(C) 1, 3, 4, 5 (D) 3, 5, 7, 1

Q.18 Which of the following is a redox reaction ***(2002)***

(A) $NaCl + KNO_3 \rightarrow NaNO_3 + KCl$

(B) $CaC_2O_4 + 2HCl \rightarrow CaCl_2 + H_2C_2O_4$

(C) $Mg(OH)_2 + 2NH_4Cl \longrightarrow MgCl_2 + 2NH_4OH$

(D) $Zn + 2AgCN \rightarrow 2Ag + Zn(CN)_2$

Q.19 The product of oxidation of I^- and MnO_4^- in alkaline medium is ***(2004)***

(A) IO_3^- (B) I_2 (C) IO^- (D) IO_4^-

Q.20 For H_3PO_3 and H_3PO_4 the correct choice is ***(2003)***

(A) H_3PO_3 is dibasic and reducing

(B) H_3PO_3 is dibasic and non-reducing

(C) H_3PO_4 is tribasic and reducing

(D) H_3PO_3 is tribasic and non-reducing

Q.21 Consider the following reaction:

$$XMnO_4^- + YC_2O_4^{2-} + ZH^+ \rightarrow xMn^{2+} + 2yCO_2 + \frac{Z}{2}H_2O$$

The values of X, Y and Z in the reaction are, respectively: ***(2013)***

(A) 5, 2 and 16 (B) 2, 5 and 8

(C) 2, 5 and 16 (D) 5, 2 and 8

Q.22 In which of the following reaction H_2O_2 acts as a reducing agent? ***(2014)***

(A) $H_2O_2 + 2H^+ + 2e^- \rightarrow 2H_2O$

(B) $H_2O_2 - 2e^- \rightarrow O_2 + 2H^+$

(C) $H_2O_2 - 2e^- \rightarrow 2OH^-$

(D) $H_2O_2 + 2OH^- - 2e^- \rightarrow O_2 + 2H_2O$

(A) (a), (b) (B) (c), (d) (C) (a), (c) (D) (b), (d)

Q.23 The equation which is balanced and represents the correct product(s) is ***(2014)***

(A) $Li_2O + 2KCl \rightarrow 2LiCl + K_2O$

(B) $\left[CoCl(NH_3)_5\right]^+ + 5H^+ \rightarrow Co^{2+} + 5NH_4^+ + Cl^-$

(C) $\left[Mg(H_2O)_6\right]^{2+}(EDTA)^{4-} \xrightarrow{\text{excess NaOH}} \left[Mg(EDTA)\right]^{2+} + 6H_2O$

(D) $CuSO_4 + 4KCN \rightarrow K_2\left[Cu(CN)_4\right] + K_2SO_4$

Q.24 From the following statements regarding H_2O_2, choose the incorrect statement: ***(2015)***

(A) It can act only as an oxidizing agent

(B) It decomposed on exposure to light

(C) It has to be stored in plastic or wax lined glass bottles in dark

(D) It has to be kept away from dust

Exercise 1

Mole Concept

Q.1 How many gm of HCl is needed for complete reaction with 69.6 gm MnO_2?

$HCl + MnO_2 \rightarrow MnCl_2 + H_2O + Cl_2$

Q.2 Titanium, which is used to make air plane engines and frames, can be obtained from titanium tetrachloride, which in turn is obtained from titanium oxide by the following process:

$$3TiO_2(s) + 4C(s) + 6Cl_2(g) \rightarrow 3TiCl_4(g) + 2CO_2(g) + 2CO(g)$$

A vessel contains 4.32 g TiO_2 5.76 g C and 6.82 g Cl_2, suppose the reaction goes to completion as written, how many gram of $TiCl_4$ can be produced? (Ti = 48).

Q.3 Sulphuric acid is produced when sulphur dioxide reacts with oxygen and water in the presence of a catalyst:

$2SO_2(g) + O_2(g) + 2H_2O(l) \rightarrow 2H_2SO_4$.

If 5.6 mol of SO_2 reacts with 4.8 mole of O_2 and a large excess of water, what is the maximum number of moles of H_2SO_4 that can be obtained?

Q.4 What weight of Na_2CO_3 of 95% purity would be required to neutralize 45.6 mL of 0.235 N acid?

Q.5 How much $BaCl_2.2H_2O$ and pure water to be mixed to prepare 50g of 12.0% (by wt.) $BaCl_2$ solution.

Q.6 To 50 litre of 0.2 N NaOH, 5 litre of 1N HCl and 15 litre of 0.1 N $FeCl_3$ solution are added. What weight of Fe_2O_3 can be obtained from the precipitate? Also report the normality of NaOH left in the resultant solution.

Q.7 0.5 g fuming H_2SO_4 (oleum) is diluted with water. The solution requires 26.7 mL of 0.4 N NaOH for complete neutralization. Find the percentage of free SO_3 in the sample of oleum.

Q.8 200 mL of a solution of mixture of NaOH and Na_2CO_3 was first titrated with phenolphthalein and N/10 HCl. 17.5 mL of HCl was required for the end point. After this methyl orange was added and 2.5 mL of same HCl was again required for next end point. Find out amount of NaOH and Na_2CO_3 in mixture.

Q.9 Potassium superoxide, KO_2, is used in rebreathing gas masks to generate oxygen:

$KO_2(s) + H_2O(l) \rightarrow KOH(s) + O_2(g)$

If a reaction vessel contains 0.158 mol KO_2 and 0.10 mol H_2O, how many moles of O_2 can be produced?

Q.10 A sample of mixture of $CaCl_2$ and NaCl weighing 4.22 gm was treated to precipitate all the Ca as $CaCO_3$ which was then heated and quantitatively converted to 0.959 gm of CaO. Calculate the percentage of $CaCl_2$ in the mixture.

Q.11 Cyclohexanol is dehyrated to cyclohexene on heating with conc. H_2SO_4. If the yield of this reaction is 75%, how much cyclohexene will be obtained from 100 g of cyclohexanol? $C_6H_{12}O \xrightarrow{\text{con. } H_2SO_4} C_6H_{10}$

Q.12 How many grams of 90% pure Na_2SO_4 can be produced from 250 gm of 95% pure NaCl?

Q.13 A precipitate of AgCl and AgBr weighs 0.4066 g. On heating in a current of chlorine, the AgBr is converted to AgCl and the mixture loses 0.0725 g in weight. Find the percentage of Cl in original mixture.

Q.14 How many milli-litre of 0.5 M H_2SO_4 are needed to dissolve 0.5 g of copper II carbonate?

Q.15 What is the strength in g per litre of a solution of H_2SO_4, 12 mL of which neutralized 15 mL of N/10 NaOH solution.

Q.16 n-butane is produced by the monobromination of ethane followed by Wurtz reaction. Calculate the volume of ethane at NTP required to produce 55 g n-butane if the bromination takes place with 90% yield and the Wurtz reaction with 85% yield.

Q.17 0.50 g of a mixture of K_2CO_3 and Li_2CO_3 required 30 mL of 0.25 N HCl solution for neutralization. What is percentage composition of mixture?

Q.18 Sodium chlorate, $NaClO_3$, can be prepared by the following series of reactions:

$2KMnO_4 + 16HCl \rightarrow 2KCl + 2MnCl_2 + 8H_2O + 5Cl_2$

$6Cl_2 + 6Ca(OH)_2 \rightarrow Ca(ClO_3)_2 + 5CaCl_2 + 6H_2O$

$Ca(ClO_3)_2 + Na_2SO_4 \rightarrow CaSO_4 + 2NaClO_3$

What mass of $NaClO_3$ can be prepared from 100 mL of concentrated HCl (density 1.18 gm/mL and 36% by mass)? Assume all other substance are present in excess amounts.

Q.19 In a determination of P an aqueous solution of NaH_2PO_4 is treated with a mixture of ammonium and magnesium ions to precipitate magnesium ammonium phosphate $Mg(NH_4)PO_4.6H_2O$. This is heated and decomposed to magnesium pyrophosphate, $Mg_2P_2O_7$ which is weighed. A solution of NaH_2PO_4 yielded 1.054g of $Mg_2P_2O_7$. What weight of NaH_2PO_4 was present originally?

Q.20 5 mL of 8 N HNO_3, 4.8 mL of 5 N HCl and a certain volume of 17 M H_2SO_4 are mixed together and made upto 2 litre. 30 mL of this acid mixture exactly neutralizes 42.9 mL of Na_2CO_3 solution containing 1 g $Na_2CO_3.10H_2O$ in 100 mL of water. Calculate the amount of sulphate ions in g present in solution.

Q.21 A sample of Mg was burnt in air to give a mixture of MgO and Mg_3N_2. The ash was dissolved in 60 Meq of HCl and the resulting solution was back titrated with NaOH. 12 Meq of NaOH were required to reach the end point. As excess of NaOH was then added and the solution distilled. The ammonia released was then trapped in 10 Meq of second acid solution. Back titration of this solution required 6 Meq of the base. Calculate the percentage of Mg burnt to the nitride.

Q.22 A mixture of ethane (C_2H_6) and ethene occupies 40 litre at 1.00 atm and at 400 K. The mixture reacts completely with 130 g of O_2 to produce CO_2 and H_2O. Assuming ideal gas behaviour, calculate the mole fractions of C_2H_4 and C_2H_6 in the mixture.

Q.23 A solid mixture 5 g consists of lead nitrate and sodium nitrate was heated below 600°C until weight of residue was constant. If the loss in weight is 28%, find the amount of lead nitrate and sodium nitrate in mixture.

Q.24 Upon mixing 45.0 mL of 0.25 M lead nitrate solution with 25 mL of 0.10 M chromic sulphate, precipitation of lead sulphate takes place. How many moles of lead sulphate are formed? Also calculate the molar concentration of the species left behind in final solution. Assume that lead sulphate is completely insoluble.

Q.25 A 10 g sample of a mixture of calcium chloride and sodium chloride is treated with Na_2CO_3 to precipitate calcium as calcium carbonate. This $CaCO_3$ is heated to convert all the calcium to CaO and the final mass of CaO is 1.12 gm. Calculate % by mass of NaCl in the original mixture.

Q.26 A mixture of Ferric oxide (Fe_2O_3) and Al is used as solid rocket fuel which reacts to give Al_2O_3 and Fe. No other reactants and products are involved. On complete reaction of 1 mole of Fe_2O_3, 200 units of energy is released?

(i) Write a balance reaction representing the above change.

(ii) What should be the ratio of masses of Fe_2O_3 and Al taken so that maximum energy per unit mass of fuel is released.

(iii) What would be energy released if 16 kg of Fe_2O_3 reacts with 2.7 kg of Al.

Q.27 A mixture of nitrogen and hydrogen. In the ratio of one mole of nitrogen to three moles of hydrogen, was partially converted into so that the final product was a mixture of all these three gases. The mixture was to have a density of 0.497 g per litre at 25°C and 1.00 atm. What would be the mass of gas in 22.4 litres at 1 atm and 273 K? Calculate the % composition of this gaseous mixture by volume.

Q.28 In one process for waterproofing, a fabric is exposed to $(CH_2)_3SiCl_2$ vapour. The vapour reacts with hydroxyl groups on the surface of the fabric or with traces of water to form the waterproofing film $[(CH_3)_2SiO]_n$ by the reaction

$$n(CH_3)_2SiCl_2 + 2nOH^- \rightarrow 2nCl^- + nH_2O + [(CH_3)_2SiO]_n$$

where n stands for a large integer. The waterproofing film is deposited on the fabric layer upon layer. Each layer is 6.0 Å thick [the thickness of the $(CH_3)_2SiO$ group]. How much $(CH_2)_3SiCl_2$ is needed to waterproof one side of a piece of fabric, 1.00 m by 3.00 m, with a film 300 layers thick? The density of the film is 1.0 g/cm³.

Q.29 Two substance P_4 and O_2 are allowed to react completely to form mixture of P_4O_6 and P_4O_{10} leaving none of the reactants. Using this information calculate the composition of final mixture when mentioned amount of P_4 and O_2 are taken.

$$P_4 + 3O_2 \longrightarrow P_4O_6$$

$$P_4 + 5O_2 \longrightarrow P_4O_{10}$$

(i) If 1 mole P_4 & 4 mole of O_2

(ii) If 3 mole P_4 & 11 mole of O_2

(iii) If 3 mole P_4 & 13 mole of O_2

Q.30 Chloride samples are prepared for analysis by using NaCl, KCl and NH_4Cl seperately or as a mixture. What minimum volume of 5% by weight $AgNO_3$ solution (sp. gr., 1.04 g mL^{-1}) must be added to a sample of 0.3 g in order to ensure complete precipitation of chloride in every possible case?

Q.31 124 gm of mixture containing $NaHCO_3$, $AlCl_3$, and KNO_3 requires 500 mL, 8% w/w NaOH solution [d_{NaOH} = 1.8 gm/mL] for complete neutralisation. On heating same amount of mixture, it known loss in weight of 18.6 gm. Calculate % composition of mixture by moles. Weak base formed doesn't interfere in reaction. Assume KNO_3 does not decompose under given conditions.

Q.32 If the yield of chloroform obtainable from acetone and bleaching powder is 75%. What is the weight of acetone required for producing 30 gm of chloroform?

Q.33 A sample of impure Cu_2O contains 66.67% of Cu. What is the percentage of pure Cu_2O in the sample?

Q.34 Equal weights of mercury and iodine are allowed to react completely to form a mixture of mercurous and mercuric iodide leaving none of the reactants. Calculate the ratio by weight of Hg_2I_2 and HgI_2 and formed. (Hg = 200, I = 127)

Redox Reactions

Q.1 Indicate the oxidation state of underlined in each case:

(i) Na<u>N</u>O_2 (b) <u>H</u>$_2$ (c) <u>Cl</u>$_2O_7$

(ii) K<u>Cr</u>O_3Cl (e) <u>Ba</u>Cl_2 (f) <u>I</u>Cl_3

(iii) K_2<u>Cr</u>$_2O_7$ (h) <u>C</u>H_2O (i) <u>Ni</u>$(CO)_4$

(iv) <u>N</u>H_2OH

Q.2 Indicate the each reaction which of the reactant is oxidized or reduced if any:

(i) $CuSO_4 + 4KI \longrightarrow 2CuI + I_2 + 2K_2SO_4$

(ii) $2Na_2S + 4HCl + SO_2 \longrightarrow 4NaCl + 3S + 2H_2O$

(iii) $NH_4NO_2 \xrightarrow{\Delta} N_2 + 2H_2O$

Q.3 Calculate the number of electrons lost or gained during the changes:

(i) $3Fe + 4H_2O \longrightarrow Fe_3O_4 + 4H_2$

(ii) $AlCl_3 + 3K \longrightarrow Al + 3KCl$

Q.4 Explain, why?

(i) H_2S acts as reductant whereas, SO_2 acts as reductant and oxidant both.

(ii) H_2O_2 acts as reductant and oxidant both.

Q.5 MnO_4^- can oxidize NO_2^- to NO_3^- in basic medium. How many mol of NO_2^- are oxidized by 1 mol of MnO_4^-?

Q.6 Which is stronger base in each pair?

(i) HSO_4^-; HSO_3^-; (ii) NO_2^-; NO_3^-;

(iii) Cl^-; ClO^-

Q.7 Fill in the blanks and balance the following equations:

(i) Zn + HNO_3 → .... + N_2O + ...

(ii) HI + HNO_3 → .... + NO + H_2O ....

Q.8 What volume of 0.20 M H_2SO_4 is required to produce 34.0 g of H_2S by the reaction:

$8KI + 5H_2SO_4 \longrightarrow 4K_2SO_4 + 4I_2 + H_2S + 4H_2O$

Q.9 20 mL of 0.2 M $MnSO_4$ are completely oxidized by 16 mL of $KMnO_4$ of unknown normality, each forming Mn^{4+} oxidation state. Find out the normality and molarity of $KMnO_4$ solution.

Q.10 $KMnO_4$ solution is to be standardized by titration against As_2O_3(s). A 0.1097 g sample of As_2O_3 requires 26.10 mL of the $KMnO_4$ solution for its titration. What are the molarity and normality of the $KMnO_4$ solution?

Q.11 0.518 g sample of limestone is dissolved and then Ca is precipitated as CaC_2O_4. After filtering and washing the precipitate, it requires 40 mL of 0.25 N $KMnO_4$ solution to equivalence point. What is percentage of CaO in limestone?

Q.12 20 mL of a solution containing 0.2 g of impure sample of H_2O_2 reacts with 0.316 g of $KMnO_4$ (acidic). Calculate:

(i) Purity of H_2O_2,

(ii) Volume of dry O_2 evolved at 27°C and 750 mm P.

Q.13 5.7 g of bleaching powder was suspended in 500 mL of water. 25 mL of this suspended on treatment with KI and HCl liberated iodine which reacted with 24.35 mL of N/10 $Na_2S_2O_3$. Calculate percentage of available Cl_2 in bleaching powder.

Q.14 Balance the following equation:

(i) $C_2H_5OH + K_2Cr_2O_7 + H_2SO_4 \longrightarrow$

$$C_2H_4O_2 + Cr_2(SO_4)_3 + K_2SO_4 + H_2O$$

(ii) $As_2S_5 + HNO_3 \longrightarrow$

$$NO_2 + H_2O + H_3AsO_4 + H_2SO_4$$

(iii) $CrI_3 + Cl_3 + KOH \longrightarrow$

$$KIO_4 + K_2CrO_4 + KCl + H_2O$$

(iv) $As_2S_3 + HClO_3 + H_2O \longrightarrow$

$$HCl + H_3AsO_4 + H_2SO_4$$

Q.15 Balance the following equations:

(i) $As_2S_3 + OH^- + H_2O_2 \longrightarrow AsO_4^{2-} + SO_4^{2-} + H_2O$

(ii) $CrI_3 + H_2O_2 + OH^- \longrightarrow CrO_4^{2-} + 3IO_4^- + H_2O$

(iii) $P_4 + OH^- + H_2O_2 \longrightarrow H_2PO_2^- + PH_3$

(iv) $As_2S_3 + NO_3^- + H^+ \xrightarrow{+H_2O} H_3AsO_4 + NO + S$

Q.16 Mg can reduce NO_3^- to NH_3 in basic solution:

$NO_3^- + Mg(s) + H_2O \longrightarrow$

$Mg(OH)_2(s) + OH^-(aq) + NH_3(g)$

A 25.0 mL sample of NO_3^- solution was treated with Mg. The $NH_3(g)$ was passed into 50 mL of 0.15 N HCl. The excess HCl required 32.10 mL of 0.10 M NaOH for its neutralization. What was the molarity of NO_3^- ions in the original sample?

Q.17 An acid solution of $KReO_4$ sample containing 26.83 mg of combined rhenium was reduced by passage through a column of granulated zinc. The effluent solution including the washing from the column, was then titrated with 0.05 N $KMnO_4$. 11.45 mL of the standard $KMnO_4$ was required for the reoxidation of all the rhenium to the perrhenate ion ReO_4^-. Assuming that rhenium was the only element reduced, what is the oxidation state to which rhenium was reduced by the zinc column.

Q.18 100 mL solution of FeC_2O_4 and $FeSO_4$ is completely oxidized by 60 mL of 0.02 M in acid medium. The resulting solution is then reduced by Zn and dil.HCl. The reduced solution is again oxidized completely by 40 mL of 0.02 M $KMnO_4$. Calculate normality of FeC_2O_4 and $FeSO_4$ in mixture.

Q.19 1 g of most sample of KCl and $KClO_3$ was dissolved in water to make 250 mL solution, 25 mL of this solution was treated with SO_2 to reduce chlorate to chloride and excess of SO_2 was removed by boiling. The total chloride was precipitated as silver chloride. The weight of precipitate was 0.1435 g. In another experiment, 25 mL of original solution was heated with 30 mL of 0.2 N ferrous sulphate solution and unreacted ferrous sulphate required 37.5 mL of 0.08 N solution of an oxidant for complete oxidation. Calculate the molar ratio of chlorate to chloride in the given mixture. Fe^{2+} reacts with ClO_3^- according to equation.

$ClO_3^- + 6Fe^{2+} + 6H^+ \longrightarrow Cl^- + 6Fe^{3+} + 3H_2O$

Q.20 (i) $CuSO_4$ reacts with KI in acidic medium to liberate I_2

$2CuSO_4 + 4KI \longrightarrow Cu_2I_2 + 2K_2SO_4 + I_2$

(ii) Mercuric per iodiate $Hg_5(IO_6)_2$ reacts with a mixture of KI and HCl following the equation:

$Hg_5(IO_6)_2 + 34KI + 24HCl \longrightarrow$

$$5K_2HgI_4 + 8I_2 + 24KCl + 12H_2O$$

(iii) The liberated iodine is titrated against $Na_2S_2O_3$ solution. One mL of which is equivalent to 0.0499 g of $CuSO_4.5H_2O$. What volume in mL of $Na_2S_2O_3$ solution will be required to react with I_2 liberated from 0.7245 g of $Hg_5(IO_6)_2$? M. wt. of $Hg_5(IO_6)_2$ = 1448.5 and M. wt. of $CuSO_4.5H_2O$ = 249.5.

Q.21 1.249 g of a sample of pure $BaCO_3$ and impure $CaCO_3$ containing some CaO was treated with dil. HCl and it evolved 168 mL of CO_2 at NTP. From this solution $BaCrO_4$ was precipitated, filtered an washed. The dry precipitate was dissolved in dilute H_2SO_4 and dilute to 100 mL. 10 mL of this solution when treated with KI solution liberated iodine which required exactly 20 mL of 0.05 N $Na_2S_2O_3$. Calculate percentage of CaO in the sample.

Q.22 A 10 g mixture of Cu_2S and CuS was treated with 200 mL of 0.75 M MnO_4^- in acid solution producing SO_2, Cu^{2+} and Mn^{2+}. The SO_2 was boiled off and the excess of MnO_4^- was treated with 175 mL of 1 M Fe^{2+} solution. Calculate percentage of CuS in original mixture.

Q.23 For estimating ozone in the air, a certain volume of air is passed through an acidified or neutral KI solution when oxygen is evolved and iodide is oxidised to give iodine. When such a solution is acidified, free iodine is evolved which can be titrated with standard $Na_2S_2O_3$ solution. In an experiment 10 litre of air at 1 atm and 27°C were passed through an alkaline KI solution, at the end, the iodine entrapped in a solution on titration as above required 1.5 mL of 0.01 N $Na_2S_2O_3$ solution. Calculate volume percentage of O_3 in sample.

Q.24 30 mL of an acidified solution of 1.5 N MnO_4^- ions, 15 mL of 0.5 N oxalic acid and 15 mL of 0.4 N ferrous salt solution are added together. Find the molarities of MnO_4^- and Fe_3^+ ions in the final solution?

Q.25 (i) 25 mL of H_2O_2 solution were added to excess of acidified solution of KI. The iodine so liberated required 20 mL of 0.1 N $Na_2S_2O_3$ for titration. Calculate the strength of H_2O_2 in terms of normality, percentage and volume.

(ii) To a 25 mL H_2O_2 solution, excess of acidified solution of KI was added. The iodine liberated required 20 mL of 0.3 N sodium thiosulphate solution. Calculate the volume strength of H_2O_2 solution.

Q.26 An aqueous solution containing 0.10 g KIO_3 (formula weight = 214.0) was treated with an excess of KI solution. The solution was acidified with HCl. The liberated I_2 consumed 45 mL of thiosulphate solution to decolorise the blue starch-iodine complex. Calculate the molarity of the sodium thiosulphate solution.

Q.27 A sample of $MnSO_4.4H_2O$ is strongly heated in air. The residue (Mn_3O_4) left was dissolved in 100 mL of 0.1 N $FeSO_4$ containing dil. H_2SO_4. This solution was completely reacted with 50 mL of $KMnO_4$ solution. 25 mL of this $KMnO_4$ solution was completely reduced by 30 mL of 0.1 N $FeSO_4$ solution. Calculate the amount of $MnSO_4.4H_2O$ in sample.

Q.28 Write complete balanced equation for the following in acidic medium by ion-electron method:

(i) $ClO_3^- + Fe^{2+} \rightarrow Cl^- + Fe^{+3} + H_2O$

(ii) $CuS + NO_3^- \rightarrow Cu^{+2} + S_8 + NO + H_2O$

(iii) $S_2O_3^{-2} + Sb_2O_3 \rightarrow SbO + H_2SO_3$

(iv) $HCl + KMnO_4 \longrightarrow Cl_2 + KCl + MnCl_2 + H_2O$

(v) $KClO_3 + H_2SO_4 \longrightarrow KHSO_4 + HClO_4 + ClO_3 + H_2O$

(vi) $HNO_3 + HBr \longrightarrow NO + Br_2 + H_2O$

(vii) $IO_4^- + I^- + H^+ \longrightarrow I_2 + H_2O$

Q.29 Balance the following equations by oxidation method:

(i) $Cu + NO_3^- + ... \longrightarrow Cu^{+2} + NO_2 + ...$ (Acid medium)

(ii) $Cl_2 + IO_3^- + OH^- \longrightarrow IO_4^- + ... + H_2O$ (Basic medium)

(iii) $H_2S + K_2CrO_4 + H_2SO_4 \longrightarrow ......$ (Acid medium)

(iv) $Fe^{+2} + MnO_4^- \longrightarrow Fe^{+3} + Mn^{+2} + ...$ (Acid medium)

(v) $KMnO_4 + H_2SO_4 + H_2O_2 \longrightarrow K_2SO_4 + MnSO_4 + H_2O + ...$ (Acid medium)

(vi) $MnO_2 + H_2O_2 \longrightarrow MnO_4^- + H_2O$ (Basic medium)

Q.30 Write complete balance equation for the following in basic medium by ion-electron method:

(i) $S_2O_4^{-2} + Ag_2O \rightarrow Ag + SO_3^{-2}$

(ii) $Cl_2 + OH^- \rightarrow Cl^- + ClO^-$

(iii) $H_2 + ReO_4^- \rightarrow ClO_2^- + Sb(OH)_6^-$

(iv) $I_2 + OH^- \rightarrow I^- + IO_3^-$

(v) $MnO_4^- + Fe^{+2} \rightarrow Mn^{+2} + Fe^{+3}$

Exercise 2

Mole Concept

Single Correct Choice Type

Q.1 'x' gms of an element 'A' on heating in a jar of chlorine give 'y' gms of ACl_2 the atomic weight of element A is

(A) $\left(\frac{x}{y-71}\right) \times \frac{35.5}{2}$ (B) $\frac{71x}{y-71}$

(C) $\frac{35.5x}{y-71}$ (D) None of these

Q.2 The amount of H_2SO_4 present in 1200 mL of 0.2 N solution is

(A) 10.76 gms (B) 11.76 gms

(C) 12.76 gms (D) 14.76 gms

Q.3 An iodized salt contains 0.5% of NaI. A person consumes 3 gm of salt everyday. The number of iodide ions going into his body everyday is

(A) 10^{-4} (B) 6.02×10^{-4}

(C) 6.02×10^{19} (D) 6.02×10^{23}

Assertion Reasoning Type

(A) If both statement-I and statement-II are true and statement-II is the correct explanation of statement-I, the mark (A).

(B) If both statement-I and statement-II are true and statement-II is not the correct explanation of statement -I, the mark (B).

(C) If statement-I is true but statement-II is false, then mark (C).

(D) If both statement-I and statement-II are false, then mark (D).

Q.4 Statement-I: 0.28 g of N_2 has equal volume as 0.44 g of another gas at same conditions of temperature and pressure.

Statement-II: Molecular mass of another gas is 44 g mol^{-1}.

Q.5 Statement-I: Boron has relative atomic mass 10.81.

Statement-II: Borons two isotopes, $^{10}_{5}B$ and $^{11}_{5}B$ and their relative abundance is 19% and 81%.

Q.6 Statement-I: The percentage of nitrogen in urea is 46%.

Statement-II: Urea is ionic compound.

Q.7 Statement-I: The oxidation state of central sulfur of $Na_2S_2O_3$ is +6.

Statement-II: Oxidation state of an element should be determined form structure.

Q.8 Statement-I: Molarity of a solution and molality of a solution both change with density.

Statement-II: Density of the solution changes when percentage by mass of solution changes.

Q.9 Statement-I: $2A + 3B \rightarrow C$, 4/3 moles of 'C' are always produced when 3 moles of 'A' and 4 moles of 'B' are added.

Statement-II: 'B' is the liming reactant for the given data.

Multiple Correct Choice Type

Q.10 Given following series of reactions:

(i) $NH_3 + O_2 \rightarrow NO + H_2O$

(ii) $NO + O_2 \rightarrow NO_2$

(iii) $NO_2 + H_2O \rightarrow HNO_3 + HNO_2$

(iv) $HNO_2 \rightarrow HNO_3 + NO + H_2O$

Select the correct option(s):

(A) Moles of HNO_3 obtained is half of moles of Ammonia used if HNO_2 is not used to produce HNO_3 by equation (iv)

(B) 100/6% more HNO_3 will be produced if HNO_2 is used to produce HNO_3 by reaction (iv) than if HNO_2 is not used to produce HNO_3 by reaction (iv)

(C) If HNO_2 is used to produce HNO_3 then 1/4th of total is produced by reaction (iv)

(D) Moles of NO produced in reaction (iv) is 50% of moles of total HNO_3 produced.

Comprehension Type

Paragraph 1: Normality is number of gram equivalents dissolved per litre of solution. It changes with change in temperature. In case of monobasic acid, normality and molarity are equal but in case of dibasic acid, normality is twice the molarity. In neutralization and redox reactions, number of mill equivalents of reactants as well as products are always equal.

Q.11 On heating a litre of a $\frac{N}{2}$ HCl solution, 2.750 g of HCl is lost and the volume of solution becomes 750 mL. The normality of resulting solution will be

(A) 0.58 (B) 0.75 (C) 0.057 (D) 5.7

Q.12 The volume of 0.1 M Ca(OH) required to neutralize 10 mL of 0.1 N HCl will be

(A) 10 mL (B) 20 mL (C) 5 mL (D) 40 mL

Q.13 Molarity of 0.5 N Na_2CO_3 is

(A) 0.25 (B) 1.0 (C) 0.5 (D) 0.125

Q.14 6.90 N KOH solution in water contains 30% by weight of KOH. The density of solution will be

(A) 1.288 (B) 2.88 (C) 0.1288 (D) 12.88

Q.15 The amount of ferrous ammonium sulphate required to prepare 250 mL of 0.1 N solution is

(A) 1.96 g (B) 1.8 g (C) 9.8 g (D) 0.196 g

Paragraph 2: A 4.925 g sample of a mixture of $CuCl_2$ and $CuBr_2$ was dissolved in water and mixed thoroughly with a 5.74 g portion of AgCl. After the reaction and solid, a mixture of AgCl and AgBr, was filtered, washed, and dried. Its mass was found to be 6.63 g.

Q.16

(1) % By mass of $CuBr_2$ in original mixture is

(A) 2.24 (B) 74.5 (C) 45.3 (D) None

(2) % By mass of Cu in original mixture is

(A) 38.68 (B) 19.05 (C) 3.86 (D) None

(3) % by mole of AgBr in dried precipitate is

(A) 25 (B) 50 (C) 75 (D) 60

(4) No. of moles of Cl^- ion present in the solution after precipitate ion are

(A) 0.06 (B) 0.02 (C) 0.04 (D) None

Paragraph 3: Water is added to 3.52 grams of UF_6. The products are 3.08 grams of a solid [containing only U, O and F] and 0.8 gram of a gas only. The gas [containing fluorine and hydrogen only], contains 95% by mass fluorine.

[Assume that the empirical formula is same as molecular formula.]

Q.17

(1) The empirical formula of the gas is

(A) HF_2 (B) H_2F (C) HF (D) HF_3

(2) The empirical formula of the solid product is

(A) UF_2O_2 (B) UFO_2 (C) UF_2O (D) UFO

(3) The percentage of fluorine of the original compound which is converted into gaseous compound is

(A) 66.66% (B) 33.33% (C) 50% (D) 89.9%

Match the Columns

Q.18 One type of artificial diamond (commonly called YAG for yttrium aluminium garnet) can be represented by the formula $Y_3Al_5O_{12}$. [Y = 89, Al = 27]

Column I	Column II
(A) Y	(p) 22.73%
(B) Al	(q) 32.32%
(C) O	(r) 44.95%

Q.19 The recommended daily does is 17.6 milligrams of vitamin C (ascorbic acid) having formula $C_6H_8O_6$. Match the following. Given: $N_A = 6 \times 10^{23}$

Column I	Column II
(A) O-atoms present	(p) 10^{-4} mole
(B) Moles of vitamin C in 1 gm of vitamin C	(q) 5.68×10^{-3}
(C) Moles of vitamin C in 1 gm should be consumed daily	(r) 3.6×10^{20}

Q.20 If volume strength of H_2O_2 solution is 'X-V' then its

Column I	Column II
(i) Strength in g/L	(p) $\frac{X}{11.2}$
(ii) Volume strength X	(q) $\frac{X}{5.6}$
(iii) Molarity	(r) $\frac{17X}{5.6}$
(iv) Normality	(s) $5.6 \times N$

(A) (i) - r, (ii) - p, (iii) - s, (iv) - q

(B) (i) - s, (ii) - p, (iii) - q, (iv) - p

(C) (i) - r, (ii) - s, (iii) - p, (iv) - q

(D) (i) - r, (ii) - q, (iii) - s, (iv) - p

Q.21 Match the entries in column I with entries in column II and then pick out correct options:

Column I	Column II
(i) M_R on mixing two acidic solutions	(p) $\frac{x \times d \times 10}{M_{solute}}$
(ii) M_R on mixing two basic solutions	(q) $n \times M \times V$ mL
(iii) M_R on mixing two acidic and basic solutions	(r) $\frac{M_1V_1 - M_2V_2}{V_1 + V_2}$
(iv) Milliequivalent	(s) $\frac{M_1V_1}{V_2}$
(v) Molarity	(t) $\frac{M_1V_1 + M_2V_2}{V_1 + V_2}$

(A) (i) - p, (ii) - r, (ii) - p, (iv) - q, (v) - s

(B) (i) - t, (ii) - t, (ii) -r, (iv) -q, (v) - p, s

(C) (i) - q, (ii) - p, (ii) -q, (iv) - r, (v) -q

(D) (i) - p, (ii) - q, (ii) - q, (iv) - r, (v) - r

Redox Reactions

Single Correct Choice Type

Q.1 One mole of N_2H_4 loses ten moles of electrons to form a new compound Y. Assuming that all the nitrogen appears in the new compound, what is the oxidation state of nitrogen in Y? (There is no change in the oxidation state of hydrogen)

(A) −1 (B) −3 (C) +3 (D) +5

Q.2 Which is best reducing agent

(A) F^- (B) Cl^- (C) Br^- (D) I^-

Q.3 In the alumino thermite process, aluminium acts as

(A) An oxidizing agent (B) A flux

(C) Reducing agent (D) A solder

Q.4 Zinc-copper couple that can be used as a reducing agent is obtained by

(A) Mixing zinc dust and copper gas

(B) Zinc coated with copper

(C) Copper coated with zinc

(D) Zinc and copper wires welded together

Q.5 In the following equations value of X is

$$ClO_3^- + 6H^+ + X \rightarrow Cl^- + 3H_2O$$

(A) $4e^-$ (B) $5e^-$ (C) $6e^-$ (D) $7e^-$

Q.6 The brown ring complex compound is formulated as $[Fe(H_2O)_5(NO)^+]SO_4$. The oxidation state of iron is

(A) 1 (B) 2 (C) 3 (D) 0

Q.7 Oxidation state of oxygen atom in potassium superoxide is

(A) –1/2 (B) –1 (C) –2 (D) 0

Q.8 In the following reaction

$3Br_2 + 6CO_3^{-2} + 3H_2O \rightarrow 5Br^- + 6HCO_3^- + BrO_3^-$

(A) Bromine is both reduced and oxidised

(B) Bromine is neither reduced nor oxidised

(C) Bromine is oxidised and carbonate is reduced

(D) Bromine is reduced and water is oxidised

Comprehension Type

Paragraph 1: The redox titration involve the chemical reaction between the oxidising agent and reducing agent in aqueous solutions under suitable conditions. Titrations involve the direct use of iodine as oxidising agent are known as iodimetric titrations while those titrations involving indirect use of iodine are known as iodometric titrations. These titrations are used for the estimation of oxidising agents like $KMnO_4$, $K_2Cr_2O_7$, $CuSO_4$ etc.

Q.9 50 mL of an aqueous solution of H_2O_2 was treated with excess of KI solution and the iodine so liberated quantitatively required 20 mL of 0.1 N solution of hypo. This titration is known as:

(A) Iodometric titration (B) Iodimetric titration

(C) Potassium iodide titration (D) All of these

Q.10 In the above problem, concentration of H_2O_2 in gm/litre is:

(A) 6.8 (B) 0.68 (C) 0.068 (D) 0.34

Q.11 0.5 gm sample of pyrolusite (MnO_2) is treated with HCl, the Cl_2 gas evolved is treated with KI, the violet vapours evolved are absorbed in 30 mL 0.1 N $Na_2S_2O_3$ solution percentage purity of pyrolusite sample is

(A) 30% (B) 50% (C) 36% (D) 26.1%

Q.12 Arsenite gets converted into arsenate by using iodine, valency factor for Arsenite and Iodine are respectively

(A) 2 and 2 (B) 2 and 1 (C) 1 and 2 (D) 5 and 2

Paragraph 2: Oxidation and reduction process involves the transaction of electrons. Loss of electrons is oxidation and the gain of electrons is reduction. It is thus obvious that in a redox reaction, the oxidant is reduced by accepting the electrons and the reductant is oxidised by losing electrons. The reactions in which a species disproportionate into two oxidation states (lower and higher) are called disproportionation reactions. In electrochemical cells, redox reaction is involved, i.e., oxidation takes place at anode and reduction at cathode.

Q.13 The reaction: $Cl_2 \rightarrow Cl^- + ClO^-_3$ is

(A) Oxidation

(B) Reduction

(C) Disproportionation

(D) Neither oxidation nor reduction

Q.14 In the reaction: $I_2 + 2S_2O_3^{-2} \rightarrow 2I^- + S_4O_6^{-2}$

(A) I_2 is reducing agent

(B) I_2 is oxidising agent

(C) $S_2O_3^-$ is reducing agent

(D) $S_2O_3^{-2}$ is oxidising agent

Q.15 Determine the change in oxidation number of sulphur is H_2S and SO_2 respectively in the following reaction: $2H_2S + SO_2 \rightarrow 2H_2O + 3S$

(A) 0, +2 (B) +2, –4 (C) –2, +2 (D) +4, 0

Multiple Correct Choice Type

Q.16 Which of the following reactions is/are correctly indicated?

Oxidant Reductant

(A) $HNO_3 + Cu \longrightarrow Cu^{2+} + NO_2$

(B) $2Zn + O_2 \longrightarrow ZnO$

(C) $Cl_2 + 2Br^- \longrightarrow 2Cl^- + Br_2$

(D) $4Cl_2 + CH_4 \longrightarrow CCl_4 + 4HCl$

Assertion and Reasoning Type

Each of the questions given below consist of statement-I and statement-II. Use the following Key to choose the appropriate answer.

(A) If both statement-I and statement-II are true, and statement-II is the correct explanation of statement-I.

(B) If both statement-I and statement-II are true, and statement-II is not the correct explanation of statement-I.

(C) If statement-I is true but statement-II is false.

(D) If statement-I is false but statement-II is true.

Q.17 Statement-I: In CrO_5 oxidation number of Cr is +6.

Statement-II: CrO_5 has butterfly structure in which

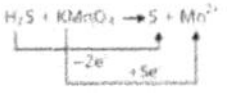

peroxide bonds are present.

Q.18 Statement-I: In PbO_4 all Pb has +8/3 oxidation number.

Statement-II: PbO_4 is mixed oxide of PbO and PbO_2

Q.19 Statement-I: $HClO_4$ is only oxidising agent.

Statement-II: Cl is most electro-negative element in H, Cl and O.

Q.20 Statement-I: In FeS_2 oxidation number of iron is +4.

Statement-II: In FeS_2 $(S^- - S^-)$ linkage is present.

Q.21 Statement-I: In given reaction H_2O_2 is oxidising & reducing agent

$$H_2O_2 \longrightarrow H_2O + \frac{1}{2}O_2$$

Statement-II: In H_2O_2 is a bleaching reagent.

Q.22 Statement-I: In basic medium colour of $K_2Cr_2O_7$ is changed from orange to yellow.

Statement-II: In basic medium $K_2Cr_2O_7$ is changed in chromate ion.

Q.23 Statement-I: $I_2 \longrightarrow IO_3^- + I^-$.

This reaction is disproportionate reaction.

Statement-II: Oxidation number of I can vary from −1 to +7.

Match the Columns

Q.24 Match the entries in column I with entries in column II and then pick out correct options:

Column I	Column II
(A) Increase in oxidation number	(p) Loss of electrons
(B) Decrease in oxidation number	(q) Redox reaction
(C) Oxidising agent	(r) Fractional oxidation number
(D) Reducing agent	(s) Zero oxidation number
(E) $2Cu^+ \rightarrow Cu^{2+} + Cu$	(t) Simple neutralisation reaction
(F) $MnO_2 + 4HCl \rightarrow MnCl_2$	(u) Gain of electrons + Cl_2 + $2H_2O$
(G) $\underline{Mn}_3O_4$	(v) Disproportion-ation
(H) $\underline{C}H_2Cl_2$	(w) Oxidation
(I) $NaOH + HCl \rightarrow NaCl + H_2O$	(x) Reduction

Q.25 Match the reactions in column I with nature of the reactions/type of the products in Column II.

Column I	Column II
(A) $O_2^- \rightarrow O_2 + O_2^{-2}$	(p) Redox reaction
(B) $CrO_4^{-2} + H^+ \rightarrow$	(q) One of the products has trigonal planar structure
(C) $MnO_4^- + NO_2^- + H^+ \rightarrow$	(r) Dimeric bridged tetrahedral metal ion
(D) $NO_3^- + H_2SO_4 + Fe^{2+} \rightarrow$	(s) Disproportionation

Previous Years' Questions

Mole Concept

Q.1 Naturally occurring boron consists of two isotopes whose atomic weights are 10.01 and 11.01. The atomic weight of natural boron is 10.81. Calculate the percentage of each isotope in natural boron. ***(1978)***

Q.2 The vapour density (hydrogen = 1) of a mixture consisting of NO_2 and N_2O_4 is 38.3 at 26.7°C. Calculate the number of moles of NO_2 in 100 g of the mixture. ***(1979)***

Q.3 A solid mixture (5.0 g) consisting of lead nitrate and sodium nitrate was heated below 600°C until the weight of the residue was constant. If the loss in weight is 28.0 percent, find the amount of lead nitrate and sodium nitrate in the mixture. **(1990)**

Q.4 'A' is a binary compound of a univalent metal. 1.422 g of A reacts completely with 0.321 g of sulphur in an evacuated and sealed tube to give 1.743 g of a white crystalline solid B, that forms a hydrated double salt, C with $Al_2(SO_4)_3$ Identify A, B and C. **(1994)**

Q.5 Calculate the molality of 1.0 L solution of 93% H_2SO_4, (weight/volume). The density of the solution is 1.84 g/mL. **(1990)**

Q.6 20% surface sites have adsorbed N_2. On heating N_2 gas evolved from sites and were collected at 0.001 atm and 298 K in a container of volume is 2.46 cm^3. Density of surface sites is 6.023×10^{14} /cm^2 and surface area is 1000 cm^2, find out the number of surface sites occupied per molecule of N_2. **(2005)**

Q.7 If 0.50 mole of $BaCl_2$ is mixed with 0.20 mole of Na_3PO_4 the maximum number of moles of $Ba_3(PO_4)_2$ that can be formed is **(1981)**

(A) 0.70 (B) 0.50 (C) 0.20 (D) 0.10

Q.8 In the standardization of $Na_2S_2O_3$ using $K_2Cr_2O_7$ is **(2001)**

(A) $\left(\dfrac{\text{Molecular Weight}}{2}\right)$

(B) $\left(\dfrac{\text{Molecular Weight}}{6}\right)$

(C) $\left(\dfrac{\text{Molecular Weight}}{3}\right)$

(D) Same as molecular weight

Q.9 The difference in the oxidation numbers of the two types of sulphur atoms in $Na_2S_4O_6$ is. **(2011)**

Read the following questions and answer as per the direction given below:

(A) Statement-I is true; statement-II is true; statement-II is the correct explanation of statement-I.

(B) Statement-I is true; statement-II is true; statement-II is not the correct explanation of statement-I.

(C) Statement-I is true; statement-II is false.

(D) Statement-I is false; statement-II is true.

Q.10 Statement-I: In the titration of Na_2CO_3 with HCl using methyl orange indicator, the volume required at the equivalent point is twice that of the acid required using phenolphthalein indicator.

Statement-II: Two moles of HCl are required for the complete neutralization of one mole of Na_2CO_3. **(1991)**

Q.11 2.68×10^{-3} moles of a solution containing an ion A^{n+} require 1.61×10^{-3} moles of MnO_4^- for the oxidation of A^{n+} to AO_3^- in acidic medium. What is the value of n? **(1984)**

Q.12 A 5.0 cm^3 solution of H_2O_2 liberates 0.508 g of iodine from an acidified KI solution. Calculate the strength of H_2O_2 solution in terms of volume strength at STP. **(1995)**

Q.13 A solution of 0.2 g of a compound containing Cu^{2+} and $C_2O_4^{2-}$ ions on titration with 0.02 M $KMnO_4$ in presence of H_2SO_4 consumes 22.6 mL of the oxidant. The resultant solution is neutralized with $NaCO_3$, acidified with dilute acetic acid and treated with excess KI. The liberated iodine requires 11.3 mL of 0.05 M $Na_2S_2O_3$ solution for complete reduction. Find out the mole ratio of Cu^{2+} to $C_2O_4^{-2}$ in the compound. Write down the balanced redox reactions involved in the above titrations. **(1991)**

Q.14 A mixture of $H_2C_2O_4$ (oxalic acid) and $NaHC_2O_4$ weighing 2.02 g was dissolved in water and the solution made up to one litre. Ten millilitres of the solution required 3.0 mL of 0.1 N sodium hydroxide solution for complete neutralization. In another experiment, 10.0 mL of the same solution, in hot dilute sulphuric acid medium, required 4.0 mL of 0.1 N potassium permanganate solution for complete reaction. Calculate the amount of $H_2C_2O_4$ and $NaHC_2O_4$ in the mixture. **(1990)**

Q.15 The unbalanced chemical reactions given in list I show missing or condition which are provided in list II. Match list I with list II and select the correct answer using the code given below the lists: **(2013)**

	List I		List II
(i)	$PbO_2 + H_2SO_4 \xrightarrow{?} PbSO_4 + O_2 + \text{other product}$	(p)	NO
(ii)	$Na_2S_2O_3 + H_2O \xrightarrow{?} NaHSO_4 + \text{other product}$	(q)	I_2
(iii)	$N_2H_4 \xrightarrow{?} N_4 + \text{other product}$	(r)	Warm
(iv)	$XeF_2 \xrightarrow{?} Xe + \text{other product}$	(s)	Cl_2

Codes:

	(i)	(ii)	(iii)	(iv)
(p)	4	2	3	1
(q)	3	2	1	4
(r)	1	4	2	3
(s)	3	4	2	1

Q.16 For the reaction $I^- + ClO_3^- + H_2SO_4 \rightarrow Cl^- + HSO_4^- + I_2$
The correct statement(s) in the balanced equation is/are: ***(2014)***

(A) Stoichiometric coefficient of HSO_4^- is 6.

(B) Iodide is oxidized.

(C) Sulphur is reduced.

(D) H_2O is one of the products

Q.17 Hydrogen peroxide in its reaction with KIO_4 and NH_2OH respectively, is acting as a ***(2014)***

(A) Reducing agent, oxidising agent

(B) Reducing agent, reducing agent

(C) Oxidising agent, oxidising agent

(D) Oxidising agent, reducing agent

Important Questions

JEE Main/Boards

Exercise 1

Mole Concept

Q.1 Q.3 Q.7
Q.11 Q.13 Q.18
Q.21 Q.29

Redox

Q.3 (C) Q.4 (F)

Exercise 2

Mole Concept

Q.1 Q.6 Q.10 Q.15
Q.19 Q.23 Q.29 Q.33

Redox

Q.1 Q.8 Q.15
Q.24 Q.25

Previous Years' Questions

Mole Concept and Redox

Q.1 Q.5 Q.14

JEE Advanced/Boards

Exercise 1

Mole Concept

Q.3 Q.7 Q.14
Q.15 Q.22 Q.26
Q.28

Redox

Q.2 Q.5 Q.13
Q.23 Q.17

Exercise 2

Mole Concept

Q.2 Q.7 Q.13
Q.16

Redox

Q.1 Q.6 Q.9
Q.16 Q.19 Q.21

Previous Years' Questions

Mole Concept and Redox

Q.3 Q.14

Answer Key

JEE Main/Boards

Exercise 1

Mole Concept

Q.1 (i) 68.125 Kg (ii) 7.63 Kg/m^2 (iii) 1.72 m

Q.2 7.818×10^{22} atoms

Q.3 4.82×10^{22} atoms

Q.4 0.437

Q.5 6.02×10^{10} g / cm^3

Q.6 0.1 M HNO_3

Q.7 14.0 tablets

Q.8 0.1 M

Q.9 $(C_3O_4) = C_{12}O_{16}$

Q.10 0.44 g

Q.11 35.5: 1, 35.5: 1, 1: 1

Q.12 0.7985, 0.798

Q.13 M_2O_3

Q.14 K_2MnO_4

Q.15 (i) 55.5 M

Q.16 (i) 3.24×10^{-22} g/molecule

(ii) 4.09×10^{22}

Q.17 55.56 moles

Q.18 7.098×10^7 g mol^{-1}

Q.29 2.03 kg

Q.20 10 mol

Q.21 260 mL

Q.22 $\frac{x}{y} = 5$

Q.23 24

Q.24 0.394 g

Q.25 79.714 gm

Q.26 0.744

Q.27 1.125, 1.99, 2.00

Q.28 0.302 M

Q.29 1800 mL

Redox Reaction

Q.1 (i) 5/2 (ii) +2 (iii) +3 (iv) +6 (v) +2 (vi) +6

(vii) +2 (viii) –2 (ix) +5/2 (x) +1 (xi) +3 (xii) +8/3

(xiii) +7 (xiv) 0 (xv) +5 (xvi) +5

Q.2 (i) $5Br^- + BrO_3^- + 6H^+ \longrightarrow 3Br_2 + 3H_2O$,

(ii) $8Cr_2O_7^{2-} + 24H_2S + 16H^+ \longrightarrow 8Cr_2O_3 + 3S_8 + 32H_2O$

(iii) $Au + 2NO_3^- + 4Cl^- + 4H^+ \longrightarrow AuCl_4^- + 2NO_2 + 2H_2O$

(iv) $3Cu_2O + 14H^+ + 2NO_3^- \longrightarrow 6Cu^{+2} + 2NO + 7H_2O$

(v) $3MnO_4^- + 4H^+ \longrightarrow 2MnO_4^{-1} + MnO_2 + 2H_2O$

(vi) $2Cu^{+2} + SO_2 + 2H_2O \longrightarrow 2Cu^+ + 4H^+ + SO_4^{-2}$

(vii) $5Cl_2 + I_2 + 3H_2O \longrightarrow 2IO_3^- + 10Cl^- + 6H^+$

(viii) $5Fe(CN)_6^{-4} + 188H^+ + 61MnO_4^- \longrightarrow 5Fe^{3+} + 30CO_2 + 30NO_3^- + 61Mn^{+2} + 94H_2O$

(ix) $6Cu_3P + 124H^+ + 11Cr_2O_7^{-2} \longrightarrow 18Cu^{+2} + 6H_3PO_4 + 22Cr^{+3} + 53H_2O$

Q.3 (a) $2Cu^{+2} + 2I^- \longrightarrow 2Cu^+ + I_2$

(i) $6Fe_3O_4 + 2MnO_4^{-1} + 8H_2O \longrightarrow 9Fe_2O_3 + 2MnO_3 + 16OH^-$

(ii) $3C_2H_5OH + 2MnO_4^- + OH^- \longrightarrow 3C_2H_3O^- + 2MnO_2(s) + 5H_2O$

(iii) $2CrI_3 + 27H_2O_2 + 10OH^- \longrightarrow 2CrO_4^{-2} + 6IO_4^- + 32H_2O$

(iv) $258KOH + K_4Fe(CN)_6 + 61Ce(NO_3)_4 \longrightarrow 61Ce(OH)_3 + Fe(OH)_3 + 36H_2O + 6K_2CO_3 + 250KNO_3$

Q.4 (i) $H_2O_2 + 2I^- + 2H^+ \longrightarrow 2H_2O + I_2$

(ii) $2Cu^{+2} + 2HI \longrightarrow 2Cu^+ + I_2 + H_2O$

(iii) $3CuO + 2NH_3 \longrightarrow 3Cu + N_2 + 3H_2O$

(iv) $3H_2SO_3 + Cr_2O_7^{-2} + 8H^+ \longrightarrow 3H_2SO_4 + 2Cr^{+3} + 4H_2O$

(v) $\therefore\ 2Cr_2O_7^{2-} + 9C_2H_4O + 16H^{\oplus} \longrightarrow 9C_2H_4O_2 + 4Cr^{+3} + 8H_2O$

(vi) $2SbCl_3 + KIO_3 + 6HCl \longrightarrow 2SbCl_5 + ICl + 3H_2O + KCl$

(vii) $As_2S_5 + 2HNO_3 \longrightarrow 5H_2SO_4 + 40NO_3 + 2H_5AsO_4 + 12H_2O$

Exercise 2

Mole Concept

Single Correct Choice Type

Q.1 B	**Q.2** C	**Q.3** B	**Q.4** B	**Q.5** B	**Q.6** B	**Q.7** D
Q.8 C	**Q.9** A	**Q.10** A	**Q.11** C	**Q.12** A	**Q.13** B	**Q.14** C
Q.15 A	**Q.16** C	**Q.17** D	**Q.18** A	**Q.19** B	**Q.20** C	**Q.21** A
Q.22 C	**Q.23** B	**Q.24** C	**Q.25** B	**Q.26** C	**Q.27** D	**Q.28** A
Q.29 C						

Redox Reaction

Single Correct Choice Type

Q.1 A	**Q.2** D	**Q.3** A	**Q.4** C	**Q.5** C	**Q.6** C	**Q.7** B
Q.8 C	**Q.9** B	**Q.10** C	**Q.11** D	**Q.12** C	**Q.13** C	**Q.14** B
Q.15 C	**Q.16** D	**Q.17** C	**Q.18** D	**Q.19** A	**Q.20** A	**Q.21** C
Q.22 A	**Q.23** D	**Q.24** B	**Q.25** C	**Q.26** A	**Q.27** A	**Q.28** D
Q.29 C	**Q.30** D	**Q.31** C	**Q.32** C			

Previous Year's Questions

Q.1 A	**Q.2** C	**Q.3** D	**Q.4** B	**Q.5** A	**Q.6** C	**Q.7** A
Q.8 D	**Q.9** B	**Q.10** B	**Q.11** D	**Q.12** A	**Q.13** C	**Q.14** A
Q.15 C	**Q.16** D	**Q.17** C	**Q.18** D	**Q.19** A	**Q.20** A	**Q21.** C
Q.22 D	**Q.23** B	**Q.24** A				

JEE Advanced/Boards

Exercise 1

Mole Concept

Q.1 116.8 gm **Q.2** 9.12 **Q.3** 5.6 **Q.4** 0.597 g

Q.5 $BaCl_2.2H_2O$ = 7.038 g, H_2O = 42.962 g **Q.6** 120 g **Q.7** 20.78%

Q.8 NaOH = 0.06 g per 200 mL, Na_2CO_3 = 0.0265 g per 200 mL

Q.9 0.1185 **Q.10** 45% **Q.11** 61.5 gm **Q.12** 320.3 gm

Q.13 6% **Q.14** 8.097 mL **Q.15** 6.125 g/litre **Q.16** 55.53 litre

Q.17 K_2CO_3 = 96%, Li_2CO_3 = 4% **Q.18** 12.9 gm **Q.19** 1.14 gm

Q.20 SO_4^{2} ion concentration = 6.528 **Q.21** 27.27% **Q.22** C_2H_6 = 0.66, C_2H_4 = 0.34

Q.23 $Pb(NO_3)_2$ = 3.32 g, $NaNO_3$ = 1.68 g

Q.24 0.0075, $[Pb^{2+}]$ = 0.0536 M, $[NO_3^-]$ = 0.32 M, $[Cr^{3+}]$ = 0.0714 M **Q.25** %NaCl = 77.8%

Q.26 (i) $Fe_2O_3 + 2Al \longrightarrow Al_2O_3 + 2Fe$; (ii) 80: 27; (iii) 10,000 units

Q.27 12.15 gm, N_2 = 14.28%, H_2 = 42.86%, NH_3 = 42.86%

Q.28 0.9413 gram **Q.29** (i) 0.5, 0.5; (ii) 0.66, 0.33; (iii) 1, 2 **Q.30** 13.4 mL

Q.31 $AlCl_3$ = 33.33%; $NaHCO_3$ = 50; KNO_3 = 16.67

Q.32 9.4 gm **Q.33** 75% **Q.34** 0.532: 1.00

Redox Reaction

Q.1 (i) +3 (ii) 0 (iii) +7 (iv) +6 (v) +2 (vi) +3 (vii) +6 (viii) 0 (ix) 0 (x) –1

Q.2 Oxidized: KI, Na_2S, NH_4^+; Reduced: $CuSO_4$, SO_2, NO_2^-

Q.3 (i) 8 electrons, (ii) electrons

Q.4 (i) Oxidation number of sulphur in H_2S and SO_2 are respectively –2 and +4.

Q.5 NO_2^- is oxidized to NO_3^- by MnO_4^- (in basic medium) which is reduced to MnO_2.

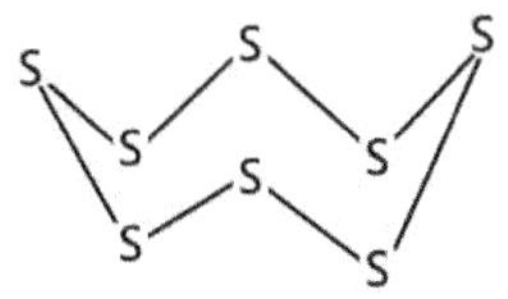

Thus, $MnO_4^- \longrightarrow MnO_3$ oxidation number decreases by 3-units

$NO_2^- \longrightarrow NO_3^-$ oxidation number increases by 2 units

Thus, $2MnO_4^- \equiv 3NO_2^-$ $\quad MnO_4^- \equiv \frac{3}{2}NO_2^- = 1.5\ \text{mol}\ NO_2^-$

Q.6 (i) HSO_3^-; (ii) NO_2^-; (ii) Cl^-

Q.7 (i) $4Zn + 10HNO_3 \longrightarrow 4Zn(NO_3)_2 + N_2O + 5H_2O$ (ii) $6HI + 2HNO_3 \longrightarrow 3I_2 + 2NO + 4H_2O$

Q.8 25 litre **Q.9** 0.5 N, 0.167 M **Q.10** 0.085 M, 0.042 N

Q.11 54% **Q.12** (i) 85%; (ii) 124.79 mL **Q.13** 30.33%

Q.14

(i)	3	2	8	3	1	2	11
(ii)	1	40	40	12	2	5	
(iii)	2	27	64	6	2	54	32
(iv)	3	14	18	14	6	9	

Q.15 (i) $As_2S_3 + 12OH^- + 14H_2O \longrightarrow 2As^{3-}O_4 + 3S^{2-}O_4 + 20H_2O$

(ii) $2CrI_3 + 10OH^- + 27H_2O_2 \longrightarrow 2Cr^{2-}O_4 + 6IO_4^- + 32H_2O$

(iii) $P_4 + 3OH^- + 3H_2O \longrightarrow 3H_2PO_2^- + PH_3$

(iv) $3As_2S_3 + 4H_2O + 10NO_3^- + 10^+ \longrightarrow 6H_3AsO_4 + 9S + 10NO$

Q.16 0.1716 **Q.17** +3 **Q.18** FeC_2O_4 = 0.03 N, $FeSO_4$ = 0.03 N

Q.19 Molar ratio = 1: 1 **Q.20** 40 mL **Q.21** 14% **Q.22** 57.4%

Q.23 1.847×10^{-3}% **Q.24** $[Fe^{3+}] = 0.1M$, $[MnO_4^-] = 0.105M$

Q.25 (i) 0.08 N, 0.136%, 0.448 volume; (ii) 1.344 **Q.26** 0.062 M **Q.27** 1.338 g

Q.28 (i) $6H^+ + ClO_3^- + 5Fe^{+2} \longrightarrow Cl^- + 5Fe^{+3} + 3H_2O$

(ii) $24CuS + 16NO_3^- + 64H^+ \longrightarrow 24Cu^{+2} + 3S_8 + 16NO + 32H_2O$

(iii) $S_2O_3^{-2} + Sb_2O_5 + 4H^+ \longrightarrow 2SbO + 2H_2SO_3$

(iv) $16HCl + 2KMnO_4 \longrightarrow 5Cl_2 + 2KCl + 2MnCl_2 + 8H_2O$

(v) $3KClO_3 + 3H_2SO_4 \longrightarrow 3KHSO_4 + HClO_4 + 2ClO_2 + H_2O$

(vi) $2HNO_3 + 6HBr \longrightarrow 2NO + 3Br_2 + 4H_2O$

(vii) $IO_4^- + 7I^- + 8H^+ \longrightarrow 4I_2 + 4H_2O$

Q.29 (i) $Cu + 4H^+ + 2NO_3^- \longrightarrow Cu^{+2} + 2NO_2 + 2H_2O$

(ii) $Cl_2 + IO_3^- + 2OH \longrightarrow IO_4^- + 2Cl^- + 2H_2O$

(iii) $3H_2S + 2K_2CrO_4 + 5H_2SO_4 \longrightarrow Cr_2(SO_4)_3 + 2K_2SO_4 + 8H_2O + 3S$

(iv) $5Fe^{2+} + MnO_4^- + 8H^+ \longrightarrow 5Fe^{+3} + Mn^{+2} + 4H_2O$

(v) $2KMnO_4 + 3H_2SO_4 + 5H_2O_2 \longrightarrow K_2SO_4 + 2MnSO_4 + 8H_2O + 5O_2$

(vi) $2H_2O_2 + 2MnO_2 + 2OH^- \longrightarrow 2MnO_4^- + 4H_2O$

Q.30 (i) $S_2O_4^{-2} + Ag_2O + 2OH^- \longrightarrow 2Ag + 2SO_3^{-2} + H_2O$

(ii) $Cl_2 + 2OH^- \longrightarrow Cl^- + ClO^- + H_2O$

(iii) $3H_2 + 2ReO_4^- \longrightarrow 2ReO_2 + 2H_2O + 2OH^-$

(iv) $2ClO_2 + SbO_2^- + 2OH^- + 2H_2O \longrightarrow 2ClO_2^- + Sb(OH)_6^{-1}$

(v) $6I_2 + 12OH^- \longrightarrow 10I^- + 2IO_3^- + 6H_2O$

(vi) $MnO_4^- + 5Fe^{+2} + 4H_2O \longrightarrow Mn^{+2} + 5Fe^{+3} + 8OH^-$

Exercise 2

Mole Concept

Single Correct Choice Type

Q.1 D **Q.2** B **Q.3** C

Assertion Reasoning Type

Q.4 B **Q.5** A **Q.6** C **Q.7** B **Q.8** A **Q.9** C

Multiple Correct Choice Type

Q.10 A, C, D

Comprehension Type

Paragraph 1: **Q.11** A **Q.12** C **Q.13** A **Q.14** A **Q.15** C

Paragraph 2: **Q.16** (1) C; (2) A; (3) B; (4) A

Paragraph 3: **Q.17** (1) C; (2) A; (3) A

Match the Columns

Q.18 A $\rightarrow$ r; B $\rightarrow$ p; C $\rightarrow$ q **Q.19** A $\rightarrow$ r; B $\rightarrow$ q; C $\rightarrow$ p **Q.20** C **Q.21** B

Redox Reaction

Single Correct Choice Type

Q.1 C **Q.2** D **Q.3** C **Q.4** D **Q.5** C **Q.6** A **Q.7** A
Q.8 B

Comprehension Type

Paragraph 1: **Q.9** D **Q.10** D **Q.11** D **Q.12** D

Paragraph 2: **Q.13** C **Q.14** B, C **Q.15** B **Q.16** A, B, D

Assertion Reasoning Type

Q.17 A **Q.18** D **Q.19** C **Q.20** D **Q.21** B **Q.22** A **Q.23** B

Match the Columns

Q.24 A → w; B → x C → u; D → p; E → v; F → q; G → r; H → s; I → t

Q.25 A → p, s; B → r; C → p, q; D → p

Previous Year's Questions

Q.1 20% **Q.2** 0.437 **Q.3** 1.7 g **Q.4** A = KO_2 **Q.5** 10.43 **Q.6** 2 **Q.7** D

Q.8 B **Q.9** 5 **Q.10** B **Q.11** 2 **Q.12** 4.48 V

Q.13 Moles of Cu^{2+}; Moles of $C_2O_4^{-2}$ = 1:2 **Q.14** 0.9 g, 1.12 g **Q.15** D **Q.16** A, B, D

Q.17 A

Solutions

JEE Main/Boards

Exercise 1

Mole Concept

Sol 1: (i) 125 pound

1 pound = lb = 545 gm

125 pound = 125 × 545 gm

= 125 × 545 × 10^{-3} kg = 68. 125 kg

(ii) 14 lb/m^2

1lb = 545 gm

In SI units = 14 × 545 × 10^{-3} kg/m^2

= 7. 63 kg/m^2

(iii) 5′8″

(1′ = 12″)

5′8″ = (12″ × 5) + 8″ = 68″

= 68 × 2. 54 cm = 1. 72 m

Sol 2: $M_{avg.}$ = (0. 932) 39 + (0. 068)41 = 39. 136

Mass of 2g-atoms = 2 ×39. 136 gm

Mass of "41K" in 2g-atoms

= 2 × 39. 136 × (0. 068)

Number of atoms = $\frac{2\times 39.136}{41}\times(0.068)\times 6.023\times 10^{23}$

= 7.818 × 10^{22}

Sol 3: Barium phosphate = $Ba_3(PO_4)_2$

No. of oxygen atoms = $\frac{6.025}{602.5}\times 8\times 6.023\times 10^{23}$

= 4.82 × 10^{22} atoms

Sol 4: Molecular weight = Vapour density × 2 = 76. 6

Let's suppose x % mole of NO_2 is there

76. 6 = x (46) + (1 – x) 92

46x = 15. 4

x = 0.3347=33. 47% = mole fraction of NO_2

Total mole = $\frac{100}{76.6}$ = 1. 305 mole

Mole of NO_2 = (0. 3347) × (1. 305) = 0.437 mole.

Sol 5: Correction: radius of fluorine

= 5 × 10^{-3} Å

Mass of nucleous = 19 × 1. 67 × 10^{-27} kg

Volume of nucleous

$= \frac{4}{3} \times \pi \times (5 \times 10{-}3)^3\ cm^3$

Density

$= \frac{3 \times 19 \times 1.67 \times 10^{-27} \times 10^3}{4 \times \pi \times (5 \times 10^{-3})^3 \times (10^{-10})^3} \frac{gm}{cm^3}$

$= \frac{95.19 \times 10^{-27} \times 10^{30}}{1570.79 \times 10^{-9}}\ gm/cm^3$

= 6.02 × 10^{13} gm/cm^3

Sol 6: mole = M_1V_1

For neutralisation $H_{HNO_3} = H_{NaOH}$

$M_1V_1 = M_2V_2$

20 × M_1 = 0. 08M × 25

$M_1 = \frac{0.08 \times 25}{20} = 0.1M$

Sol 7: HCl produced perday

= 3. 0 gm × 2. 5 = 7. 5 gm = $\frac{7.5}{36.5}$mole

Moles of $Al(OH)_3$ in an antacid tablet

$= \frac{400 \times 10^{-3}}{(27+51)} = \frac{0.4}{78}$ mole

3 × mole of $Al(OH)_3$ = moles of HCl

$3 \times n \times \frac{0.4}{78} = \frac{7.5}{36.5}$

$n \approx 14$

Sol 8: HCl + $AgNO_3 \rightarrow$ AgCl + HNO_3

$M \times 10 \times 10^{-3} = \frac{0.1435}{108+35.5}$

M = 0. 1 M.

Sol 9: Lets it is C_xO_y

x(12) + y(16) = 400

$\frac{x(12)}{400} = 0.36$

x = 12; y = 16

the formula is $C_{12}O_{16} = (C_3O_4)_4$

Sol 10: $C_xH_y + O_2 \rightarrow H_2O + CO_2$

0. 1 mole   0. 2 mole

Mole of $H_2O = \frac{0.18}{18} = 0.1$ mole

Mole of CO_2 = 0.02 mole

Mole of O_2 required = $\frac{1}{2}(0.1) + 0.02$ = 0.07 mole

y = Mole of H = (0. 1)2 = 0.2 mole

x = Mole of C = 0.02 mole

Mass of hydro is = (0.02)12 + (0.2)1

= 0.24 + 0.2 = 0.44 gm.

Sol 11: PCl_3 and PH_3

2257%   91. 18%

Mass of Cl in PCl_3 = 3 × 35. 5 = 106. 5

Mass of H in PH_3 = 3 × 1 = 3

$\left.\begin{array}{l} \text{ratio} = \frac{106.5}{3} = 35.5 \\ \text{Ratio of Cl : H} = 35.5 \end{array}\right\}$ equal.

Hence prove.

Sol 12: Exp. (I) CuO → Cu + X

Ratio of mass of

Cu : CuO = $\frac{1.098}{1.375} = 0.7985$

Exp. (II) $\underset{1.179\ gm}{Cu} \xrightarrow{HNO_3} Cu(NO_3)_2 \longrightarrow \underset{1.476\ gm}{CuO}$

Ratio of mass of Cu : CuO = $\frac{1.179}{1.479} = 0.798$

both ratio are same. Hence prove.

Sol 13: M_xO_y

x × atomic mass of M = 0. 540 .... (i)

y × 16 = 1. 020 – 0. 540

y = 0. 03

Dulong-Petit law (atomic mass of M)

× 0. 216 = 5. 83

Atomic mass of M ≅ 27 .... (ii)

Petit x × (27) = 0. 540

x = 0. 02

Formula of metal oxide = M_2O_3.

Sol 14: Let's say substance is 100 gm

Moles of K = $\frac{39.7}{39} = 1.017$

Moles of Mn = $\frac{29.9}{55} = 0.5436$

Moles of O = $\frac{100 - 39.7 - 29.9}{16} = 1.9$

so by seeing on ratio of K : Mn : O

empirical formula is K_2MnO_4.

Sol 15: Molarity = $\frac{\text{no. of moles}}{\text{volume (in litre)}}$

No. of moles = $\frac{\text{mass}}{18} = \frac{1000 \times 0.997}{18}$ per litre.

Molarity = 55. 38 M

Sol 16: (a) Mass = (8 × 12) + (10 × 1) +(4 × 14) + (2 × 16)

= 96 + 10 + 56 + 32

Mass = 194 amu

= $194 \times 1.66 \times 10^{-24}$ gm/molecule

= 3.24×10^{-22} gm/molecule

(b) Molecular mass of Cl_2 = 71

Total no. of electrons in one molecule of Cl_2 = 34

So no. of electrons = $\frac{0.142}{71} \times 34 \times 6.023 \times 10^{23}$

= 4.029×10^{22}

Sol 17: Molarity = moles per litre

$= \frac{\text{Mass}}{18 \times V} = \frac{\text{Density} \times \text{Volume}}{18 \times \text{Volume}} = \frac{1000}{18} = 55.55M$

Sol 18: Volume of plant virus = ($\pi r^2 h$)

= $\pi \times (75 \times 10^{-10})^2 \times (5000 \times 10^{-10})$

= 8.835×10^{-23} m^3

= $8.835 \times 10^{-23} \times (10^{+2})^3$ cm^3

= 8.835×10^{-17} cm^3

Mass = $\frac{8.835 \times 10^{-17}}{0.75}$ gm

= 11.78×10^{-17} gm

= $11.78 \times 6.023 \times 10^{+23} \times 10^{-17}$

= 7.098×10^{7} g mol^{-1}

Sol 19: 25% of heavy water = 0.5 litre.

Mass of heavy water

= $0.5 \times 10^3 \times 1.06$ gm/cm^3 = 530 gm

Mass of normal water

= $1.5 \times 10^3 \times 1$ gm/cm^3 = 1500 gm

Total mass = 2030 gm = 2. 030 kg

Sol 20 $SO_2Cl_2 + 2H_2O \rightarrow H_2SO_4 + 2HCl$

Initially	2. 5	0	0
after	0	2. 5	5

For 2. 5 moles of H_2SO_4, KOH = 5 mole

For 5 mole of HCl, KOH = 5 mole

Total KOH = 5 + 5 = 10 mole

Sol 21: $\underbrace{NH_4Cl + MgCl_2}_{2\,\%\text{ by mass}} + \underbrace{AgNO_3}_{5\%\text{ by mass}} \rightarrow$

$AgCl + NH_4NO_3 + Mg(NO_3)_2$

d = 1040 gm/lit.

Moles of Cl^- in

$NH_4Cl = \frac{2}{(14 + 4 + 35.5)} = 0.0373$ mole

Moles of Cl^- in

$MgCl_2 = 2 \times \frac{2}{(24 + 71)} = 0.0421$ mole

Total mole of Cl^-

= Total mole of $AgNO_3$ required

= 0.0421 + 0.0373 = 0.07940 mole

Mass of $AgNO_3$ = 0.07940 × 170 = 13.49 gm

Mass of solution of

$AgNO_3 \times 13.49 \times \frac{100}{5} = 269.97\text{gm}$

Volume required = $\frac{269.97}{1.04} cm^3 = 259.59\ cm^3$

Sol 22: Oxalic acid = $H_2C_2O_4$

Formic acid = HCOOH

$H_2C_2O_4 + H_2SO_4 \rightarrow SO_4^{2-} + H_4C_2O_4^{2+}$

$HCOOH + H_2SO_4 \xrightarrow{\Delta} SO_4^{2-} + 2H^+ + HCOOH$

$H_4C_2O_4^{2+} + KOH \rightarrow C_2O_4^{2-}$

Sol 23: CH_4 › x mole › Molecular weight − 16

(C_2H_4) → y mole → Molecular weight = 28

Mean molecular weight

$= 20 = \frac{x(16)+y(28)}{x+y}$

$20 = 16 + \frac{12y}{x+y}$

$0.33 = \frac{y}{x+y}$

$\frac{x}{x+y} = 0.66$

x : y = 2 : 1

If x : y = 1 : 2

Then, mean molecular weight = $\frac{1(16)+2(28)}{3} = 24$

Sol 24:

$2KClO_3 \rightarrow 2KCl + 3O_2$

4.369×10^{-3} 6.55×10^{-3}

$4KClO_3 \rightarrow 3KClO_4 + KCl$

3.794×10^{-3} 2.845×10^{-3}

Moles of oxygen produced

$= \frac{146.8\times10^{-3}}{22.4} = 6.55 \times 10^{-3}$ mole

Total mole of $KClO_3 = \frac{1}{39+35.5+48}$

$= 8.163 \times 10^{-3}$ mole

Moles of $KClO_3$ in IInd reaction = 3.794×10^{-3} moles

Moles of $KClO_4$ produced in IInd reaction

$= 2.8496 \times 10^{-3}$ mole

Mass of $KClO_4 = 2.8456 \times 10^{-3} \times (39 + 35.5 + 64)$

= 0.394 gm

Sol 25: Let's say we have 100 gm mix.

$Fe_3O_4 = FeO.Fe_2O_3 \rightarrow FeO + Fe_2O_3$

x gram 0.310 gram 0.680x gram

Initially FeO → (100 − x) gram

Total (FeO) → (100 − x + 0.310 x) gm

= (100 − 0.690 x) gm

$2FeO + \frac{1}{2}O_2 \rightarrow Fe_2O_3$

(105 − 0. 690x)gm

$2\times\left(\frac{100-0.690x}{72}\right) = \frac{(105-0.690x)}{160}$

$(100 \times 0.690x) \times \frac{160}{36} = 105 - 0.690x$

4000 − 27.6 x = 945 − 6.21 x 3055 = 21.39 x

x = oxygen external = 5gm = $\frac{5}{32}$mole

So moles of FeO that was present

$\frac{(100-0.690x)}{56+16} = \frac{4\times5}{32}$

Fe_3O_4 + x = 79.71 gm

FeO = 100 − x = 20.29 gm

Sol 26: $Zn + 2I \rightarrow ZnI_2$

m m

2x moles of Zn = moles of I

(to complete reaction)

$2 \times \frac{m}{65}$ = 2x moles of Zn

$\frac{m}{127}$ = moles of I

Since moles of I < 2x moles of Zn

So Zn will be left unreacted

Zn unreacted = $\frac{m}{65} - \frac{m}{127\times2}$ mole

Mass Zn unreacted = $m - \frac{65}{254}m = 0.744$ m

Sol 27: Mole of $P_4 = \frac{2}{4\times31} = \frac{1}{62}$ mole

Moles of $O_2 = \frac{2}{2\times16} = \frac{1}{16}$ mole

P_4	+ $3O_2$	→ P_4O_6	
x	3x	0	Initial
0	0	x	After

P_4	+ $5O_2$	→ P_4O_{10}	
y	5y	0	Initial
0	0	y	After

$\left.\begin{aligned} x+y &= \frac{1}{62} = 0.0161 \\ 3x+5y &= \frac{1}{16} = 0.0625 \end{aligned}\right\}$ by solving

$y = 7.056 \times 10^{-3}$

$x = 9.0435 \times 10^{-3}$

Mass of $P_4O_6 = 9.0435 \times 10^{-3}$

$[(4 \times 31) + (6 \times 16)] = 1.9895$ gm

Mass of $P_4O10 = 7.056 \times 10^{-3}$

$[(4 \times 31) + (10 \times 16)] = 2.003$ gm

Sol 28: Moles of aluminium $= \frac{2.7}{27} = 0.1$ mole

Moles of H_2SO_4 in solution

$= \frac{(1.18 \times 100) \times 0.25}{98} = 0.\ 3010$ mole

$2Al + 3H_2SO_4 \rightarrow Al_2(SO_4)_3 + 3H_2$

For consumption of Al, required mole of

$H_2SO_4 = (0.10) \times \frac{3}{2} = 0.\ 15$ mole

remaining mole of $H_2SO_4 = 0.3010 - 0.15 = 0.151$ mole

Molarity $= \frac{0.151}{500} \times 1000 = 0.302M$

Sol 29: $KMnO_4 + H_2SO_4 \rightarrow K_2SO_4 + MnSO_4 + H_2O + (O)$ .... (1)

$FeC_2O_4 + H_2SO_4 \rightarrow FeSO_4 + H_2C_2O_4$ .... (2)

$FeSI_4 + H_2C_2O_4 + H_2SO_4 + O \rightarrow Fe_2(SO_4)_3 + CO_2 + H_2$ .... (3)

$3KMnO_4 + 5FeC_2O_4 \rightarrow Fe^{3+} + 2CO_2 + Mn^{2+}$

$(0.\ 5)V = \frac{(1.5)}{5} \times 3$

$V = 1.\ 8$ lit. $= 1800$ mL

Redox Reactions

Sol 1: (a) $(N_2H_5)_2\ \underbrace{SO_4}_{2}$

$2(N_2H_5)^+ + SO_4^{2-}$

$\downarrow$

$2N^{x-} + 5H^+$

$\therefore$ Oxidation number of N $= -\frac{5}{2}$

$-2x + 5 = 0; x = \frac{5}{2}$

(b) $Mg_3N_2 \rightarrow 3Mg^{x+} + 2N^{3-}$

$3x - 6 = 0$

$x = 2$ $\quad Mg^{2+}$

(c) $\underbrace{[Co(NH_3)_5 Cl]Cl_2}$

$[Co(NH_3)_5 Cl]^{+2} + 2Cl^-$

$Co^{2+} + (NH_3)_5\ Cl^-$

$\therefore x - 1 = +2 \qquad x = +3$

Co^{+3}

(d) K_2FeO_4

$2K^+\ Fe^{x+}\ 4O^{2-}$

$+2 + x - 8 = 0; x = +6$

Fe^{+6}

(e) $Ba(H_2PO_2)_2$

$Ba^{2+}\ (H_2PO_2)^-$

$2H^+ + Px^+ + 2O^{2-}$

$2 + x - 4 = 0; x = +2$

$\therefore P^{+2}$

(f) H_2SO_4

$+2 + x - 8 = 0; x = +6$

S^{+6}

(g) CS_2

$-4 + 2x = 0; x = +2$

S^{+2}

(h) S^{-2}

(i) $Na_2S_4O_6$

$+2 + 4x - 12 = 0$

$S^{+5/2}$

$x = +\frac{5}{2}$

(j) S_2Cl_2

$+2x - 2 = 0; x = +1$

S^{+1}

(k) RNO_2

$+1 + x - 4 = 0; x = 3$

N^{+3}

(l) Pb_3O_4

$+3x - 8 = 0$

$x = +\frac{8}{3} \quad Pb^{+8/3}$

(m) $\underline{S_2}O_8^{2-}$

$2x - 16 = -2$

$2x = 14; x = +7$

S^{+7}

(n) $\underline{C_6}H_{12}O_6$

$+6x + 12 - 12 = 0; x = 0$

C^0

(o) $Mg_2P_2O_7$

$+2(2) + 2x - 14 = 0$

$x = +5$

P^{+5}

(p) $KClO_3$

$+1 + x - 6 = 0; x = +5$

Cl^{+5}

Sol 2: (a) $\overset{-1}{Br} + \overset{+5}{Br}O_3^- + H^+ \rightarrow \overset{0}{Br_2} + H_2O$

$Br^{-1} \rightarrow \frac{1}{2} Br_2 + e^-$ ......(1)

$5e^- + BrO_3^- + 6H^+ \rightarrow \frac{1}{2} Br_2 + 3H_2O$ ......(2)

((1) × 5) + (2)

$5Br^- + BrO_3^- + 6H^+ \rightarrow 3Br_2 + 3H_2O$

(b) $H_2\overset{-2}{S} + \overset{-2}{Cr_2}O_7^{2-} + H^+ \longrightarrow$

$\overset{+3}{Cr_2}O_3 + \overset{0}{S_8} + H_2O$

$H_2S \rightarrow \frac{1}{8} S_8 + 2e^- + 2H^+$ ......(1)

$8H^+ + Cr_2O_7^{2-} + 6e^- \rightarrow Cr_2O_3 + 4H_2O$ ......(2)

((1) × 3) + (2)

$3H_2S + Cr_2O_7^{2-} + 2H+ \rightarrow \frac{3}{8} S_8 + 6H^+ + Cr_2O_3 + 4H_2O$

$8Cr_2O_7^{2-} + 24H_2S + 16H^+ \rightarrow 8Cr_2O_3 + 3S_8 + 32H_2O$

(c) $Au + 4Cl^- \rightarrow AuCl_4^- + 3e^-$ ......(1)

$NO_3^- + e^- + 2H^+ \rightarrow NO_2 + H_2O$ ......(2)

(1) + (2 × (2))

$Au + 4Cl- + 2NO_3^- + 4H^+ \rightarrow AuCl_4^- + 2NO_2 + 2H_2O$

(d) $Cu_2O + 2H^+ \rightarrow 2Cu^{+2} + 2e^- + H_2O$ ......(1)

$H^+ + NO_3^- + 3e^- \rightarrow NO + 2H_2O$ ......(2)

(3 × (1)) + (2 × (2))

$3Cu_2O + 6H^+ + 8H^+ + 2NO_3^- \rightarrow 6Cu^{2+} + 3H_2O + 2NO + 4H_2O$

$3Cu_2O + 2NO_3^- + 14H^+ \rightarrow 6Cu^{2+} + 2NO + 7H_2O$

(e) $\overset{+6}{Mn}O_4^{2-} \longrightarrow \overset{+7}{Mn}O_4^- + \overset{+2}{Mn}O_2$

$4H^+ + \overset{+6}{Mn}O_4^{2-} + 2e^- \rightarrow MnO_2 + 2H_2O$ ......(1)

$MnO_4^{2-} \longrightarrow MnO_4^- + e^-$ ......(2) × 2

$3MnO_4^{2-} + 4H^+ \rightarrow MnO_2 + 2MnO_4^- + 2H_2O$

(f) $Cu^{2+} + SO_2 \rightarrow Cu^+ + SO_4^{2-}$

$Cu^{2+} + e^- \rightarrow Cu^+ \times (2)$

$H_2O + SO_2 \rightarrow SO_4^{2-} + 2e^- + 4H^+$

$2Cu^{2+} + SO_2 + 2H_2O \rightarrow 2Cu^+ + SO_4^{2-} + 4H^+$

(g) $\overset{0}{Cl_2} + \overset{0}{I_2} \longrightarrow \overset{0}{IO_3^-} + Cl^-$

$2e^- + Cl_2 \rightarrow 2Cl^- \times (5)$

$3H_2O + I_2 \rightarrow 2IO_3^- + 10e^- + 6H^+$

$3H_2O + 5Cl_2 + I_2 \rightarrow 2IO_3^- + 10Cl^- + 6H^+$

(h) $Fe^{+2} \rightarrow Fe^{+3} + e^-$

$2H_2O + 6C^{+2} \rightarrow 6CO_2^{+4} + 6(2e^-) + 24H^+$

$8H_2O + 6N^{3-} \rightarrow 6\overset{+5}{N}O_3^- + 6(8e^-) + 36H^+$

$5e^- + \overset{+7}{Mn}O_4^- + 8H^+ \rightarrow Mn^{2+} + 4H_2O \times (6H^+)$

$Fe(CN)_6^{4-} + 30H^2O \rightarrow Fe^{3+} + 6CO_2$

$6NO_3^- + 60H^+ + 61e^- \times (5)$

$5Fe(CN)_6^{4-} + 61MnO_4^- + 188H^+ \rightarrow 5Fe^{+3}$

$+30CO_2 + 30NO_3^- + 61Mn^{2+} + 94H_2O$

(i) $Cu^{+1} \rightarrow Cu^{+2} + e^- \times (3)$

$P^{3-} + 4H_2O \rightarrow \overset{+5}{H_3PO_4} + 8e^- + 5H^+$

$Cu_3P + 4H_2O \rightarrow 3Cu^{2+} + H_3PO_4 + 5H^+ + 11e^-$

$\overset{+6}{Cr_2O_7^{2-}} + 3e^- + 14H^+ \rightarrow Cr^{+3} + 7H_2O$

$6Cu_3P + 124H^+ + 11Cr_2O_7^{2-} \rightarrow 18Cu^{2+}$
$\qquad\qquad + 6H_3PO_4 + 22Cr^{+3} + 53H_2O$

Sol 3: (a) $2Cu^{2+} + 2I^- \rightarrow 2Cu + I_2$

(b) $\overset{+8/3}{Fe_3O_4} + 4H_2O \rightarrow \overset{+3}{Fe_2O_3} +$
$\qquad 8OH^- + 2e^- \times (3)$

$3OH^- + \overset{+7}{MnO_4^-} + 3e^- \rightarrow$
$\qquad\qquad MnO_2 + 2H_2O \times (2)$

$3Fe_3O_4 + 12H_2O + 8OH^- + 2MnO_4^- \rightarrow$
$\qquad 3Fe_2O_3 + 24OH^- + 4H_2O$

$6Fe_3O_4 + 2MnO_4^- + 8H_2O \rightarrow 9Fe_2O_3 + 16OH^- + 2MnO_3$

(c) $\overset{2}{C_2H_5OH} + OH^- \rightarrow \overset{-1}{C_2H_3O^-} + H2O$ Re$^-$

$3e^- + \overset{+7}{MnO_4^-} + 4H2O \rightarrow \overset{+4}{MnO_2} + 8OH^-$

$3C_2H_5OH + 2MnO_4^- + OH- \rightarrow 3C_3H_3O^- + 2MnO_2 + 5H_2O$

(d) $Cr^{+3+} 8OH^- \rightarrow \overset{+6}{CrO_4^{2-}} + 4H_2O + 3e^-$

$3I^- + 8OH^- \rightarrow \overset{+7}{3IO_4} + 24e^- + 4H_2O$

$e^- + \overset{-1}{H_2O_2} + H_2O \rightarrow H_2O^{-2} + 2OH^-$

$2CrI_3 + 27H_2O_2 + 10OH^- \rightarrow$
$\qquad 2CrO_4^{2-} + 6IO_4^- + 32H_2O$

(e) $258KOH + K4Fe(CN)_6 + 61Ce(NO_3)_4 \rightarrow 61Ce(OH)_3 +$
$Fe(OH)_3 + 36H_2O + 6K_2CO_3 + 250KNO_3$

Sol 4: (a) $I^\ominus + H_2O_2 \rightarrow H_2O + I_2$
$\qquad$ (acidic medium)

$(\overset{-1}{I^\ominus} \xrightarrow{1e^-} \overset{0}{I_2}) \times 2$

$\overset{+1\ -1}{H_2O_2} \xrightarrow{2\times1e} H_2O^{-2}$

$2I^\ominus + H_2O_2 \rightarrow H_2O + I_2 + 2H^+ + OH^\ominus$

$\therefore 2HI + H_2O_2 \rightarrow 2H_2O + I_2 + H^+$

(b) $Cu^{+2} + I^\ominus \rightarrow Cu^+ + I_2$

$(Cu^{+2} \xrightarrow{1e^-} Cu^+) \times 2$

$(I^\ominus \xrightarrow{1e^-} I) \times 2$

$2Cu^{+2} + 2I^\ominus \rightarrow 2Cu^+ + I_2$

By the oxidation number method,

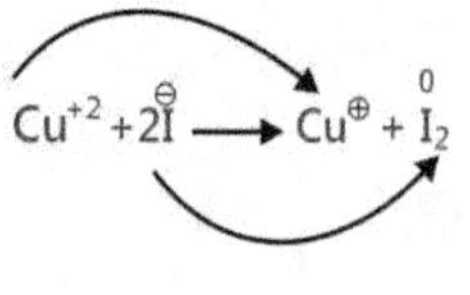

$\therefore$ To balance the electrons transferred,

$2Cu^{+2} + 2I^\ominus \rightarrow 2Cu^+ + I_2$

To balance charges on both sides,

$2Cu^{+2} + 2I^\ominus + 2H^+ \rightarrow 2Cu^+ + I_2 + H_2O$

$\therefore 2Cu^{+2} + 2HI \rightarrow 2Cu^+ + I_2 + H_2O$

(c) $CuO + NH_3 \rightarrow Cu + N_2 + H_2O$

To balance the electrons transferred to balance oxygen

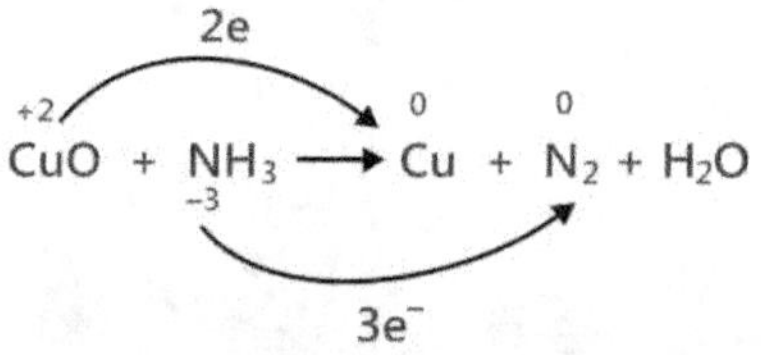

$3CuO + 2NH_3 \longrightarrow 3Cu + N_2 + 3H_2O$

(d) $\overset{+2\ -1}{H_2SO_3} + Cr_2O_7^{2-} \longrightarrow \overset{0}{H_2SO_4} + 2Cr^{+3}$
$\quad(+4)\qquad(+12)\qquad(+6)\qquad(+6)$
$\qquad\qquad(2e^-)\qquad(6e)$

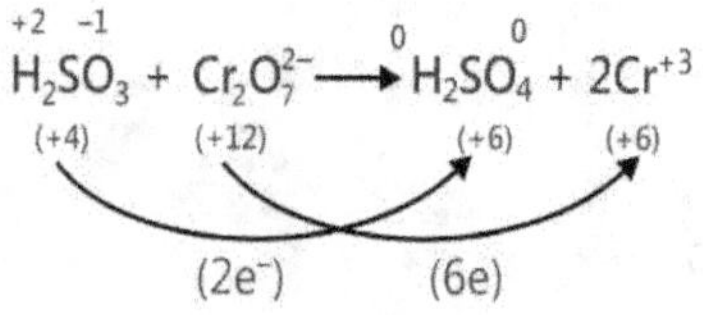

To balance the number of electrons transferred,

$9H_2SO_3 + Cr_2O_7^{2-} \rightarrow 9H_2SO_4 + 2Cr^{+3}$

To balance charges on both sides,

$3H_2SO_3 + Cr_2O_7^{2-} + 8H^+ \rightarrow 3H_2SO_4 + 2Cr^{+3} + 4H_2O$

We observe that the number of oxygen atoms are simultaneously balanced

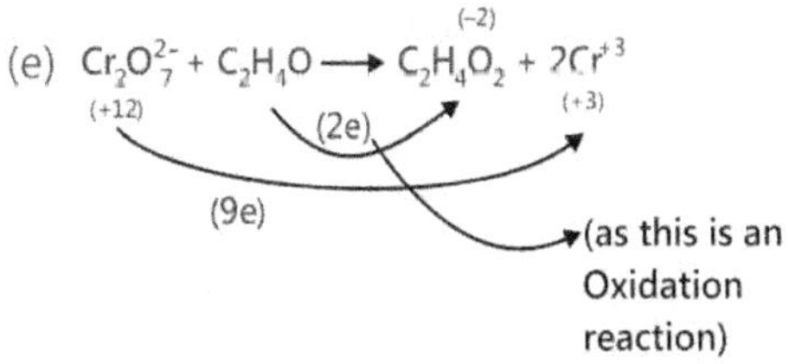

$\therefore\ Cr_2O_7^{2-} + C_2H_4O \rightarrow 9C_2H_4O_2 + 4Cr^{+3}$

To balance charges on both sides,

$\therefore\ 2Cr_2O_7^{2-} + 9C_2H_4O + 16H^+ \rightarrow 9C_2H_4O_2 + 4Cr^{+3} + 8H_2O$

(f) $3Cl^- + \overset{+3}{SbCl_3} \rightarrow \overset{+5}{SbCl_5} + 2e^- \times (2)$

$6e^- + \overset{+5}{KIO_3} + 6H^+ \rightarrow \overset{+1}{I} + 3H_2O + K^+$

$2SbCl_3 + KIO_3 + 8HCl \rightarrow 2SbCl_5 +$
$ICl + 4H_2O + KCl$

(g) $\overset{+5}{As_2}S_5^{-2}$

$As^{+5} \rightarrow \overset{+5}{H_3AsO_4}$ √ No redox charge

$5S^{2-} + 4H_2O \rightarrow 5H_2\overset{+6}{S}O_4 + 40e^-$

$e^- + \overset{+5}{HNO_3} + H^+ \rightarrow \overset{+4}{NO_2} + H_2O$

$As_2S_5 + 2HNO_3 \rightarrow 5H_2SO_4 + 40NO_2 + 2H_3ASO_4 + 12H_2O$

Sol 5: Disproportionation is a specific type of redox reaction in which a species is simultaneously reduced and oxidized to form two different product.

Eg. $\underset{+6}{MnO_4^{2-}} \longrightarrow \underset{+7}{MnO_4^-} + \underset{+4}{MnO_2}$

Sol 6: Ion-electron method :-

(I) Divide the complete equations into two half reaction.

(II) Balance the atoms in each half reaction separately according to the following steps :-

(a) Balance all atoms other then O and H.

(b) For O and H.

1. Acidic Medium:

(i) Add H_2O to the side which is oxygen deficient.

(ii) Add H^+ to the side which is hydrogen deficient.

2. Basic Medium:

(i) Add OH^- to the side which has less -ve charge.

(ii) Add H_2O to the side which is oxygen deficient.

(iii) Add H^- to the side which is hydrogen deficient.

3. Oxidation State Method: This method is based on the fact that the number of electrons gained during reduction must be equal to the number of e^-s lost during oxidation.

Sol 7: Definition of Redox Reaction: Reaction which involves change in oxidation state of their atom, generally involve the transfer of electron between species. So, the most essential conditions that must be satisfied is the exchange of electron change in oxidation state.

Sol 8: No, oxidation state term is just introduced to easily calculate the exchange of electron in redox reaction.

So, oxidation no. of an element in a particular compound represents the no. of e^-s lost or gained by an element during its change from free state into that compound or it represent the extent of oxidation or reduction of an element during its charge from free state into that compound.

Sol 9: Redox Couple: Oxidation half reaction and reduction half reaction contributes to redox couple

$M \longrightarrow M^{+n} + ne^-$ Oxidation

$A + ne^- \longrightarrow A^{-n}$ Reduction

$M + A \longrightarrow M^{+n} + A^{-n}$

Sol 10: (1) Combination of half cells (a) and (b)

$Zn(s) / Zn^{2+}(aq) || Cu(s) / Cu^{2+}(aq)$

Anode Cathode

E.M.F. of the cell, E

$= E_{right} - E_{left} = 0.34 - (-0.76)$

$= 1.10$ V

(2) Combination of half cells (b) and (d)

$Cu(s) / Cu^{2+}(aq) || Ag(s) / Ag^{2+}(aq)$

Anode Cathode

E.M.F. of the cell, E

$= E_{right} - E_{left} = 0.80 - (+0.34)$

$= 0.46$ V

Sol 11: (a)

$$MnO_{4\,(aq.)}^{-} + C_2H_2O_{4\,(aq.)} \rightarrow Mn^{2+}_{(aq.)} + CO_{2(g)} + H_2O_{(\ell)}$$

↓ ↓ ↓ ↓

Oxidant Reductant Reductant Oxidant

(i) Ion-electron method :-

$$\overset{+7}{MnO_4^-} \longrightarrow Mn^{+7} \quad \text{—— } 5e^-) \times 2$$

$$\overset{+3}{C_2}\overset{+2}{H_2}\overset{-8}{O_4} \xrightarrow{2e^-} \overset{+4}{CO_2} \quad \text{—— } 2e^-) \times 5$$

$$2MnO_4^- + 5C_2H_2O_4 \rightarrow 2Mn^{+2} + 10CO_2$$

(ii) Oxidation number method :-

$$\underset{(+7)}{MnO_4^-} + \underset{(+6)}{C_2H_2O_4} \rightarrow \underset{(+2)}{Mn^{+2}} + \underset{(+4)}{2CO_2}$$

5e⁻ 2e⁻

(b) $\overset{-2}{H_2S_{(aq.)}} + Cl_{2(aq)} \longrightarrow \overset{0}{S_{(S)}} + Cl^-_{(aq.)}$

↓ ↓

Reductant Oxidant

$H_2S^{-2} \longrightarrow S + 2e^- + 2H^+$

$Cl_2 + 2e^- \longrightarrow 2Cl^-$

$H_2S + \overset{0}{Cl_2} \rightarrow \overset{0}{S} + 2Cl^-$

$H_2S^{-2} + \overset{0}{Cl_2} \rightarrow \overset{0}{S} + 2Cl^{-2}$

-2e⁻ +2e⁻

So, no need to multiply this equation with any co-efficients.

Sol 12 (a) $Fe^{3+} + e^- \rightarrow Fe^{2+}$

$2I^- - 2e^- \rightarrow I_2$

(b) $Zn \rightarrow Zn^{2+} + 2e^-$

$2H^+ + 2e^- \rightarrow H_2(g)$

(c) $Al^{3+} + 3e^- \rightarrow Al(s)$

$Ag^+ + e \rightarrow Ag(s)$

Sol 13: Oxidation: Increase in oxidation number

$2Cl^- \longrightarrow Cl_2$

Reduction: Decrease in oxidation number

$$KMnO_4 \longrightarrow Mn^{2+}$$

Sol 14: (a) Combination reaction: Reaction in which two or more elements on compounds combine together to form a single compound

$2Mg + O_2 \longrightarrow 2MgO + \text{heat}$

(b) Decomposition reaction: Reaction is the separation of a chemical compound into elements or simpler compounds

$2H_2O_2 \longrightarrow 2H_2O + O_2$

(c) Displacement reaction:

Reaction in which on element or ion moves out of ore compound and into another

Eg. $Fe + CuSO_4 \longrightarrow Cu + FeSO_4$

Sol 15: Oxidation No.: No. of e⁻s lost or gained by an element during its change from free state into compound or represent the extent of oxidation or reduction of an element during its change from free state into that compound.

Valence: Number of valence bonds a given atom has formed or can form with one or more than one with other atoms.

Sol 16: S in SO_2 has oxidation state +4. It lies between the minimum oxidation state (-2) and maximum oxidation state (+6) of S. Thus, S in SO_2 can show an increase in its ox. no. (i.e., act as reducant) or can show a decrease in its ox. no. (i.e. acts as oxidant). On the other hand in H_2S, S is in - 2 oxidation state and can only increase its oxidation state to act as reducant.

Sol 17: Half Reaction: This is either the oxidation or reduction reaction component of a redox reaction. This is obtained by considering the change in oxidation states of individual substances involved in the redox reaction

$$Zn + CuSO_4 \longrightarrow ZnSO_4 + Cu$$

$Zn \longrightarrow Zn^{2+} + 2e^-$

oxidation half reaction

$$e^- + CuSO_4 \longrightarrow Cu + SO_4^{2-}$$

Reduction half reaction.

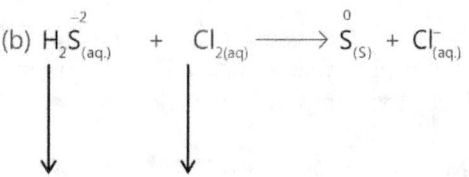

Sol 18: (i) Oxidation-Oxidation is the loss of electrons or an increase in oxidation state by a molecule, atom, or ion.

(ii) Reduction-Reduction is the gain of electrons or a decrease in oxidation state by a molecule, atom, or ion.

(iii) Oxidizing agent an oxidizing agent is a chemical species that removes an electron from another species.

(iv) Reducing agent-Reducing agent is an element or compound that loses an electron to another chemical species in a redox chemical reaction.

Exercise 2

Mole Concept

Single Correct Choice Type

Sol 1: (B) $A + O_2 \rightarrow$

$M_1V_1 = N_2V_2$

Equivalent of A = Equivalent of O_2

$$\frac{x}{\text{Equivalent weight of A}} = \left(\frac{16}{16}\right) \times 2$$

$$\frac{x}{2} = \text{equivalent weight of A}$$

Sol 2: (C) Mass O_2 in 88 gm

$$CO_2 = \frac{88 \times 32}{44} = 64\text{gm}$$

$$\text{Mole of O} = \frac{64}{16} = 4 \text{ mole}$$

So, mass of CO is = 4 × (12 + 16) = 112 gm

Sol 3: (B) $Mg + \frac{1}{2}O_2 \rightarrow MgO$

0.25 mole ↓ 0.5 mole

Mass of MgO = 0. 5 × (24 + 16) = 20 gm

Sol 4: (B) Let's diabasic acid is $C_xH_yO_z$

$$\text{Weight of C} = \frac{x(12)}{M}$$

$$\text{Weight} = H = \frac{y}{M}$$

$$\text{Weight of O} = \frac{z(16)}{M}$$

$x(12) = 8 \times y \Rightarrow 3x = 2y$

$$x(12) = \frac{1}{2} \times 16(z) \Rightarrow 3x = 2z$$

$$y = z = \frac{3}{2}x$$

$\Rightarrow$ Empirical formula $C_2H_3O_3$

$$Ag_2(C_xH_yO_z) \xrightarrow{\Delta} 2Ag$$

$$\frac{0.5934}{108}$$

$$\text{Mole of salt} = \frac{0.5934}{2 \times 108} = \frac{1}{[216 + (24 + 3 + 48)x]}$$

216 + 75x = 364

$x \sim z$

So the formula would be = $C_4H_6O_6$

Sol 5: (B) $12C(s) + 11H_2(g) + \frac{11}{2}O_2(g) \rightarrow C_{12}H_{22}O_{11}(s)$

$\frac{84}{12}$	$\frac{12}{1}$	$\frac{56}{22.4}$
7	12	2. 5

Here O_2 is limiting reagent

$$\text{Moles of } C_{12}H_{22}O_{11} \text{ formed} = \frac{2.5}{11} \times 2 = \frac{5}{11} \text{ mole}$$

$$\text{Mass} = \frac{5}{11} \times [(12 \times 12) + 22 + (11 \times 16)] = 155.45 \text{ gm}$$

Sol 6: (B) $M(CO_3) \rightarrow CO_2 + MO$

or

$M_2(CO_3) \rightarrow CO_2 + M_2O$

Mass of CO_3 = 12 + 48 = 60

$$\text{Mole of } CO_2 = \frac{12.315}{(PV)}(RT) = \frac{12.315}{1 \times (12.315)} \times 0.0821 \times$$

300 = 0.5 mole

Mole of $M(CO_3)$ or M_2CO_3 = 0.5 mole

So, mass of CO_3 in carbonate = 0.5 × 60 = 30 gram

Checking all options one by one

(B) is correct.

Sol 7: (D) Empirical formula would be NH_2 because ratio of N and H is given 1 : 2. By stability we see N_2H_4 is correct answer.

Sol 8: (C) $C_xH_y + O_2 \rightarrow CO_2 + H_2O$

5V 10V 5V ↓ 10V

By oxygen atom balance $H_2O = 10V$

By hydrogen atom balance $y(5) = 2(10)$ ∴ $y = 4$

By carbon balance $= x(5) = 5$

the molecule is CH_4

Sol 9: (A) Molecular weight of $NO_2 = 32 + 14 = 46$

Molecular weight of $NO = 16 + 14 = 30$

let's $x = NO \quad 1 - x = NO_2$

$34 = x(30) + (1 - x)\, 46$

$16x = 12$

$x = \frac{3}{4}$ so NO_2 % = 25%

Sol 10: (A) $5A_2 + 2B_4 \rightarrow 2AB_2 + 4A_2B$

4 mole

Molecular mass of $AB_2 = 250$

Molecular mass of $A_2B = 140$

$B_4 = 480$

$A_2 = 20$

Moles of AB_2 to be produced $= \frac{1000}{250} = 4$ mole

Moles of A_2B to be produced $= \frac{1000}{140} = 7.14$ mole

So, mass of A_2, B_4 would to according to AB_2

So, mass of A_2 required $= \frac{5}{2} \times 4 \times 20 = 10 \times 20 = 200$ gm

Mass of B_4 required $= 4 \times 480 = 2 \times 960$ gm $= 1920$ gm

Total mass $= 1920 + 200 = 2120$ gm

Sol 11: (C) $C_xH_yO_z + O_2 \rightarrow CO_2 + H_2O$

132 gm 54 gm

⇓ ⇓

3 mole 3 mole

$y = 6$ mole

$x = 3$ mole

Sol 12: (A) $Zn + \frac{1}{2}O_2 \rightarrow ZnO$

x gm

$$\frac{1}{2} \times \frac{x}{65} = \frac{v}{22.4}$$

$$v = \frac{x}{65} \times 11.2 = \frac{2x}{65} \times 5.6 \text{ lit.}$$

Sol 13: (B) Let's say 100 gm of clay is given initially

12 gm water

x gm silica

y gm other

After that (100 – A)gm of clay

(12 – A) gm water $\left(\frac{100 - A}{2}\right)$ silica

$$(12 - A) = \frac{(100 - A) \times 7}{100}$$

$1200 - 100A = 700 - 7A$

$93A = 500$

$$A = \frac{500}{93}$$

By conservation of silica

$$\frac{100 - \frac{500}{93}}{2} = x$$

$x = 47.31$

Sol 14: (C) $C_2H_4O_2 + 2O_2 \rightarrow 2CO_2 + 2H_2O$

x gm 620–x gm

11 11

$\frac{x}{60}$ mole $\frac{620 - x}{32}$

To produce maximum energy $C_2H_4O_2$ and O_2 will be fully consumed.

$$x\frac{x}{60} = \frac{620 - x}{32} \times \frac{1}{2}$$

$64x = 37200 - 60x$

$x = 300$ gm

Weight of $CO_2 = 2 \times \frac{300 \times 44}{60} = 440$ gm

Sol 15: (A) (Organic compound) + $H_2O \rightarrow N_2$
0.42 gm

Moles of N_2 = $\frac{PV}{RT} = \frac{860}{760} \times \frac{100}{11} \times \frac{10^{-3}}{0.08 \times 250}$

$= \frac{86}{167200} = 5.143 \times 10^{-4}$

Mass of N_2 = $5.143 \times 10^{-4} \times 28 = 0.0144$ gm

Fraction = $\frac{0.0144}{0.42} = 0.034 = \frac{10}{3}\%$

Sol 16: (C) Moles of H^+ = (0. 1) (0. 3) + (0. 2) (0. 3) × 2 = 0. 15 mole

Normality = $\frac{0.15}{500} \times 1000 = 0.3N$

Sol 17: (D) Moles of NaOH = (0. 300) (0. 5) = 0. 15 moles

For molarity = 0. 2 M = $\frac{0.15}{V}$

V = 750 mL

Volume to be added = 750 – 300 = 450 mL

Sol 18: (A) Moles of water = $\frac{250}{18} = 13.888$ mole

urea = $NH_2{-}\overset{\overset{\large O}{||}}{C}{-}NH_2$

Moles urea = $\frac{3}{60}$ = 0. 05 mole

Mole fraction = 0.0036

Sol 19: (B) $P4S_3 + 8O_2 \rightarrow P_4O_{10} + 3SO_2$

Moles of O_2 = $\frac{384}{32} = 12$mole

Moles of P_4S_3 = $\frac{440}{124 + 96} = 2$ mole

L. R. = O_2

So mass of P_4O_{10} produced

$= \frac{12}{8} \times [124 + 160]$ = 426 gm

Sol 20: (C) $PCl_5 \rightarrow PCl_3 + Cl_2$

Initially 1 mole

After $\frac{1}{2}$mole $\quad \frac{1}{2}$mole $\quad \frac{1}{2}$mole

Initially $M_{avg.} = M_{PCl_5}$ = 31 + 5(35. 5) = 208. 5

After $M_{avg.} = \frac{M}{3/2} = \frac{208.5}{3} \times 2 = 208.5 \times \frac{2}{3}$

So change in $M_{avg.}$ = 33. 33%

Sol 21: (A) $3Mg + 2NH_3 \rightarrow Mg_3N_2 + 3H_2$

2 mole 2 mole

L. R. = Mg

Mass of Mg_3N_2 produced = $\frac{2}{3} \times (72 + 28) = \frac{200}{3}$gm

Sol 22: (C) Let's say solution is in 100 gm.

HCl mole = 1 mole

Molality = $\frac{\text{moles of solute}}{\text{mass of solvent (in kg)}}$

Molality = $\frac{1}{(100 - 36.5)} \times 1000$ = 15. 75

Sol 23: (B) Weight of $Na_2CO_3.xH_2O$ in 10 mL solution

$= \frac{0.025}{5} = 0.07$

$2 \times \frac{0.07}{46 + 12 + 48 + x(18)} = \frac{9.9}{10} \times 10^{-3}$

$\frac{0.14}{106 + 18x} = 0.99 \times 10^{-3}$

0.07 = [104.94 + 17.82 x] × 10^{-3}

35.06 = 17.82 x

x ~ 2

Sol 24: (C) Washing soda (Na_2CO_3) in 25 cc = 0. 12 gm

$2 \times \frac{0.12}{106 + 18x} = 1.7 \times 10^{-3}$

240 = 180. 2 + 30. 6 x

x ~ 2

Percentage of carbonate = $\frac{106}{106 + 36}$ ~ 76%

Sol 25: (B) No. of carbon atoms

$= \frac{1.2 \times 10^{-3}}{12} \times 6.023 \times 10^{23}$ = 6. 02 × 10^{19}

Sol 26: (C) $21.31 = (0.79)(24) + (0.21 - x)(25) + (x)(26)$

$24.31 = 18.96 + 5.25 + x$

$x = 0.1 = 10\%$

Sol 27: (D) Using HPh

$$Na_2CO_3 \xrightarrow{HCl} NaHCO_3^- + H^+$$

$$NaOH \xrightarrow{HCl} NaCl + H^+$$

Using

MeOH

$$Na_2CO_3 \xrightarrow{HCl} H_2CO_3 + NaCl$$

$$NaOH \xrightarrow{HCl} NaCl + H^+$$

Moles of HCl used in HPh = 4m mole

Moles of HCl used in MeOH = 4.5 m mole

It means that for $NaHCO_3 \rightarrow H_2CO_3$ required mole of HCl 0.5 m mole

Moles of Na_2CO_3 = 0.5 m mole

$= 0.5 \times 10^{-3} \times 106$ gm/250 mL

= 5.3 mg/25 mL

= 2.12 g/L

NaOH moles = 3.5 m mole

Mass of NaOH = 40 × 3.5

= 140 mg/25 mL

= 5.6 g/L

None of these

Sol 28: (A) Moles of HCl = 0.25 × 30 m mole = 7.5 m mole

$$\frac{7.5}{2} = 10^{-3} = \frac{x(0.5)}{138} + \frac{(1-x)0.5}{74}$$

7.5×10^{-3}

Solving this x ~ 96%

K_2CO_3 ~ 96%

Li_2CO_3 ~ 4%

Sol 29: (C) $KMnO_4 + 5FeSO_4 \rightarrow Mn^{2+} + Fe^{3+}$

$\frac{2.0}{152}$ moles of $KMnO_4$

$$= \frac{2.0}{5 \times 152} = 2.631 \times 10^{-3}$$

Volume × molarity = moles

$$\text{Volume} = \frac{2.631 \times 10^{-3}}{0.05} = 52.63 \text{ mL}$$

Redox Reactions

Single Correct Choice Type

Sol 1: (A) Equivalent weight = Molecular weight × n_{factor}

$= (M_0)_{FeSO_4} \times 1$

$Fe^{+2} \rightarrow Fe^{+3}$

Sol 2: (D) Equivalent weight = Molecular weight × n_{factor}

$= (M)_{K_2Cr_2O_7} \times 6$

$Cr_2O_7 \rightarrow Cr^{+3}$

Sol 3: (A) $H_2S + KMnO_4 \rightarrow S + Mn^{2+}$

$-2e^-$ $+5e^-$

$$\frac{m}{34} \times 2 = \frac{1.58}{158} \times 5 = 0.85$$

Sol 4: (C) $\overset{+5}{HNO_3} + \overset{0}{I_2} \rightarrow \overset{+5}{I_2O_5} + \overset{+4}{NO_2}$

$$\frac{m}{63} \times 1 = \frac{127}{127 \times 2} \times 10 = 315$$

Sol 5: (C) $N_1V_1 = N_2V_2$

Oxalic acid $KMnO_4$

$10 \times 10^{-3} \times N = 20 \times 10^{-3} \times 0.02 \times (5)$

n-factor =0.2N

Sol 6: (C) $\frac{m}{34} \times 2 = 10 \times 10^{-3} \times 1$; m = 0.17

$$\therefore \text{Purity} = \frac{0.17}{0.2} \times 100 = 85\%$$

Sol 7: (B) $KMnO_4 + FeC_2O_4 \rightarrow Mn^{2+} + Fe^{3+} + CO_2$

$$n_{KMnO_4} \times 5 = 1 \times (1 + 2) = \frac{3}{5}$$

Sol 8: (C) $M_2x_2 + xH_2 \rightarrow 2M + xH_2O$

1 mole M_2x_2 gives 2 moles M

$$\therefore \frac{3.15g}{(MW)} \rightarrow \frac{1.05}{M}$$

$\therefore 6M = 2M + x(16)$

$\therefore 4M = x(16)$

$\therefore$ M = 4x

Now, $(EW)_M = \dfrac{(MW)_M}{x} = \dfrac{M}{X}$

$\therefore (EW)_M = \dfrac{M}{X} = 4.$

Sol 9: (B) Oxidation means increase of oxidation number

$\therefore$ So, there is loss of electrons.

Sol 10: (C) $K_2Cr_2O_7$

$2K^+ \quad 2Cr^{x+} \quad 7O^{2-}$

$+2 + 2x - 14 = 0$

$x = +6$

Sol 11: (D) $K_2\overset{+6}{Cr_2}O_7 \rightarrow K_2\overset{+3}{Cr_2}O_7$

$+6 \rightarrow +3$

$\therefore$ Change in oxidation No. = [3].

Sol 12: (C) $P + NaOH \rightarrow \overset{-3}{PH_3} + NaH_2\overset{+1}{PO_2}$

P is getting oxidised and also reduced

$\therefore$ It is oxidation and reduction

(D is proportionation)

Sol 13: (C) CH_2O

$C^{x+} \quad 2H^+ \quad O^{2-}$

$x + 2 - 2 = 0$

$x = 0$

Sol 14: (B) $CH_4 \rightarrow \quad C^{x+} + 4H^+ \quad x = -4$

$CH_3Cl \quad C^{x+} + 3H^+ + Cl^- \quad x = -2$

$CH_2Cl_2 \quad x + 2 - 2 = 0 \quad x = 0$

$CHCl_3 \quad x + 1 - 3 = 0 \quad x = 2$

$CCl_4 \quad x - 4 = 0 \quad x = +4$

Sol 15: (C) Redox: Exchange of electrons

$\therefore$ Change in oxidation state.

(c) $Ba^{2+}\ O_2^{2-} + H_2\overset{+6}{S}O_4 \rightarrow Ba^{2+}\ \overset{+6}{S}O_4^{2-} + H_2\overset{-1}{O_2}$

No change in oxidation

Sol 16: (D) (a) $+4 - 6 + x = 0; \; x = +2$

(b) $+2 + x - 8 = 0; \; x = +6$

(c) $+2x - 2 = 0; \; x = +1$

(d) $x + 5(0) = 0; \; x = 0$

Sol 17 (C) (a) $H\overset{+1}{Cl}O^{2} \quad +1 + x - 2 = 0; \; x = +1$

(b) $HClO_2 \quad +1 + x - 4 = 0; \; x = +3$

(c) $HClO_3 \quad +1 + x - 6 = 0; \; x = 5$

(d) $HClO_4 \quad +1 + x - 8 = 0; \; x = +7$

Sol 18: (D) $M^{3+} \rightarrow M^{6+} + 3e^-$

Sol 19: (A) $\overset{+x}{Mn}O_4^- \quad +x - 8 = -1; x = +7$

Sol 20: (A) The oxidation number of carbon in $CHCl_3$ is +4

Sol 21: (C) $Pb^{2+} \rightarrow Pb^{4+} + 2e^-$

Sol 22: (A) $C_{12}H_{22}O_{11}$

$12x + 22 + 11(-2) = 0; \; x = 0$

Sol 23: (D) SO_4^{2-}

$S^{x+} + 4O^{2-}$

$x - 8 = -2; \; x = +6$

Sol 24: (B)

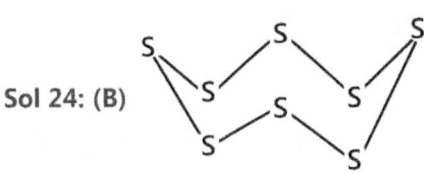

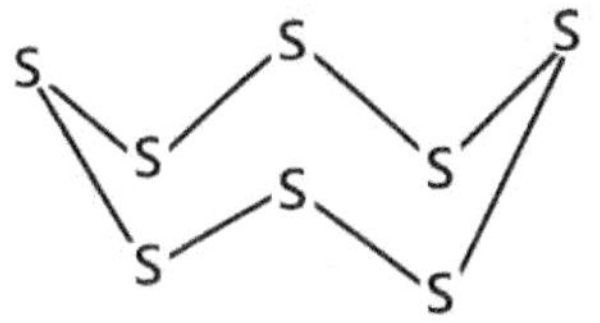

Sol 25: (C) $SO_3^{2-} \longrightarrow \overset{+6}{S}O_4^{2-}$

$-2e^-$

Sol 26: (A) $S_2O_7^{2-} \quad +2x - 14 = -2$

$2x = +12; x = +6$

Sol 27: (A) $\overset{+4}{Mn}O_2 + 4H^+ + X \longrightarrow \overset{2+}{Mn} + H_2O$

$-2e^-$

S_8

$8x = 0; x = 0$

Covalency = 2

Sol 28: (D) N_3H

$3x + 1 = 0; \; x = x = -\dfrac{1}{3}$

Sol 29: (C) $K_3 [Fe(CN)_6]$

$+3 - (x - 6) = 0; \; x = +3$

Sol 30: (D) MH_2

$+ x + 2 = 0; \; x = -2$

Sol 31: (C) $\overset{-3}{PH_3}$ to $\overset{+5}{H_3PO_4}$

Sol 32: (C) In chlorine atom had +5 oxidation state.

$HClO_3 + 1 + x - 6 = 0; \; x = 5$

Previous Years' Questions

Mole Concept

Sol 1: (A) If we consider that $\frac{1}{6}$, in place of $\frac{1}{12}$, mass of carbon atom is taken to be the relative atomic mass unit, the mass of one mole of a substance will decrease twice.

Sol 2: (C) According to definition of molar solution is one that contains one mole of a solute in one litre of the solution.

Sol 3: (D) $2Al_{(s)} + 6HCl_{(s)} \longrightarrow$

$$2Al^{3+}_{(aq)} + 6Cl^{-}_{(aq)} + 3H_{2(g)}$$

For each mole of HCl reacted, 0.5 mole of is formed at STP.

1 mole of an ideal gas occupies 22.4 lit at STP.

Volume of H_2 gas formed at STP per mole of HCl reacted is 22.4 × 0.5 litre.

Sol 4: (B) From molecular formula of magnesium phosphate, it is evident that each mole of $Mg(PO_4)_2$ contains 8 mole of oxygen atoms.

Therefore, 0.25 mole of oxygen atom will remain present in $\frac{0.25}{8}$ mole i.e. 3.125×10^{-2} mole of $Mg(PO_4)_2$.

Sol 5: (A) 200 mg of CO_2 = 200 $\times 10^{-3}$ = 0.2 g

44g of CO_2 = 6 × 10^{23} molecules (approx.)

$0.2g \text{ of } CO_2 = \frac{6 \times 10^{23}}{44} \times 0.2$

$= 0.0272 \times 10^{23} = 2.72 \times 10^{21}$ molecule

Now, 10^{21} molecule are removed.

So remaining molecules $= 2.72 \times 10^{21} - 10^{21}$

$= 10^{21}(2.72 - 1) = 1.72 \times 10^{21}$ molecule

Now, 6.023×10^{23} molecules = 1 mol

$1.72 \times 10^{21} \text{ molecules} = \frac{1 \times 1.72 \times 10^{21}}{6.023 \times 10^{23}}$

$= 0.285 \times 10^{-2} = 2.85 \times 10^{-3}$

Sol 6: (C) $\underset{+12/\text{two atom}}{K_2Cr_2O_7} + 4H_2SO_4 \rightarrow K_2SO_4 + \underset{+6/\text{two atom}}{Cr_2(SO_4)_3}$

Change by 6

$$\text{Eq. wt.} = \frac{\text{Mol. wt.}}{6}$$

Sol 7: (A) $M_f = \frac{M_1V_1 + M_2V_2}{V_1 + V_2}$

$$= \frac{0.5 \times \frac{3}{4} + 2 \times \frac{1}{4}}{1} = 0.875 \text{ M}$$

Sol 8: (D) 18 g H_2O contains 2 g H

$\therefore$ 0.72 g H_2O contains 0.08 g H.

44 g CO_2 contains 12 g C

$\therefore$ 3.08 g CO_2 contains 0.84 g C

$\therefore \; C : H = \frac{0.84}{12} : \frac{0.08}{1} = 0.07 : 0.08 = 7 : 8$

$\therefore$ Empirical formula = C_7H_8

Sol 9: (B) $M_{0.98}O$

Consider one mole of the oxide.

Moles of M = 0.98, Moles of O^{2-} = 1

Let moles of M^{3+} = x

$\Rightarrow$ Mole of M^{2+} = 0.98 – X

$\Rightarrow$ Doing charge balance

(0.98 - x) × 2 + 3x – 2 = 0

$\Rightarrow$ 1.96 – 2x + 3x – 2 = 0

$\Rightarrow$ x = 0.04

$\Rightarrow$ % of $M^{3+} = \frac{0.04}{0.98} \times 100 = 4.08\%$

Sol 10: (B) Let the mass of O_2 = x

Mass of N_2 = 4x

Number of moles of $O_2 = \frac{x}{32}$

Number of moles of $N_2 = \frac{4x}{28} = \frac{x}{7}$

$\therefore$ Ratio $= \frac{x}{32} : \frac{x}{7} = 7 : 32$

Sol 11: (D) 1 g of $C_8H_7SO_3Na = \frac{1}{206}$ mole

$$\underset{\frac{1}{206}\text{ mole}}{2C_8H_7SO_3Na} + \underset{\frac{1}{412}\text{ mole}}{Ca^{2+}} \rightarrow (C_8H_7SO_3)_2Ca + 2Na^+$$

Sol 12: (A) $C_xH_y(g) + \left(x + \frac{y}{4}\right)O_2(g) \rightarrow xCO_2(g) + \frac{y}{2}H_2O(\ell)$

15 mL

Volume of O_2 used $= \frac{20}{100} \times 375 = 75$ ml.

Volume of air remaining = 300 mL

Total volume of gas left after combustion = 330 mL

Volume of CO_2 gases after combustion = 330 − 300 = 30 mL.

$$\underset{15\text{ ml}}{C_xH_y(g)} + \underset{75\text{ ml}}{\left(x + \frac{y}{4}\right)O_2(g)} \rightarrow \underset{30\text{ ml}}{xCO_2(g)} + \frac{y}{2}H_2O(\ell)$$

$\frac{x}{1} = \frac{30}{15} \Rightarrow x = 2$

$\frac{x + \frac{y}{4}}{1} = \frac{75}{15} \Rightarrow x + \frac{y}{4} = 5$

$\Rightarrow y = 12$

$\Rightarrow C_2H_{12}$

Redox Reactions

Sol 13: (C) Prevent action of water and salt.

Sol 14: (A) $2H\overset{-1}{I} + H_2\overset{+6}{S}O_4 \longrightarrow \overset{0}{I_2} + \overset{+4}{S}O_2 + 2H_2O$

Sol 15: (C) CH_2O

x + 2 − 2 = 0; x = 0

Sol 16: (D) $K_2Cr_2O_7 + 6KI + 7H_2SO_4 \longrightarrow$

$4K_2SO_4 + Cr_2(SO_4)_3 + 7H_2O + 3I_2$

$\overset{+3}{Cr_2}(SO_4)_3 \longrightarrow 2\overset{+3}{Cr} + 3SO_4^{-2}$

Sol 17: (C) Number of e^- transferred in each case is 1, 3, 4, 5.

Sol 18: (D) $\overset{0}{Zn} + 2\overset{+1}{Ag}CN \longrightarrow 2\overset{0}{Ag} + \overset{+2}{Zn}(CN)_2$

Oxidation: Zn → Zn(CN)₂ ; Reduction: AgCN → Ag

Sol 19: (A) $6MnO_4^- + I^- + 6OH^- \longrightarrow$

$6MnO_4^{2-} + IO_3^- + 2H_2O$

Sol 20: (A) $H-O-\underset{\downarrow \atop O}{\overset{H \atop |}{P}}-OH$, hence it is dibasic. It acts as a reducing agent also.

Sol 21: (C)

$\underset{vf = 1(7-2) = 5}{MnO_4^-} + \underset{vf = 2(3-2) = 2}{C_2O_4^{2-}} + H^+ \rightarrow Mn^2 + CO_2 + H_2O$

$\therefore$ Balanced Equation:

$2MnO_4^- + 5C_2O_4^{2-} + 16H^+ \rightarrow 2Mn^2 + 10CO_2 + 8H_2O$

So, x = 2, y = 5 & z = 16.

Sol 22: (D) The reducing agent oxidises itself:

(A) $H_2O_2^{-1} + 2H^+ + 2e^- \rightarrow 2H_2O^{-2}$

(B) $H_2O_2^{-1} - 2e^- \rightarrow O_2^0 + 2H^+$

(C) $H_2O_2^{-1} + 2e^- \rightarrow 2\overset{-2}{O}H^-$

(D) $H_2O_2^{-1} + 2OH^- - 2e^- \rightarrow O_2^0 + H_2O$

Note: Powers of 'O' are oxidation number of 'O' in the compound.

Sol 23: (B)

The complex $\left[CoCl(NH_3)_5\right]^+$ decomposes under acidic medium, so $\left[CoCl(NH_3)_5\right]^+ + 5H^+ \rightarrow Co^{2+} + 5NH_4^+ + Cl^-$

Sol 24: (A) H_2O_2 can undergo reduction as well as oxidation because oxidation number of oxygen in H_2O_2 is -1. So, it can act both as reducing agent and oxidising agent.

JEE Advanced/Boards

Exercise 1

Mole Concept

Sol 1: $4HCl + MnO_2 \rightarrow MnCl_2 + 2H_2O + Cl_2$

69.6 gm

69.6 gm of MnO_2 = $\frac{69.6}{87}$ mole

Mole of HCl = $\frac{69.6}{87} \times 4$ mole

Weight of HCl = $\frac{69.6}{87} \times 4 \times 35.5 = 116$ gm

Sol 2: $3TiO_{2(s)} + 4C_{(s)} + 6Cl_{2(g)}$

4.32 gm 5.76 gm 6.82 gm

0.054 mole 0.48 mole 0.0960 mole

$\rightarrow 3TiCl_{4(g)} + 2CO_{2(g)} + 2CO_{(g)}$

L. R. = Cl_2

So $TiCl_4$ mole produced

$= \frac{1}{2} \times 0.0960 = 0.048$ mole

Weight of $TiCl_4$ produced = 0.048 × 190 = 9.12 gm

Sol 3: $2SO_{2(g)} + O_{2(g)} + 2H_2O(\ell) \rightarrow 2H_2SO_4$

5.6 moles 4.8 moles

L. R. = SO_2

So H_2SO_4 mole obtained in maximum = 5.6 mole

Sol 4: Na_2CO_3 = x gram

Pure Na_2CO_3 = (0.95)x gm

$Na_2CO_3 + 2HCl(acid) \rightarrow H_2CO_3 + 2NaCl$

Mole of acid = (45.6 mL) × (0.235) = 10.716 m mole

Moles of Na_2CO_3 required = 5.358 m mole

Weight of Na_2CO_3 required = (0.95)x 5.358 (106) × 10^{-3}

x = 0.597 gm

Sol 5: $BaCl_2$ = 12%

Molecular weight of $BaCl_2.2H_2O$ = 208 + 36 = 244

$BaCl_2$ = 6 gm

$BaCl_2.\ 2H_2O = 6 \times \frac{244}{208} = 7.038$ gm

H_2O = 42.962 gm.

Sol 6: NaOH mole = 50(0.2) = 10 mole

HCl mole = 5 mole

$FeCl_3$ mole = 1.5 mole (acidic)

$NaOH + HCl \rightarrow NaCl + H_2O$

After this reaction NaOH left = 5 mole

$FeCl_3 + 3NaOH \rightarrow Fe(OH)_3 + 3NaCl$

1. 5

After this reaction NaOH left = 5 – (1. 5)3 = 0. 5 mole

Volume after reaction = 15 + 5 + 50 = 70 litre

Normality = $\frac{0.5}{70} = 7.142 \times 10^{-3}$N

$2Fe(OH)_3 \rightarrow Fe_2O_3 + 3H_2O$

Weight of Fe_2O_3 = $\frac{1.5}{2} \times 160$ = 120 gm

Sol 7: Oleum = $H_2S_2O_7 = H_2SO_4 + SO_3$

$H_2SO_4 + 2NaOH \rightarrow Na_2SO_4 + 2H_2O$

Mole of NaOH = (26.7)x (0.4) m mole = 10.68 m mole

Mole of H_2SO_4 = 5.34 m mole

Weight of H_2SO_4 = 0.523 gm

$H_2S_2O_7 = H_2SO_4 + SO_3$

x gram (0.5 – x) gm

$SO_3 + H_2O \rightarrow H_2SO_4$

$= \frac{(0.5-x)}{80} \times 98$ gm

Total H_2SO_4 = x + $\frac{(0.5-x)98}{80} \times 0.523$

x = $\frac{0.0895}{0.225}$ ~ 0.3977 gm

$\% SO_3 = \frac{0.5 - 0.3977}{0.5} \sim 20.4\%$

Sol 8: HPh: $NaOH + HCl \rightarrow NaCl + H_2O$ ...(1)

$Na_2CO_3 + HCl \rightarrow NaHCO_3 + NaCl$ ...(2)

after MeOH:

$NaHCO_3 + HCl \rightarrow H_2CO_3 + NaCl$ ...(3)

Mole of HCl (when HPh) = 1.75 m mole

Mole of HCl (when MeOH) = 0.25 m mole (extra added)

Amount of $NaHCO_3$ = 0.25 m mole

Amount of HCl required in (2) and (3)

= $(0.25)_2$ = 0.5 m mole

Amount of Na_2CO_3 = 0.25 m mole

Amount of NaOH = 1.75 – (0.25) = 1.5 m mole

NaOH (in gram) = $1.5 \times 10^{-3} \times 40$ = 0.06 gm per 200 mL

Na_2CO_3 (in gram) = $0.25 \times 10^{-3} \times 106$

= 0.0265 gm/200 mL

Sol 9: $2KO_{2(s)} + H_2O_{(l)} \rightarrow 2KOH_{(S)} + \frac{3}{2}O_{2(g)}$

0. 158 mole 0. 1 mole

L. R = KO_2

Moles of O_2 formed = $\frac{3}{4} \times 0.158 = 0.1185$

Sol 10: $CaCl_2 + H_2CO_3 \rightarrow CaCO_3 + 2HCl$

$CaCO_3 \rightarrow CaO + CO_2$

0.959 gm

Moles of CaO = 0.017125 mole

Moles of $CaCl_2$ = 0.017125 mole

Mass of $CaCl_2$ = (0.017125) × 111

= 1.9 gm

% of $CaCl_2$ = $\frac{1.9}{4.22} = 45\%$

Sol 11: $C_6H_{12}O \xrightarrow{\text{Conc. } H_2SO_4} C_6H_{10}$

100 gm

Moles of cyclohexanol = $\frac{100}{100}$ = 1 mole

Mole of cyclohexene = 0. 75 mole

Mass of cyclohexene = (0. 75) × 89 = 66. 75 gm

Sol 12: $2NaCl \rightarrow Na_2SO_4$ (By Na = atom balance)

Pure NaCl mole = $\frac{(0.95)250}{23 + 35.5}$ = 4.059 mole

Pure Na_2SO_4 = $\frac{4.059}{2} \times (46 + 96)$ = 288. 24 gm

Na_2SO_4 (90% pure) = $\frac{288.24 \times 100}{90}$ = 320.27 gm

Sol 13: $\underset{(0.466-x)gm}{AgCl} \longrightarrow$ unreacted

$\underset{x\ gm}{AgBr} \rightarrow AgCl$

AgCl formed = $\frac{x}{188} \times (1435)$ gm = 0.763 x

Total weight after reaction = 0.4066 – x + 0.763 x

Weight lost = (1 – 0.763)x = 0.0725

x = 0.306 gm = 30.6%

Weight of Cl in initial mixture

= $(0.4066 - 0.306) \times \frac{35.5}{143.5}$ = 0.0248 gm

% of Cl = $\frac{0.0248}{0.4066} = 6.1\%$

Sol 14: $CaCO_3 + H_2SO_4 \rightarrow CaSO_4 + H_2CO_3$

0. 5 gm

Moles of $CaCO_3$ = moles of H_2SO_4

required = $\frac{0.5}{63.5 + 60}$ = 4.048×10^{-3} mole

m litre of 0.5M H_2SO_4 required $\frac{4.048}{0.5}$ = mL = 8. 096 mL

Sol 15: $H_2SO_4 + 2NaOH \rightarrow Na_2SO_4 + 2H_2O$

Moles of NaOH = $15 \times \frac{1}{10} = 1.5$ m mole

Moles of H_2SO_4 required = $\frac{1.5}{2}$ = 0.75 m mole

In 12 mL, mole of H_2SO_4 = 0.75 m mole

In 1 L, mole of H_2SO_4 = $\frac{0.75}{12}$ mole

In 1 L, weight of H_2SO_4 required

= $\frac{0.75}{12} \times 98$ gram = 6.125 gm/L

Sol 16: Ethane (C_2H_6) $\xrightarrow{\text{monobromination}}$

$\xrightarrow[\text{reaction}]{\text{wurtz}}$ n-butane (C_4H_{10})

$2\,C_2H_6 \rightarrow C_4H_{10}$ (by carbon balance)

XV

Let's say volume of ethane = x l

Weight of $C_4H_{10} = \dfrac{x}{2\times 22.4}\times\dfrac{90}{100}\times\dfrac{85}{100}\times 58 = 55.53\ l$

x = 55. 53 l

Sol 17: Mole of HCl = 30 × 0.25 m mole = 7.5 m mole

let's say x fraction is K_2CO_3 so

$$\frac{7.5}{2}\times 10^{-3} = \frac{x(0.5)}{138}+\frac{(1-x)(0.5)}{74}$$

x ~ 96%

Sol 18: Mass of solution of HCl

= 100 × 1.18 gm = 118 gm

Mass of HCl in solution = (0.36) (118) = 42. 48 gm

n_{HCl} = mole of HCl = $\dfrac{42.48}{365}$ = 1.163 mole

$$2KMnO_4 + \underset{1.163\text{ mole}}{16HCl} \rightarrow 2KCl + 2MnCl_2 + 8H_2O + \underset{0.363\text{ mole}}{5Cl_2}$$

$$\underset{0.363\text{ mole}}{6Cl_2} + 6Ca(OH)_2 \rightarrow \underset{0.0606\text{ mole}}{Ca(ClO_3)_2} + 5CaCl_2 + 6H_2O$$

$$\underset{0.0606\text{ mole}}{Ca(ClO_3)_2} + Na_2SO_4 \rightarrow CaSO_4 + \underset{0.1212\text{ mole}}{2NaClO_3}$$

Mass prepared of $NaClO_3$

= 0.1212 x molecular weight = 12.911 gm

Sol 19: $NaH_2PO_4 \xrightarrow[NH_4^+]{Mg^{2+}} Mg(NH_4)PO_4 \cdot 6H_2O$

$\xrightarrow{\Delta} \dfrac{1}{2}\, Mg_2P_2O_7$ (by P-balance)

Mole of $Mg_2P_2O_7 = \dfrac{1.054}{224} = 4.747 \times 10^{-3}$

Weight of $NaH_2PO_4 = 2 \times 4.747 \times 10^{-3} \times 119.98$

(Molecular weight) = 1.139 gm

Sol 20: Moles of HNO_3 = 8 × 5 m mole = 40 m mole

Mole of HCl = 4. 8 × 5 m mole = 24 m mole

Let's say volume of H_2SO_4 is V mL

So mole of H_2SO_4 = 17 V m mole

Moles of HNO_3 in 30 mL (picked up from 2 l sol)

$= \dfrac{40}{2000}\times 30 = \dfrac{120}{200}$ m mole

Moles of HCl in 30 mL (picked up from 2 l sol)

$= \dfrac{24}{2000}\times 30 = \dfrac{72}{200}$ m mole

Moles of H_2SO_4 is 30 mL (picked up from 2 lt. sol)

$= \dfrac{17V}{2000}\times 30 = \dfrac{51V}{200}$ m mole

Total moles of H^+ from 30 mL solution $= \dfrac{120}{200}+\dfrac{72}{200}+\dfrac{102V}{200}$

$= \left(\dfrac{192+102V}{200}\right)$ m mole

Mole $Na_2CO_3 \cdot 10H_2O = \dfrac{1}{286}$ mole

Mole of OH = $\dfrac{2}{286}$ mole (in 100 mL)

Mole of OH in 42.9 mL = $\dfrac{2\times 0.429}{2.86}$

$= \dfrac{0.858}{286} = 0.003$ mole

$10^{-3}\times\left(\dfrac{192+102V}{200}\right) = 0.003$; V = 4 mL

Amount of sulphate ion in gm = $\dfrac{51\times 4}{200}\times(96)\times 10^{-3}$

= 0.097.92 gm/30 mL

= 6.528 gm/L

Sol 21: $Mg \xrightarrow[O_2]{N_2} MgO + Mg_3N_2$

$$MgO + \underset{x\text{ Meq.}}{2HCl} \rightarrow \underset{\frac{x}{2}\text{ mole}}{MgCl_2} + H_2O$$

$$\underset{(60-x)\text{Meq.}}{Mg_3N_2} + \underset{(60-x)\text{Meq.}}{HCl} \rightarrow \underset{\frac{60-x}{2}\text{mole}}{3MgCl_2} + \underset{\left(\frac{60-x}{3}\right)\text{mole}}{2NH_3}$$

$$\underset{\left(x+\frac{60-x}{2}\right)}{MgCl_2} + \underset{12\text{Meq.}}{2NaOH} \rightarrow Mg(OH)_2 + 2NaCl$$

$NH_3 + HCl \rightarrow NH_4^+ + Cl^-$

initially	10 Meq.	x mole
after	(10 – x) m mole	

$$HCl + \underset{6\text{ Meq.}}{NH_4^+} + \underset{6\text{ Meq.}}{\overline{OH}} \rightarrow NH_4OH + Cl^-$$

$$\frac{x}{2}+\frac{60-x}{2}=\frac{12}{2}$$

x = 27.27%

Sol 22: $PV = n_T RT$

(1) (40) = n_T (0.0821) (400)

n_T = Total mole = 1.2180

$$\underset{x\ mole}{C_2H_6}+\frac{7}{2}O_2 \rightarrow 2CO_2 + 3H_2O$$

$$\underset{(1.218-x)\ mole}{C_2H_4} + 3O_2 \rightarrow 2CO_2 + 2H_2O$$

Mole of O_2 required

$$= \frac{7}{2}x + 3(1.218 - x) = \frac{130}{32}$$

x = 0.817 mole

Mole fraction of C_2H_4 = $\frac{1.218-0.817}{1.218} = 0.33$

Mole fraction of C_2H_6 = 0.67

Sol 23: $\underset{x\ gm}{Pb(NO_3)_2} \xrightarrow{\Delta} PbO + 2NO_2 + \frac{1}{2}O_2$

$$\underset{(5-x)gm}{2NaNO_3} \xrightarrow{\Delta} Na_2O + 2NO_2 + \frac{1}{2}O_2$$

$$\left[\frac{(5-x)}{85}\right]+\frac{1}{4}\left(\frac{5-x}{85}\right)+2(x)$$

Sol 24: $\underset{11.25\ m\ mole}{3Pb(NO_3)_2}+\underset{2.5\ m\ mole}{Cr_2(SO_4)_3} \rightarrow 3PbSO_4 + 2Cr(NO_3)_3$

L. R. = $CrSO_4$

So moles of $PbSO_4$ formed

= 2. 5 m mole × 3

= 7. 5 m mole

Molar conc. of $[Pb^{2+}] = \frac{11.25-7.5}{70}$ = 0. 0536 M

Molar conc. of $[NO_3^-] = \frac{(2\times 11.25)}{70}$ = 0. 32 M

$$[Cr^{3+}] = \frac{2\times 2.5}{70} = 0.0714M$$

Sol 25: NaCl

$CaCl_2 + Na_2CO_3 \rightarrow CaCO_3 + 2NaCl$

$$CaCO_3 \xrightarrow{\Delta} \underset{1.12gm}{CaO} + CO_2$$

Mole of CaO = $\frac{1.12}{56}$ = 0.02 mole

Moles of $CaCl_2$ = 0.02 mole

Weight of $CaCl_2$ = 2.22 gm

NaCl = 10 – 2.22 = 7.78 gm

% NaCl = 77.8%

Sol 26: (i) $Fe_2O_3 + 2Al \rightarrow Al_2O_3 + 2Fe$

(ii) Mole ratio (to complete reaction) = 1 : 2

mass ratio = 1 × (112 + 48) (2 × 27) = 80 : 27

(iii) 2.7 kg of Al = $\frac{2700}{27}$ mole = 100 mole

16 kg of Fe_2O_3 = $\frac{16000}{160}$ mole = 1000 mole

L . R. = Al

So energy released = 200 × $\frac{100}{2}$ = 10000 unit

Sol 27: $N_2 : H_2$ (mole) = 1 : 3

$N_2 + 3H_2 \rightarrow 2NH_3$

Initially 1 3a

after 1–x 3–3x 2x

P(Molecular weight) = SRT

1(M.W.) = (0.497) × (0.0821) (298)

Molecular weight = 12. 15 gm

$$\frac{(2x)(17)+(3-3x)2+(1-x)28}{4-2x} = 12.15$$

34x + 6 – 6x + 28 – 28x = 48.63 – 24.31 x

24.31x = 14.63

x = 0.602

% composition by volume

$$N_2 = \frac{1-0.602}{4-2(0.602)} = \frac{0.398}{2.795} = 14.21\%$$

H_2 = 3(N_2%) = 42.86%

$$NH_3 = \frac{2(0.602)}{2.795} = 42.86\%$$

We know average molecular weight = 12.15

So (1) (12.15)

$$= \frac{Mass}{22.4} \times (0.0821) \times (273)$$

Mass = 12.14 gm.

Sol 28: $x(CH_3)_2SiCl_2 + Zn\overline{O}H \rightarrow ZnCl^- + nH_2O + [(CH_3)_2SiO]_n$

Volume of film = $6 \times 10^{-10} \times 300 \times 1 \times 3$ m^3

= 54×10^{-8} m3 = 0.54 cm^3

Mass of the film = 0. 54 gm

Mole of $[(CH_3)_2SiO]_n = \frac{0.54}{n[30+28+16]}$

Mass of $[(CH_3)_2SiCl_2] = \frac{0.54}{71} \times (58+71) = 0.941$ gm.

Sol 29 $P_4 + 3O_2 \rightarrow P_4O_6$ ....(i)

$P_4 + 5O_2 \rightarrow P_4O_{10}$ ....(ii)

(i) $P_4 + 3O_2 \rightarrow P_4O_6$

1–x 4–y

$P_4 + 5O_2 \rightarrow P_4O_{10}$

x y

5x = y

3(1 – x) = 4 – y

3 – 3x = 4 – 5x

$x = \frac{1}{2}$

$y = \frac{5}{2}$

$P_4O_6 = P_4O_{10} = 50\%$

(ii) $P_4 + 3O_2 \rightarrow P_4O_6$

3–x 11–y

$P_4 + 5O_2 \rightarrow P_4O_{10}$

x y

5x = y

3(3 – x) = 11 – y

9 – 3x = 11 – 5x

x = 1

$P_4O_{10} = \frac{1}{3}$; $P_4O_6 = \frac{2}{3}$

(iii) $P_4 + 3O_2 \rightarrow P_4O_6$

3–x 13–y

$P_4 + 5O_2 \rightarrow P_4O_{10}$

x y

5x = y

3(3 – x) = 13 – y

9 – 3x = 13 – 5x

x = 2

$P_4O_6 = 2$

$P_4O_{10} = 1$

Sol 30: $Cl^- + AgNO_3 \rightarrow AgCl + NO_3^-$

Let's say V mL must be added

Weight of solution = (1. 04 V) gm

Weight of $AgNO_3$ = 0.05 × (1.04 V)gm

Moles of $AgNO_3 = \frac{(0.05)(1.04V)}{173}$

Minimum moles of Cl^- (it will be case of more molecular weight i. e. KCl)

$$= \frac{0.3}{39+35.5} = \frac{0.3}{74.5}$$

$$\frac{(0.05)(1.04V)}{173} = \frac{0.3}{74.5}$$

V = 13. 4 mL.

Sol 31: In 500 mL of NaOH

Weight of solution = 1.8 × 500 = 900 gm

So, weight of NaOH = (0.08) (900) = 72 gm

Mole of NaOH = $\frac{72}{40}$ = 1.8 mole

Moles of H^+ = 18 mole

On heating $NaHCO_3 \rightarrow CO_2 + H_2O$

On C-balance $n_{CO_2} = n_{NaHCO_3}$

Mass of $NaHCO_3 = \frac{18.6}{44} \times (84) = 33.50$ gm

$$H^+ = 1.8 = \frac{18.6}{44} + 3\left[\frac{x}{27+3(35.5)}\right] + 0$$

$1.8 = \frac{x}{44.5} + 0.418 \Rightarrow x = 61.5$ gm = mass of $AlCl_3$

Mass of KNO_3 = 124 – 97 = 27 gm

Total mole = 0.267 + 0.460 + 0.422 = 1.149 mole

Sol 32: $\frac{1}{2}\underset{\text{(acetone)}}{CH_3COCH_3} + \frac{3}{2}CaOCl_2 \rightarrow \underset{30gm}{CHCl_3} + x$

Mole of $CHCl_3 = \dfrac{30}{119.5}$

By carbon balance

Mole of acetone (ideally)

$= \dfrac{1}{2} \times \dfrac{30}{119.5}$

As the yield is 75%

So, weight required

$= \dfrac{30}{2 \times 119.5} \times \dfrac{100}{75} \times (58) = 9.7 \text{ gm}$

Sol 33: $Cu_2O + x$

Let's assume total 100 gm is given

Cu = 66.67 gm

$0 = \dfrac{66.67}{63.5} \times \dfrac{1}{2} \times 16$

Oxygen (O) = 8.4

% Cu_2O = 66.67 + 8.4 = 75%

Sol 34: $Hg \quad + \quad I_2 \rightarrow \quad HgI_2$

$\left(\dfrac{M}{200} - x\right) \quad \left(\dfrac{M}{254} - \dfrac{x}{2}\right)$

$2Hg \; + \; I_2 \; \rightarrow \; Hg_2I_2$

x mole $\dfrac{x}{2}$

Let's say M gm is initially taken

$\dfrac{M}{200} - x = \dfrac{M}{254} - \dfrac{x}{2}$

$M\left(\dfrac{54}{200}\right)\dfrac{1}{254} = \dfrac{x}{2}$

$M = \left(\dfrac{254}{0.54}\right)x$

(gm) $Hg_2I_2 = \dfrac{x}{2} \times (200 + 127) \times 2 = 327 \, x \, HgI_2$(gm)

$= \left(\dfrac{M}{200} - x\right) = \left[\dfrac{254}{(0.54)(200)} - 1\right]$

x Molecular weight = (1.351 × 454) × HgI_2

$HgI_2 : Hg_2I_2$ = 0.532 : 1.

Redox Reactions

Sol 1: (a) $NaNO_2$

$Na^{+} N^{x+} 2O_2^{-}$ + 1 + x − 4 = 0; x = +3

(b) H_2

2x = 0; x = 0

(c) Cl_2O_7

2x − 14 = 0; x = +7

(d) $KCrO_3Cl$

$K^{+} Cr^{x+} 3O^{2-} Cl^{-}$; + 1 + x − 6 − 1 = 0; x = +6

(e) Ba Cl_2

+ x − 2 = 0; x = +2

(f) ICl_3

+ x − 3 = 0; x = +3

(g) $K_2Cr_2O_7$

+ 2 + 2x − 14 = 0; x = +6

(h) CH_2O

+ x + 2 − 2 = 0; x = 0

(i) Ni $(CO)_4$

+ x + 0 = 0; x = 0

(j) NH_2OH

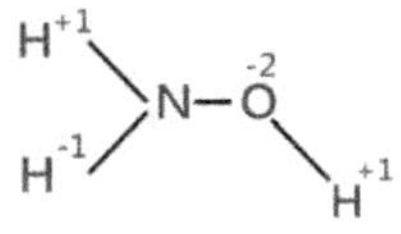

+ 3 + x − 2 = 0; x = -1

Sol 2: (a) $\overset{+1}{CuSO_4} + 4KI \longrightarrow 2\overset{+1}{CuI} + \underset{\text{Reduced}}{\overset{0}{I_2}} + \underset{\text{Oxidised}}{\overset{+2}{2K_2SO_4}}$

(b) $2Na_2S + 4HCl + SO_2 \longrightarrow 4NaCl + \underset{\text{Oxidised}}{3S} + \underset{\text{Reduced}}{2\,H_2O}$

(c) $NH_4NO_2 \xrightarrow{\Delta} \underset{\text{Reduced}}{N_2} + \underset{\text{Oxidised}}{2H_2O}$

Sol 3: (a) $3Fe + 4H_2O \longrightarrow Fe_3O_4 + 4H_2$ = 8 electrons

(b) $AlCl_3 + 3K \longrightarrow Al + 3KCl$ = 3electrons

Sol 4: (a) Sulphur shows various oxidation states such as −2, 0, +2, +4, +6

In H_2S, oxidation no. of S is '−2'

So now it can only act as a reducing agent because it can't get more electrons since S^{2-} is in its lowest oxidation state.

But in SO_2, oxidation state of S is +4 which is an intermediate oxidation state. So, it can gain or lose electrons and can go to –2, 0, +2 or higher +6 oxidation state.

(b) Oxidation state of oxygen in H_2O_2 is '–1' so it can get oxidised or reduced because O have multiple oxidation state like –2, –1, 0. By losing electrons it can form O_2 and act as a reducing agent and by gaining e^-s, it can form H_2O and behave as an oxidising agent.

Sol 5: NO_2^- is oxidized to NO_3^- by MnO_4^- (in basic medium) which is reduced to MnO_2

$$\underset{+7}{MnO_4^-} + \underset{+3}{NO_2^-} \longrightarrow \underset{+5}{NO_3^-} + \underset{+4}{MnO_2}$$

oxidation ($NO_2^- \to NO_3^-$)

reduction ($MnO_4^- \to MnO_2$)

Thus, $MnO_4^- \longrightarrow MnO_2$ oxidation number decreases by 3-units

$NO_2^- \longrightarrow NO_3^-$ oxidation number increases by 2 units

Thus, $2MnO_4^- \equiv 3NO_2^-$

$$MnO_4^- \equiv \frac{3}{2}NO_2^- = 1.5 \text{ mol } NO_2^-$$

Sol 6: (a) HSO_3^- (b) NO_2^- (c) Cl^-

Sol 7: (a) $4Zn + 10HNO_3 \longrightarrow 4Zn(NO_3)_2 + N_2O + 5H_2O$

(b) $6HI + 2HNO_3 \longrightarrow 3I_2 + 2NO + 4H_2O$

Sol 8: 5 moles of H_2SO_4 can produce

1 mole of H_2S

$0.2 \times V \times 10^{-3} = nH_2SO_4$

(equating equivalents)

∴ Volume = 25 lit.

Sol 9: ∴ $20 \times 0.2 \times 2 = 0.167$ M

Normality = $n_f \times M = 0.5$ N

Sol 10: mole of $As_2O_3 = 5.54 \times 10^{-4}$

equating equivalents,

$(5.54 \times 10^{-4}) \times (2) = (26.1 \times 10^{-3}) \times M \times 5$

∴ Molarity = 8.49×10^{-3},

Normality = molarity × n-factor

$= (8.49 \times 10^{-3}) \times (5) = 4.24 \times 10^{-2}$

Sol 11: $CaO \longrightarrow CaC_2O_4$

$$\overset{+3}{CaC_2O_4} + K\overset{}{MnO_4} \longrightarrow \overset{+4}{CO_2} + Mn^{2+}$$

+5e⁻ (Mn), –2e⁻ ($C_2O_4^{2-}$)

Equating equivalents

$(\text{equivalent})_{CaC_2O_4} = (\text{equivalent})_{KMnO_4}$

$n_{CaC_2O_4} \times 2 = 40 \times 0.25 \times 10^{-3}$

Moles of $CaC_2O_4 = 5 \times 10^{-3}$

∴ Mole of CaO = $CaC_2O_4 = 5 \times 10^{-3}$

∴ Mass of CaO = 0.28

∴ % composition

$$= \frac{0.28}{0.518} \times 100 = 54\%$$

Sol 12: Reaction

$$KMnO_4 + \overset{-1}{H_2O_2} \longrightarrow Mn^{2+} + O_2$$

5e⁻ (Mn), +2e ($H_2O_2 \to O_2$)

Assume mass of H_2O_2 = x gm

∴ Molarity of

$$H_2O_2 = \frac{\frac{x}{34} \times 1000}{20} = 147 \text{ x M}$$

$$\text{Moles of } KMnO_4 = \frac{0.316}{158} = 2 \times 10^{-3}$$

Now equating equivalents,

Equivalents of H_2O_2 = Equivalents of $KMnO_4$

$1.47x \times 20 \times 10^{-3} \times 2 = 2 \times 10^{-3} \times 5$

∴ x = 0.17 gm

∴ Purity of H_2O_2 = 85% (i)

moles O_2 evolved = moles of H_2O_2 consumed.

∴ Moles of $O_2 = 5 \times 10^{-3}$

$$\therefore \text{Volume} = \frac{nRT}{P} = 124.8 \text{ ml} \quad \text{(ii)}$$

Sol 13: (CaOCl)+ Cl^- 5.7 gm is taken

∴ Lets take moles of $CaOCl_2$ = x

molarity of

$$CaOCl_2 = \frac{x}{500} \times 1000 = 2x$$

Now on treatment with KI + HCl

$(CaOCl)^+Cl^- + KI + HCl \rightarrow I_2$

$\downarrow$ treated with $Na_2S_2O_3$

$\therefore$ Equivalents of I_2 = equivalents of $Na_2S_2O_3$

$= 24.35 \times \frac{1}{10} \times 10^{-3} = 2.435$ milliequ.

Now, equiv. of I_2 = equiv. of Bleaching powder

$2.435 \times 10^{-3} = 2x \times 25 \times 10^{-3}$

$\therefore x = 4.87 \times 10^{-2}$

$\therefore$ Mass of bleaching powder = 1.73 gm

$\therefore$ % availability $= \frac{1.73}{5.7} \times 100 = 30.33\%$

Sol 14: (i) $3C_2H_5OH + 2K_2Cr_2O_7 + 8H_2SO_4 \longrightarrow$

$3\ C_2H_4O_2 + 1Cr_2(SO_4)_3 + 2\ K_2SO_4 + 11H_2O$

(ii) $1As_2S_5 + 40HNO_3 \longrightarrow$

$40NO_2 + 12H_2O + 2H_3AsO_4 + 5H_2SO_4$

(iii) $2CrI_3 + 27Cl_3 + 64KOH \longrightarrow$

$6KIO_4 + 2K_2CrO_4 + 54KCl + 32H_2O$

(iv) $3As_2S_3 + 14HClO_3 + 18H_2O \longrightarrow$

$14HCl + 6H_3AsO_4 + 9H_2SO_4$

Sol 15:

(i)

$As_2S_3 + 12OH^- + 14H_2O \longrightarrow 2As^{3-}O_4 + 3S^{2-}O_4 + 20H_2O$

(ii)

$2CrI_3 + 10OH^- + 27H_2O_2 \longrightarrow 2Cr^{2-}O_4 + 6IO_4^- + 32H_2O$

(iii) $P_4 + 3OH^- + 3H_2O \longrightarrow 3H_2PO_2^- + PH_3$

(iv) $3As_2S_3 + 4H_2O + 10NO_3^- + 10^+ \longrightarrow$

$6H_3AsO_4 + 9S + 10NO$

Sol 16:

$\overset{+5}{NO_3^-} + Mg(s) + H_2O \rightarrow Mg(OH)_{2(g)} + OH^-_{(aq.)} + \overset{-3}{NH_{3(g)}}$

+8e

2e⁻

$NH_3 + HCl \longrightarrow NH_4Cl$

Say molarity of NO_3^- ions = x M

$\therefore$ Moles of $NO_3^- = x \times 25 \times 10^{-3}$

Equivalents of NO_3^-

= Equivalents of $NH_3 = 8 \times x + 25 \times 10^{-3} = 0.2\ x$

$\therefore$ Moles of $NH_3 = 0.2\ x$

Moles of NaOH $= 32.10 \times 10^{-3} \times 0.1 = 3.21 \times 10^{-3}$

Now, moles of HCl = (moles of NH_3) + (moles of NaOH)

$50 \times 0.15 \times 10^{-3} = 0.2x + 3.21 \times 10^{-3}$

$x = 2.145 \times 10^{-2}$

$\therefore$ Molarity = 8x = 0.1716 M

Sol 17: $K\overset{+7}{Re}O_4 + Zn \longrightarrow Zn^{2+} + Re^{+x}$

$\downarrow \overset{+7}{KMnO_4}$

$\overset{+7}{ReO_4^-} + Mn^{2+}$

$\therefore$ Moles of $KReO_4 = 9.28 \times 10^{-5}$

$\therefore$ Moles of $KMnO_4 = 0.05 \times 11.45 \times 10^{-3} \times 5$

Now equating equivalents of

$Re^{x+} = KMnO_4$

$(x_{Re^{x+}})(7 - x) = 5.725 \times 10^{-4}$ .... (1)

Now equating equivalents of

$KReO_4 = Re^{x+}$

$9.28 \times 10^{-5} = Re^{x+} = x_{Re^{x+}}$

$\therefore 7 - x = \frac{5.725 \times 10^{-4}}{9.28 \times 10^{-5}}$; x = +1

Sol 18: Let moles of FeC_2O_4 = x

$FeSO_4$ = y

$FeC_2O_4 + FeSO_4 + KMnO_4 \rightarrow Fe^{2+} + CO_2\uparrow$

$\downarrow$ Zn-dil HCl

Fe^{+3}

$\downarrow$ $KMnO_4$

Fe^{+2}

Now, (2x + x + y) 0.1 = 60 × 0.02 × 5

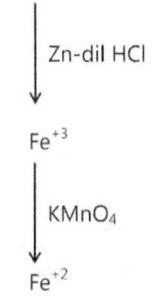

$3x + y = 60$

$x + y = 40 \times 0.02 \times 5 = 4$

$\therefore 4 - x = 6 - 3x$

$2x = 2$

$x = 1$

$y = 3$

$\therefore$ Normality $= 1 \times 3 \times 10^{-2} = 0.03N$ of FeC_2O_4

$= 3 \times 10^{-2} = 0.03$ M of $FeSO_4$

Sol 19: Mass of KCl = x gm

$H_2O = 1 - x - y$ gm

$KClO_3 = y$ gm

Treating with SO_2

$$\overset{+5}{ClO_3^-} + SO_2 \longrightarrow SO_4^{2-} + Cl^-$$

with $+6e^-$ and $-2e^-$ electron transfers marked

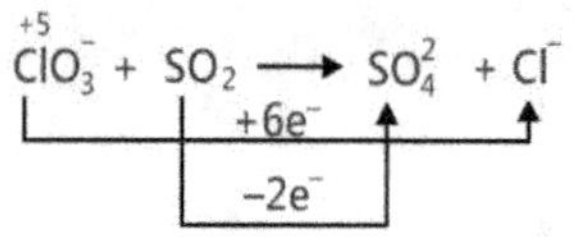

Then silver chloride formed

$\therefore$ Total moles of chloride $= 10^{-3} = \dfrac{x}{74.5} + \dfrac{y}{122.5}$

Now for another experiment

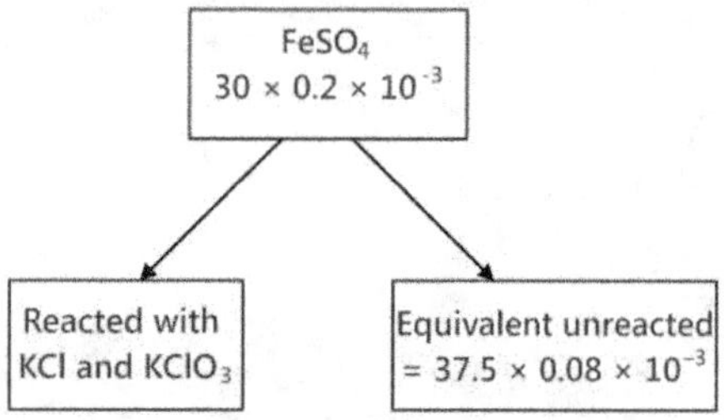

$\therefore$ Equivalents reacted $= 3 \times 10^{-3}$

$$\overset{+5}{ClO_3^-} + 6Fe^{2+} + 6H^+ \rightarrow Cl^- + 6Fe^{3+} + 3H_2O$$

with $+6e^-$ and $+e^-$ electron transfers marked

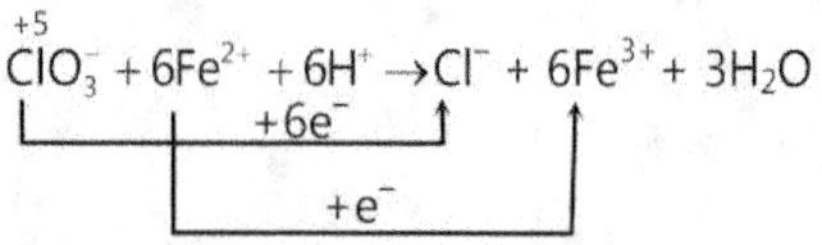

Equivalents of Fe^{2+} = Equivalent of ClO_3^-

3×10^{-3}

$= \dfrac{\frac{y}{122.5}}{25010} \times 10^{+3} \times 25 \times 10^{-3} \times 6$

$y = 0.6125$ gm

Moles of ClO_3^- = 0.005

Molar ratio = 1 : 1

Sol 20: (iii) One mL of $Na_2S_2O_3$ is equivalent to

0.0499 gm of $CuSO_4$

0.2 millimole of $CuSO_4$

Since reaction is balanced

$Hg_5(IO_6)_2 \longrightarrow 8I_2$

| 1 | 8 |

5×10^{-4} 4 millimole

$2\,CuSO_4 \longrightarrow I_2$

| 2 | 1 |

0.2 millimole   0.1 millimole

$\therefore$ 0.1 millimole of I_2 equivalent to 0.2 millimole of $CuSO_4$

But we have 4 millimole

So, 1 mL $\longrightarrow$ 0.1 millimole

40 mL $\longleftarrow$ 4 millimole

Sol 21: $BaCO_3 + CaCO_3 + CaO$

with x, y labels and dil. HCl arrow

$CO_2\uparrow \quad H_{CO_2} = 7.5 \times 10^{-3}$

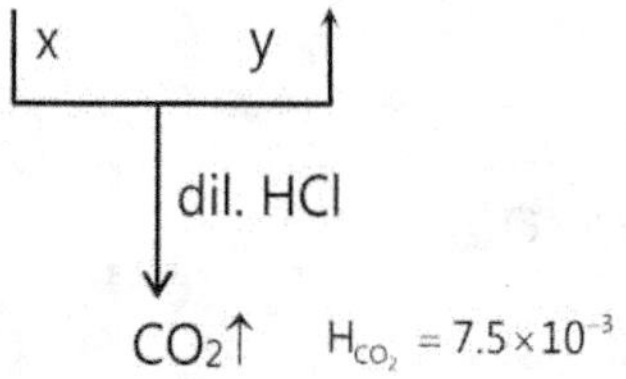

$x + y = 7.5 \times 10^{-3}$

$BaCrO_4 + I^- \longrightarrow I_2 + Cr^{3+}$

$0.1\,x$

Equating equivalents of $BaCrO_4 = I^-$

$0.1\,x - 3 = 20 \times 0.05 \times 10^{-3}$

$x = \dfrac{10^{-2}}{3} = 3.33 \times 10^{-3}$

$\therefore y = 4.167 \times 10^{-3}$

Mass of $BaCO_3 = 0.659$ gm

$CaCO_3 = 0.4167$ gm

$\therefore$ Mass of CaO = 0.1733

% CaO = 13.87%

Sol 22: $Cu_2S + CuS + MnO_4^- \longrightarrow$

      x   y

$Mn^{2+} + Cu^{2+} + SO_2$

$200 \times 0.75 \times 10^{-3} = \dfrac{175 \times 10^{-3}}{5} +$ moles of required

$\therefore$ Moles required of MnO_4^- = 0.115

Let moles of $Cu_2S^{-2} = x$

           CuS = y

$\therefore 0.115 \times 5 = 8x + 6y$

Let m = mass of Cu_2S

8x + 6y = 0. 575

$$\frac{8m}{159} + \frac{6(10-m)}{95.5} = 0.575$$

$[(5.03 \times 10^{-2}) - (6.28 \times 10^{-2})]\ m = -5.327 \times 10^{-2}$

m = 4.26 gm

$$\%\ CuS = \frac{5.74}{10} \times 100 = 57.4\%$$

Sol 23: $2H^+ + \overset{0}{O_3} + I^- \rightarrow O_2 + I_2 + H_2O^{-2}$

moles of air

$= 0.406 = \frac{PV}{RT}$ ↓ $Na_2S_2O_3$

Equivalents of $I_2 = 1.5 \times 10^{-3} \times 0.01 = 1.5 \times 10^{-5}$

Moles of O_3 = x

∴ x × 6 = 1.5×10^{-5}

x = 2.5×10^{-6}

∴ Volume of $O_3 = 1.847 \times 10^{-4}$ lit.

$$\%\ O_3 = 1.847 \times 10^{-3} = \frac{V_{O_3}}{10} \times 100$$

Sol 24: $MnO_4^- + H_2C_2O_4 + Fe^{2+} \longrightarrow$

30×1.5   1.5×0.5   15×0.4

$Fe^{3+} + CO_2 + Mn^{2+}$

(30 × 1.5) – (15 × 0.5 + 15 × 0.4) = 31.5 milliequivalents

$$\therefore \text{Final of } MnO_4^- = \frac{31.5 \times 10^{-3}}{60 \times 10^{-3}} = 0.525\ M$$

Final molarity of

$$Fe^{3+} = \frac{15 \times 0.4 \times 10^{-3}}{60 \times 10^{-3}} = 0.1\ M$$

∴ Final normalities $MnO_4^- = 0.105M$

$Fe^{3+} = 0.1M$

Sol 25: (i) Equivalents of $I^- = 20 \times 0.1 \times 10^{-3}$

$$H_2\overset{-1}{O_2} + I^- \longrightarrow H_2\overset{-2}{O} + I_2$$

∴ Equivalents of $H_2O_2 = 2 \times 10^{-3}$

$$\therefore \text{Moles of } H_2O_2 = \frac{2 \times 10^{-3}}{2} = 10^{-3}$$

$$\text{Now, molarity} = \frac{10^{-3}}{25} \times 10^3 = 0.04\ M$$

∴ Normality = 0.04 × 2 = 0.08 N

(ii) $20 \times 0.3 \times 10^{-3}$ = equivalent of I^-

$$\therefore \text{Normality of } H_2O_2 = \frac{20 \times 0.3 \times 10^{-3}}{25 \times 10^{-3}} = 0.24$$

∴ Strength = 1.344

Sol 26: Let molarity of $Na_2S_2O_3$ solution = xM

∴ Equivalent of thiosulphate

= Eq. of I^- = Eq. of I_2 = $x \times 45 \times 10^{-3}$

$$\frac{0.1}{214} \times 5 = x \times 45 \times 10^{-3}$$

x = 0.062 M

Sol 27: $\overset{+2}{Mn}SO_4 \cdot 5H_2O \xrightarrow{\Delta} \overset{+8/3}{Mn_3}O_4$

Now $\overset{+8/3}{Mn_3}O_4 + FeSO_4 \rightarrow Fe^{3+} + Mn^{+2} + KMnO_4$

Let assume no. of moles of

$MnSO_4 \cdot 5H_2O$ = x

∴ Moles of Mn_3O_4 = 3x

$3x(6) + 100 \times 0.1 \times 10^{-3} = 0.12 \times 100 \times 10^{-3}$

25 × N = 30 × 0.1

N = 0.12

∴ x = 1.11×10^{-4}

∴ Mass of $MnSO_4 \cdot 4H_2O$ = 1.338 gm.

Sol 28:

(i) $\overset{+5}{Cl}O_3^- + Fe^{+2} \longrightarrow Cl^- + Fe^{+3} + H_2O$

$6H^+ + \overset{+5}{Cl}O_3 + 5e^- \longrightarrow Cl^- + 3H_2O$

$Fe^{+2} \longrightarrow Fe^{+3} + e^- \times (5)$

$6H^+ + ClO_3^- + 5Fe^{2+} \rightarrow 5Fe^{3+} + Cl^- + 3H_2O$

(ii) $8CuS^{-2} \rightarrow S_8 + 16e^- + 8Cu^{2+}$ ×(3)

$4H^+ + \overset{+5}{N}O_3^- + 3e^- \rightarrow \overset{+2}{N}O + 2H_2O$ ×(16)

$24CuS + 16NO_3^- + 64H^+ \longrightarrow 24Cu^{2+} + 3S_8 +$

$16NO + 32H_2O$

(iii) $S_2O_3^{2-} + Sb_2O_5 \longrightarrow SbO + H_2SO_3$

$6H^+ + Sb_2O_5 + 6e^- \rightarrow 2SbO + 3H_2O$

$H_2O + \overset{+2}{S_2}O_3^{2-} \rightarrow 2H_2\overset{+4}{S}O_3 + 4e^- + 2H^+$

$Sb_2O_5 + S_2O_3^{2-} + 4H^+ \rightarrow 2SbO + 2H_2SO_3$

(iv) $2HCl^{-1} \rightarrow Cl_2^{o} + 2H^+ + 2e^- \quad \times (5)$

$5e^- + K\overset{+7}{M}NO_4 \rightarrow KCl + \overset{+2}{Mn}Cl_2 + 4H_2O \times (2) + 8H^+$

$10HCl + 2KMnO_4 \rightarrow 5Cl_2 + 2KCl + 2MnCl_2 + 8H_2O$

(v) $H_2SO_4 + K\overset{+5}{Cl}O_3 \rightarrow H\overset{+7}{Cl}O_4 + 2e^- + KHSO_4$

$3SO_4 + K\overset{+5}{Cl}O_3 \rightarrow \overset{+4}{Cl}O_2 + H_2O + KHSO_4$

$3KClO_3 + 3H_2SO_4 \rightarrow 3KHSO_4 + HClO_4 + 2ClO_2 + H_2O$

(vi) $4H^+ + H\overset{+5}{N}O_3 + 3e^- \longrightarrow \overset{+2}{N}O + 2H_2O \times (2)$

$2H\overset{+1}{B}r \longrightarrow Br_2 + 2e^- + 2H^+ \quad \times (3)$

$2HNO_3 + 6HBr \rightarrow 2NO + 3Br_2 + 4H_2O$

(vii) $H^+ + 2\overset{+7}{I}O_4^- + 14e^- \rightarrow \overset{0}{I_2} + 4H_2O$

$2I^- \rightarrow I_2 + 2e^- \times (7)$

$IO_4^- + 7I^- + 8H^+ \rightarrow 4I_2 + 4H_2O$

Sol 29: $P_4 + 3O_2 \rightarrow P_4O_6$ .... (i)

$P_4 + 5O_2 \rightarrow P_4O_{10}$ .... (ii)

(i) $P_4 + 3O_2 \rightarrow P_4O_6$

1–x   4–y

$P_4 + 5O_2 \rightarrow P_4O_{10}$

x   y

$5x = y$

$3(1 - x) = 4 - y$

$3 - 3x = 4 - 5x$

$x = \frac{1}{2}$

$y = \frac{5}{2}$

$P_4O_6 = P_4O_{10} = 50\%$

(ii) $P_4 + 3O_2 \rightarrow P_4O_6$

3–x   11–y

$P_4 + 5O_2 \rightarrow P_4O_{10}$

x   y

$5x = y$

$3(3 - x) = 11 - y$

$9 - 3x = 11 - 5x$

$x = 1$

$P_4O_{10} = \frac{1}{3}$; $P_4O_6 = \frac{2}{3}$

(iii) $P_4 + 3O_2 \rightarrow P_4O_6$

3–x   13–y

$P_4 + 5O_2 \rightarrow P_4O_{10}$

x   y

$5x = y$

$3(3 - x) = 13 - y$

$9 - 3x = 13 - 5x$

$x = 2$

$P_4O_6 = 2$

$P_4O_{10} = 1$

Sol 30:

(i) $H_2O + 2e^- + \overset{+1}{Ag_2}O \rightarrow 2Ag + 2OH^-$

$4OH^- + S_2O_4^{2-} \rightarrow 2SO_3^{2-} + 2e^- + H_2O$

$S_2O_4^{2-} + Ag_2O + 2OH^- \rightarrow 2Ag + 2SO_3^{2-} + H_2O$

(ii) $\overset{o}{Cl_2} + 2e^- \rightarrow 2Cl^-$

$\overset{o}{Cl_2} + 2OH^- \rightarrow 2\overset{+1}{Cl}O^- + 2e^- + 2H_2O$

$Cl_2 + 2OH \rightarrow Cl^- + ClO^- + H_2O$

(iii) $2OH^- + \overset{0}{H_2} \rightarrow \overset{+1}{H_2}O + 2e^- + H_2O \times (3)$

$3e^- + \overset{+7}{Re}O_4 \rightarrow \overset{+4}{Re}O_2 + 4OH^- \quad \times(2)$

$3H_2 + 2ReO_4 \rightarrow 2ReO_2 + 2H_2O + 2OH^-$

(iv) $\overset{+4}{Cl}O_2 + e^- \rightarrow \overset{+3}{Cl}O_2^- \quad \times(2)$

$2H_2O + \overset{+3}{Sb}O_2^- \rightarrow \overset{+5}{Sb}(OH)_6^- + 2e^- + 2OH^- + H_2O$

$2ClO_2 + SbO_2^- + 2OH^- + 2H_2O \rightarrow 2ClO_2^- + Sb(OH)_6^-$

(v) $4H_2O + \overset{+7}{MnO_4^-} + 5e^- \rightarrow Mn^{+2} + 8OH^-$

$Fe^{+2} \rightarrow Fe^{+3} + e^- \qquad \times(5)$

$MnO_4^- + 5Fe^{+2} + 4H_2O \rightarrow Mn^{2+} + 5Fe^{3+} + 8OH^-$

Exercise 2

Mole Concept

Single Correct Choice Type

Sol 1: (D) $A + Cl_2 \rightarrow ACl_2$

$$\frac{x}{M} \quad \frac{y}{71+M}$$

$$\frac{x}{M} = \frac{y}{71+M}$$

$71x + Mx = My$

$$M = \frac{71x}{x-y}$$

Sol 2: (B) Equivalents of H_2SO_4 = 1.200 × 0.2 = 0.24

Moles of H_2SO_4 = 0.12

Mass of H_2SO_4 = 0.12 × 98 = 11.76 gm

Sol 3: (C) NaI consumption per day

$$= \frac{0.5}{100} \times 3 \text{ gm} = 0.015 \text{ gm}$$

Number of I^- = $\frac{0.015}{127+23} \times 6.023 \times 10^{23} = 6.023 \times 10^{19}$

Assertion Reasoning Type

Sol 4: (B) Statement-I: moles of $N_2 = \frac{0.28}{28} = 0.01$ mole

PV = nRT

At same P and T, $V \propto n$

If M. W. = 44 gm of gas

n = 0. 01 mole

$V \propto n$

So, volume will be same as moles are also same.

Sol 5: (A) We know that for isotopes

$M_{avg.} = x(M_1) + (1-x)M_2$

So, statement-II is explaining statement-I and both are correct.

Sol 6: (C) Statement-I: Mass of urea = 60

```
H2N–C–NH2
    ||
    O
```

Mass of nitrogen = 28

$$\% = \frac{28}{60} = 46.66$$

Statement-II: Urea not ionic.

Sol 7: (B) Statement-I: $S_2O_3^{2-}$

2x + 3(–2) = –2

x = +2

Statement: Yes, Because these may be per-oxide bond.

Sol 8: (A) Statement-I: Molarity = $\frac{n}{v}$ density increases $\Rightarrow n \uparrow$ (at const. V)

$$= \text{molality} = \frac{\text{moles of solute}}{\text{mass of solvent}}$$

Density increases = moles of solute ↑

Molality and molarity both changes.

Statement-II: Density results in change in mass thus increases moles.

Sol 9: (C) Statement-I: Incorrect because it depends in extent of reaction

Statement-II: Correct.

Multiple Correct Choice Type

Sol 10: (A, C, D) (A) $NH_3 \rightarrow HNO_3 + HNO_2$ (till reaction III)

by nitrogen balance

$$n_{HNO_3} = \frac{1}{2} n_{NH_3}$$

(B) $3HNO_2 \rightarrow HNO_3 + 2NO + H_2O$

Let's say 1 mole of NH_3 is initially taken.

It makes $\frac{1}{2} - \frac{1}{2}$ mole of HNO_2 and HNO_3 till reaction-III $\frac{1}{2}$ mole HNO_2 make $\frac{1}{6}$ mole of HNO_3 in reaction-IV so HNO_3 made

$= \frac{1}{2}+\frac{1}{6}$ mole $= \frac{2}{3}$ mole

% increase $= \dfrac{\frac{1}{6}}{\frac{1}{2}} = \frac{100}{3}\%$

(C) By above data, it is correct

(D) Mole of NO produced $= \frac{1}{2}\times\frac{2}{3}$ = 50 % of HNO_3

Comprehension Type

Paragraph 1

Sol 11: (A) Initially mole of HCl $= \frac{1}{2}$ mole

$= \frac{1}{2} \times 36.5$ gm = 18.25 gm

So, after heating mole of HCl

$= \frac{18.25-2.75}{36.5} = \frac{15.5}{36.5} = 0.424$ mole

Normality $= \frac{0.424}{0.750} = 0.5662 \sim 0.58$

Sol 12: (C) Please note that, there is a small hypo in questions,

Instead of Ca(OH), it should be $Ca(OH)_2$

$Ca(OH)_2 + 2HCl \rightarrow CaCl_2 + 2H_2O$

Moles of HCl = 0.1 × 10 = 1m mole

Moles of $Ca(OH)_2$ required = 0.5 m mole

Volume $= \frac{0.5}{0.1}$ mL = 5 mL

Sol 13: (A) We know valency factor for Na_2CO_3 is 2

So, molarity will be $= \frac{0.5}{2} = 0.25M$

Sol 14: (A) 6.90 N means in 1 lit. solution

KOH = 6.90 moles

Weight of KOH = 6.90 × (56) = 386. 4 gm

given 30% by weight is KOH

So, weight of solution = 12.88 gm

Density $= \frac{12.88}{1} = 12.88$

Sol 15: (C) Ferrous ammonium sulphate

$= FeSO_4(NH_4)_2SO_4.6H_2O$

Molecular weight = 390

Moles in 0.1 N, 250 mL $= \frac{(0.1)(0.250)}{\text{Valency factor}}$

$Fe^{2+} \rightarrow Fe^{3+}$ Valency factor = 1

Mass of ferrous ammonium sulphate required

= (0.1) (0.250) (390) = 9.8 gm

Paragraph 2

Sol 16: $\underset{(4.925-x)gm}{CuCl_2} + \underset{(5.74-y)gm}{AgCl} \rightarrow$ unreacted

$\underset{x\,gm}{CuBr_2} + \underset{y\,gm}{2AgCl} \rightarrow 2AgBr + CuCl_2$

Let's say initially $CuBr_2$ = x gm

$CuCl_2$ = 4. 925 – x gm

AgCl = y gram (reacts with reacted)

AgCl = 5. 74 – y gram (in reacted)

Finally same AgCl → AgBr and $CuBr_2$ → $CuCl_2$ (completely)

Moles of AgCl in reaction $= \frac{y}{143.5}$

= Mole of AgBr produced

Finally AgCl = (5.74 – y) gm

AgBr $= \frac{y}{143.5} \times (80+108) = y(1.310)$

AgCl + AgBr = 6.63 = 5.74 + y(0.310)

y = 2.87 gm

So moles of $CuBr_2$ $= \frac{2.87}{2\times143.5} = \frac{x}{223.5}$

x = 2.235 gm

(1) **(C)** $CuBr_2$ mass % $= \frac{2.235}{4.925} = 45.38\%$

(2) **(B)**% mass of Cu =

$\left[\left(\frac{2.235}{223.5}\right)+\left(\frac{4.925-2.235}{63.5+71}\right)\right]\times\frac{63.5}{4.925}$

$= \frac{0.03\times63.5}{4.925} = 38.68\%$

(3) **(B)** Mole % of AgBr =

$$\frac{\left(\frac{2.87}{143.5}\right)}{\left(\frac{2.87}{143.5}\right)+\left(\frac{9.74-2.87}{143.5}\right)} = 50\%$$

(4) **(A)** Moles of $CuBr_2$ = Moles of $CuCl_2$ produced

= 0.01 mole

Moles of $CuCl_2$ initially take

$$= \frac{4.925-2.235}{134.5} = 0.02 \text{ mole}$$

Mole of Cl^- in final solution = (0.01 + 0.02) × 2 = 0.06

Paragraph 3

Sol 17: $UF_6 + xH_2O \rightarrow UO_xF_y + \text{gas } (F_{6-y}, H_{2x})$

3.52gm 3.08 gm 0.8 gm

0.01 mole

Gas contains 95% fluorine by mass

$$= 0.8 \times \frac{95}{100} = (6 - y) \times 19$$

y = 5. 96

$$0.8 \times \frac{5}{100} = (2x)$$

x = 0. 02

(1) **(C)** So empirical formula $F_{6-596}\ H_{2(0.02)}$

$= F_{0.04}\ H_{0.04}$

= HF

(2) **(A)** Empirical formula of solid = UO_xF_y

final reaction

$UF_6 + BH_2O \rightarrow UO_xF_y + A(HF)$

0. 01 0. 01

A = 2B (H-balance)

6 = Y + A (F - balance)

$B = X = \frac{A}{2}$ (O - balance)

Y = 6 – A

$UO_{\frac{A}{2}}F_{6-A}$ molecular weight $= \frac{3.08}{0.01} = 308\text{gm}$

$238 + \frac{A}{2}(16) + (6 - A) \times 19 = 308$

8A + 114 – 19A = 70

11A = 44

A = 4

So UO_2F_2

(3) **(A)** % of F converted $= \frac{A}{6} = 66.66\%$

Match the Columns

Sol 18: A → r; B → p; C → q

$\frac{1}{3}$ Al_5O1_2 molecular weight = 267 + 135 + 196 = 598

(A) $Y = \frac{267}{598} = 44.95\%$

(B) $Al = \frac{135}{598} = 22.57\%$

(C) $O = \frac{196}{598} = 32.32\%$

Sol 19: A → r; B → q; C → p

$C_6H_8O_6$ molecular weight = 72 + 8 + 96 = 176

Moles of $C_6H_8O_6 = \frac{17.6 \text{ mg}}{176} = 0.1$ m mole

(A) O – atom $= 6 \times n_{C_6H_8O_6} \times N_A = 3.6 \times 10^{20}$

(B) Mole $= \frac{1}{176} = 5.68 \times 10^{-3}$

(C) Moles of $C_6H_8O_6$ = 0.1 m mole

Sol 20: **(C)** Volume strength

$2H_2O_2(\ell) \rightarrow O_2(g) + 2H_2O(\ell)$

1 lt. of H_2O_2 gives x lt. of O_2 gas then X is said to be volume strength of H_2O_2

It X – V is given at S. T. P. then

Mole of O_2 produced $= \frac{X}{22.4}$

Mole of H_2O_2 required $= \frac{X}{11.2}$ (in litre)

Molarity $= \frac{X}{11.2}$

Normality $= \frac{X}{11.2} \times \text{(valency factor)} = \frac{X}{5.6}$

Strength in g/L $= \frac{X}{11.2} \times 34 = \frac{17x}{5.6}$

Volume strength = Normality × 5.6

Sol 21: (B) (A) acid + acid → No reaction

$$M_{avg.} = \frac{M_1V_1 - M_2V_2}{V_1 + V_2} = \frac{\text{Total no. of moles}}{\text{Total volume}}$$

Similarly (B)

(C) acid + basic →

$$M_{avg.} = \frac{M_1V_1 - M_2V_2}{V_1 + V_2}$$

(D) Mili equivalent = x × M × VmL

(E) Molarity = $\frac{\text{moles}}{\text{volume}} = \frac{M_1V_1}{\text{Volume}} = \frac{M_1V_1}{V_2}$

or $= \frac{\text{moles}}{\text{volume}} = \frac{\text{mass (gm)}}{M_{solute} \times \text{Volume (lt.)}}$

Redox Reactions

Single Correct Choice Type

Sol 1: (C) $\overset{-2}{N_2H_4} \rightarrow y = 10e^-$

Each nitrogen coses $5e^-$

∴ Oxidation no. of N in

y = −2 + 5 = +3

Sol 2: (D) The ore which get easily oxidised is best reducing agent

$I^- \rightarrow \frac{1}{2} I_2$ is most feasible because.

Sol 3: (C) Alumino thermite process :-

$Al + Mn_3O_4 \rightarrow Al_2O_3 + Mn$

↓

Reducing agent

Sol 4: (D) (a) Oxidation number of S in $H_2S = +2$

Oxidation number of S in SO_2 = +4

(b) H_2O_2 can undergo reduction as well as oxidation because oxidation number of oxygen in H_2O_2 is -1. So, it can act both as reducing agent and oxidising agent.

Sol 5: (C) $\overset{+5}{ClO_3^-} + 6H^+ + x \rightarrow Cl^- + 3H_2O$

↓

$6e^- + 5 \rightarrow -1$

Sol 6: (A) $[Fe(H_2O)_5(NO)^+]^{-1}\ SO_4^{2-}$

$Fe^{x+}\ 5(H_2O)^0\ (NO)^+$

x + 1 = +2; x = +1

Sol 7: (A) KO_2^-

$K^+\ O_2^-\quad 2x = -1$

$\therefore x = -\frac{1}{2}$

Sol 8: (B) $\overset{0}{3Br_2} + 6CO_3^{2-} + 3H_2O \rightarrow 5Br^- + 6HCO_3^- + \overset{+5}{+5BrO_3^-}$

$BrO \rightarrow Br^{-1}$ Reduction

$BrO \rightarrow Br^{+5}$ Oxidation

Comprehension Type

Paragraph 1

Sol 9: (D) $H_2O_2 + KI \longrightarrow I_2$

(×5) ↓

hypo

Sol 10: (D) Eq. of hypo solution eq. of I_2

$20 \times 0.1 \times 10^{-3} = 50 \times 10^{-3} \times N_{H_2O_2}$

$\therefore N_{H_2O_2} = 0.04$

∴ Concentration of H_2O_2

in gm/lit. = $\frac{0.04}{4} \times 34 = 0.34$

Sol 11: (D) ∴ Eq. of MnO_2 + Eq. of hypo solution

$\frac{m}{87} \times 2 = 30 \times 0.1 \times 10^{-3}$

m = 0.1305

$\therefore \% = \frac{0.1305}{0.5} \times 100 = 26.1\%$

Sol 12: (D) $As^{5+} + 2I^- \longrightarrow As^{3+} + I_2$

($As^{5+} \rightarrow As^{3+}$: $+2e^-$; $2I^- \rightarrow I_2$: $-2e^-$)

∴ Valence factor = 5 for As

2 for I

Paragraph 2

Sol 13: (C) $\overset{0}{Cl_2} \longrightarrow Cl^- + \overset{+5}{ClO_3^-}$

$+e^-$; $-5e^-$

Disproportionation

(oxidation as well as reduction)

Sol 14: (B, C) $\overset{0}{I_2} + 2\overset{+2}{S_2}O_3^{2-} \longrightarrow 2\overset{-1}{I^-} + \overset{+5}{S_4}O_6^{2-}$

$+e^-$; $-5e^-$

Sol 15: (B) $2H_2\overset{-2}{S} + \overset{+4}{SO_2} \longrightarrow 2H_2O + 3S$

$-2e^-$; $+4e^-$

$H_2S \rightarrow S \quad 0-(-2) = +2$

$SO_2 \rightarrow S \quad 0-(4) = -4$

Multiple Correct Choice Type

Sol 16: (A, B, D) Meq. of formed = Meq. of HCl used for NH_3

$= 50 \times 0.15 - 32.10 \times 0.10$

$= 4.29$

These Meq. of NH_3 are derived using valance factor of $NH_3 = 1$ (an acid base reaction)

In redox change valence factor of NH_3 is 8;

$8e + N^{5+} \rightarrow N^{3-}$

$\therefore$ Meq. of NH_3 for valence factor

Also, Meq. of NO_3^- = Meq. of NH_3

$= 8 \times 4.29 = 34.32$

$\therefore N_{NO_3^-} = \dfrac{34.32}{25} = 1.37$

$(N \times V \text{ in mL} = \text{Meq.})$

Assertion Reasoning Type

Sol 17: (A)

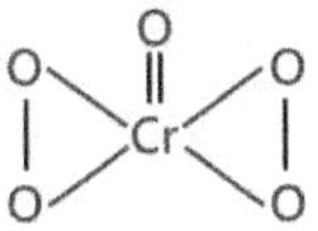

Valency of Cr is 6 all O have higher electronegativity than Cr

$\therefore$ Cr's oxidation no. = +6

$K_2Cr_2O_7 \underset{H^+}{\overset{OH^-}{\rightleftharpoons}} K_2CrO_4$

Orange dichromate — Yellow chromate

	$BaCl_2$	$+ Al_2(SO_4)_3$	$\longrightarrow BaSO_4$	$+ AlCl_3$
Meq. before reaction	$30 \times 0.1 = 3$	$40 \times 0.2 = 8$	$0 = 0$	$0 = 0$
Meq. after reaction	0	5	3	3

Sol 18: (D) Avg. oxidation no. of Pb_3O_4 is $+\frac{8}{3}$. But in reality, Pb_3O_4 is made up of $PbO + PbO_2$ .So, actively, Pb have oxidation state +2, +4.

Sol 19: (C) Oxidation no. of Cl = +7 it can not be greater than this

$\therefore$ It can get only reduced

$\therefore$ $HClO_4$ is an oxidising agent

$\therefore$ In $HClO_3$, oxidation no. of chlorine = + 5

E. N. order O > Cl > H

Sol 20: (D) Since S_2^{2-} has S – S⁻ linkage structure

$\therefore$ $FeS_2^{2-} \rightarrow Fe^{2+}$ (S – S) oxidation no. = +2

Sol 21: (B) Yes, the given reaction is an example of disproportionation

$\therefore$ H_2O_2 is a reducing as well as an oxidising agent

So it is not only bleaching (oxidising agent)

$H_2\overset{-1}{O_2} \longrightarrow H_2\overset{-2}{O} + \frac{1}{2}\overset{0}{O_2}$

$+e^-$; $-e^-$

Sol 22: (A) $K_2Cr_2O_7 \underset{H^+}{\overset{OH^-}{\rightleftharpoons}} K_2CrO_4$

Orange dichromate — Yellow chromate

Sol 23: (B) $\overset{0}{I_2} \longrightarrow \overset{+5}{IO_3^-} + \overset{-1}{I^-}$

$-5e^-$; $+e^-$

These reactions show E° > 0

∴ It is not feasible because iodine can show multiple oxidation state.

Match the Columns

Sol 24: $A \rightarrow w$; $B \rightarrow x$; $C \rightarrow u$; $D \rightarrow p$; $E \rightarrow v$; $F \rightarrow q$; $G \rightarrow r$; $H \rightarrow s$; $I \rightarrow t$

(1) Increase in oxidation no:- Loss of electrons (oxidation)

(2) Decrease in oxidation no:- Gain of only e^-s (reduction)

(3) Oxidation agent:- Gain of e^-s

(4) Reducing agent:- Loss of e^-s

(5) $2Cu^+ \longrightarrow Cu^{2+} + Cu$

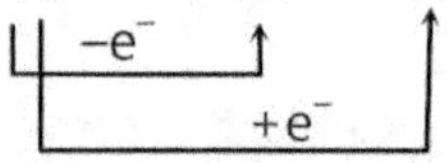

Disproportionation reaction

(6) Redox reaction

(7) Mn_3O_4 oxidation no:-

$+\frac{8}{3}$ fractional

(8) CH_2Cl_2

$x + 2 - 2 = 0$ zero oxidation no.

$x = 0$

(9) NaOH + HCl → NaCl + H2O

Simple neutralisation reaction

Sol 25: $A \rightarrow p, s$; $B \rightarrow r$; $C \rightarrow p, q$; $D \rightarrow p$

(a) $\overset{-\frac{1}{2}}{O_2^-} \longrightarrow \overset{0}{O_2} + \overset{-1}{O_2^{2-}}$

Disproportionation

Redox reaction

(b) $\underset{\text{tetrahedral}}{CrO_4^{2-}} + H^+ \rightarrow \underset{\text{dimeric bridged tetrahedral ion}}{Cr_2O_7^{2-}}$

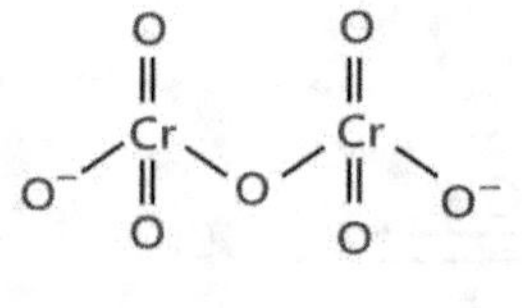

(c) $\underset{\text{tetrahedral}}{MnO_4^-} + NO_2^- + H^+ \rightarrow Mn^{2+} + \underset{\text{trigonal plonar}}{NO_3^-}$

Redox Reaction

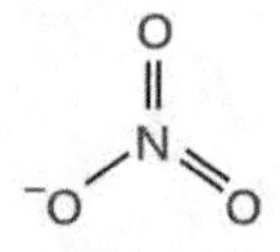

(d) $NO_3^- + H_2SO_4 + Fe^{2+} \rightarrow$

$Fe^{3+} + NO_2 + H_2O$

Redox reaction

Previous Years' Questions

Mole Concept

Sol 1: Average atomic weight

$$= \frac{\text{Percentage of an isotope} \times \text{atomic weight}}{100}$$

$$\Rightarrow 10.81 = \frac{10.01x + 11.01(100 - x)}{100}$$

$\Rightarrow x = 20\%$

Therefore, natural boron contain 20% (10.01) isotope and 80% other isotope.

Sol 2: From the vapour density information,

Molar mass = Vapour density × 2

(∵ Molar mass of H_2 = 2)

= 38.3 × 2 = 76.6

Now, let us consider 1.0 mole of mixture and it contains x mole of N_2.

$\Rightarrow 46x + 92(1 - x) = 76.6$

$\Rightarrow \quad x = 0.3348$

Also, in 100 g mixture, number of moles $= \frac{100}{76.6}$

⇒ Moles of in mixture

$$= \frac{100}{76.6} \times 0.3348 = 0.437$$

Sol 3: Heating below 600°C converts $Pb(NO_3)_2$ into PbO but to $NaNO_3$ into $NaNO_2$ as:

$$Pb(NO_3)_2 \xrightarrow{\Delta} PbO(s) + 2NO_2 \uparrow + \frac{1}{2}O_2 \uparrow$$

MW. 330 222

$$NaNO_3 \xrightarrow{\Delta} NaNO_2(s) + \frac{1}{2}O_2 \uparrow$$

MW.85 69

Weight loss $= 5 \times \frac{28}{100} = 1.4$ g

⇒ Weight of residue left = 5 − 1.4 = 3.6 g

Now, let the original mixture contain x g of $Pb(NO_3)_2$

$\because$ 330 g gives 222 g PbO

$\therefore$ x g $Pb(NO_3)_2$ will give $\frac{222x}{330}$ g PbO

Similarly, 85 g $NaNO_3$ gives 69 g

$\Rightarrow$ (5 – x)g will give

$\frac{69(5-x)}{85}$ g $NaNO_2$

$\Rightarrow$ Residue: $\frac{222x}{330}+\frac{69(5-x)}{85}=3.6g$

x = 3.3 g $Pb(NO_3)_2$

$\Rightarrow$ $NaNO_3$ = 1.7 g

Sol 4: Compound B forms hydrated crystals with $Al_2(SO_4)_3$ Also, B is formed with univalent metal on heating with sulphur. Hence, compound B must have the molecular formula M_2SO_4 and compound A must be an oxide of M which reacts with sulphur to give metal sulphate as

$$A+S\longrightarrow \underset{B}{M_2SO_4}$$

$\because$ 0.321 g sulphur gives 1.743 g of M_2SO_4

$\therefore$ 32.1 g S (one mole) will give 174.3 g M_2SO_4

Therefore, molar mass of M_2SO_4 = 174.3 g

$\Rightarrow$ 174.3 = 2 × Atomic weight of M + 32.1 + 64

$\Rightarrow$ Atomic weight of M = 39, metal is potassium (K)

K_2SO_4 on treatment with aqueous $Al_2(SO_4)_3$ gives potash-alum.

$$\underset{B}{K_2SO_4}+Al_2(SO_4)_3+24H_2O\longrightarrow \underset{C}{K_2SO_4Al_2(SO_4)_3\cdot 24H_2O}$$

If the metal oxide A has molecular formula MO_x, two moles of it combine with one mole of sulphur to give one mole of metal sulphate as

$$2KO_x+S\longrightarrow K_2SO_4$$

$\Rightarrow$ x = 2, i.e., A is KO_2.

Sol 5: 93% H_2SO_4 solution weight by volume indicates that there is 93 g H_2SO_4 in 100 mL of solution.

If we consider 100 mL solution, weight of solution = 184 g

Weight of H_2O in 100 mL solution

= 184 – 93 = 91 g

$$\Rightarrow \text{Molality}=\frac{\text{Moles of solute}}{\text{Weight of solvent(g)}}\times 1000$$

$$=\frac{93}{98}\times\frac{1000}{91}=10.43$$

Sol 6: Partial pressure of N_2 = 0.001 atm,

T = 298 K, V= 2.46 dm^3.

From Ideal Gas law : pV = nRT

$$n(N_2)=\frac{pV}{RT}=\frac{0.001\times 2.46}{0.082\times 298}=10^{-7}$$

$\Rightarrow$ No. of molecules of

$=6.023\times 10^{23}\times 10^{-7}$

$=6.023\times 10^{17}$

Surface sites used in adsorption

$$=\frac{20}{100}\times 6.023\times 10^{17}=2\times 6.023\times 10^{16}$$

$\Rightarrow$ Sites occupied per molecules

$$=\frac{\text{Number of sites}}{\text{Number of molecules}}=\frac{2\times 6.023\times 10^{16}}{6.023\times 10^{16}}=2$$

Sol 7: (D) The balanced chemical reaction is

$$3BaCl_2+2Na_3PO_4\longrightarrow Ba_3(PO_4)_2+6NaCl$$

In this reaction, 3 moles of $BaCl_2$ combined with 2 moles of Na_3PO_4 Hence, 0.5 mole of $BaCl_2$ requires

$\frac{2}{3}\times 0.5=0.33$ mole of Na_3PO_4.

Since available Na_3PO_4 (0.2 mole) is less than required mole (0.33), it is the limiting reactant and would determine the amount of product $Ba_3(PO_4)_2$.

$\because$ 2 moles of Na_3PO_4 gives 1 mole $Ba_3(PO_4)_2$

$\therefore$ 0.2 mole of Na_3PO_4 would give

$\frac{1}{2}\times 0.2$ = 0.1 mole $Ba_3(PO_4)_2$

Sol 8: (B) The following reaction occur between

$S_2O_3^{-2}$ and $Cr_2O_7^{-2}$:

$$26H^++3S_2O_3^{-2}+4Cr_2O_7^{-2}\longrightarrow 6SO_4^{-2}+8Cr^{3+}+13H_2O$$

Change in oxidation number of $Cr_2O_7^{-2}$ per formula unit is 6 (it is always fixed for $Cr_2O_7^{-2}$).

Hence, equivalent weight of $K_2Cr_2O_7$

$$=\frac{\text{Molecular weight}}{6}$$

Sol 9: $Na_2S_4O_6$ is a salt of $H_2S_4O_6$ which has the following structure

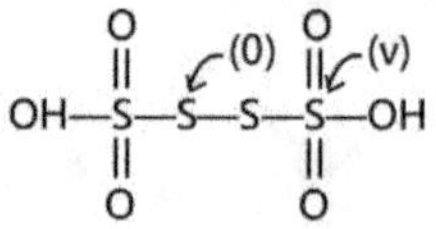

$\Rightarrow$ Difference in oxidation number of two types of sulphur = 5.

Sol 10: (B) 1. Both assertion and reason are factually true but the reason does not exactly explain the assertion. The correct explanation is, methyl orange and phenolphthalein changes their colours at different pH.

Sol 11: For the oxidation of A^{n+} as:

$A^{n+} \longrightarrow AO_3^-$ n-factor = 5 – n

Gram equivalent of $A^{n+} = 2.68 \times 10^{-3}$ (5 – n)

Now equating the above gram equivalent with gram equivalent of $KMnO_4$:

$2.68 \times 10^{-3}(5-n) = 1.61 \times 10^{-3} \times 5$

$\Rightarrow n = +2$

Sol 12: The redox reaction involved is :

$H_2O_2 + 2I^- + 2H^+ \longrightarrow 2H_2O + I_2$

If M is molarity of H_2O_2 solution, then

$5M = \dfrac{0.508 \times 1000}{254}$

($\because$ 1 mole $H_2O_2 \equiv$ 1 mole I_2)

$\Rightarrow M = 0.4$

Also, n-factor of H_2O_2 is 2, therefore normality of H_2O_2 solution is 0.8 N.

$\Rightarrow$ Volume strength = Normality × 5.6

= 0.8 × 5.6 = 4.48 V

Sol 13: With $KMnO_4$ oxalate ion is oxidized only as:

$5C_2O_4^{-2} + 2MnO_4^- + 16H^+ \longrightarrow 2Mn^{2+} + 10CO_2 + 8H_2O$

Let, in the given mass of compound, x millimol of $C_2O_4^{-2}$ ion is present, then

Meq. of $C_2O_4^{-2}$ = Meq of MnO_4^-

$\Rightarrow 2x = 0.02 \times 5 \times 22.6; \Rightarrow x = 1.13$

At the later stage, with I^-, Cu^{2+} is reduced as :

$2Cu^{2+} + 4I^- \longrightarrow 2CuI + I_2$

and $I_2 + 2S_2O_3^{-2} \longrightarrow 2I^- + S_4O_6^{-2}$

Let there be x millimole of Cu^{2+}

$\Rightarrow$ Meq of Cu^{2+} = Meq of I_2 = meq of hypo

$\Rightarrow x = 11.3 + 0.05 = 0.565$

$\Rightarrow$ Meq of Cu^{2+} : Meq of $C_2O_4^{-2}$ = 0.565 : 1.13 = 1 : 2

Sol 14: Let us consider 10 mL of the stock solution contain x millimol oxalic acid $H_2C_2O_4$ and y millimol of $NaHC_2O_4$. When titrated against NaOH, basicity of oxalic acid is 2 while that of $NaHC_2O_4$ is 1.

$\Rightarrow 2x + y = 3 \times 0.1 = 0.3$ ...(i)

When titrated against acidic $KMnO_4$, n-factor of both oxalic acid and $NaHC_2O_4$ would be 2.

$2x + 2y = 4 \times 0.1 = 0.4$ ...(ii)

Solving equations (i) and (ii) gives y = 0.1, x = 0.1.

$\Rightarrow$ In 1.0 L solution, mole of $= \dfrac{0.1}{1000} \times 100 = 0.01$

Mole of $NaHC_2O_4 = \dfrac{0.1}{1000} \times 100 = 0.01$

$\Rightarrow$ Mass of $H_2C_2O_4$ = 90 × 0.01 = 0.9 g

Mass of $NaHC_2O_4$ = 112×0.01 = 1.12 g

Sol 15: (D) (p) $PbO_2 + H_2SO_4 \xrightarrow{\Delta} PbSO_4 + O_2 + \frac{1}{2}O_2$

(q) $2Na_2S_2O_3 + Cl_2 + 2H_2O \longrightarrow 2NaCl + 2NaHSO_4 + 2S$

(r) $N_2H_4 + 2I_2 \longrightarrow N_2 + 4HI$

(s) $XeF_2 + 2NO \longrightarrow Xe + 2NOF$

Sol 16: (A, B, D) The balanced equation is

$ClO_3^- + 6I^- + 6H_2SO_4 \rightarrow 3I_2 + Cl^- + 6HSO_4^- + 3H_2O$

Sol 17: (A) $KIO_4 + H_2O_2 \rightarrow KIO_3 + H_2O + O_2$

$NH_2OH + 3H_2O_2 \rightarrow HNO_3 + 4H_2O$

2. ATOMIC STRUCTURE

1. INTRODUCTION

The word atom in Greek means indivisible, i.e. an ultimate particle which cannot be further subdivided. This idea of all matter ultimately consisting of extremely small particles was conceived by ancient Indian and Greek philosophers. The old concept was put on firm footing by John Dalton with his atomic theory that was developed by him during the years 1803–1808.

1.1 Dalton's Atomic Theory

Dalton's atomic theory (1808) is based on the following two laws: law of conservation of mass and law of definite proportions. He also proposed law of multiple proportions, as a logical consequence of his theory.

(a) Every element is composed of extremely small indestructible particles called atoms ('atom' in Greek means indivisible).

(b) Atoms of any one element are all similar but they differ from atoms of another element.

(c) Atoms of each element are fundamental particles, have a characteristic mass but do not have any structure.

(d) Atoms of various elements take part in a chemical reaction to form compound (which is called molecule).

(e) In any compound, the relative number and kinds of atoms are constant.

2. DISCOVERY OF FUNDAMENTAL PARTICLES

2.1 Cathode Rays – Discovery of Electron

(a) In mid-1800s, scientists (William Crookes 1879, Julius Plucker 1889) started to study the discharge of electricity through partially evacuated tubes. Gases are normally poor conductor of electricity and they do not conduct electricity under normal pressure even with an applied potential of 1000 volt. However when the pressure was reduced to 10^{-2} mm at a potential of 1000 volt, Crookes and Plucker noticed that:

(i) From the cathode surface a glow surrounding the cathode began and the space left between the glow and cathode was named Crookes dark space. Under this condition electric current starts to flow from one electrode to other.

(ii) When the pressure is sufficiently low, the glow fills whole of the tube.

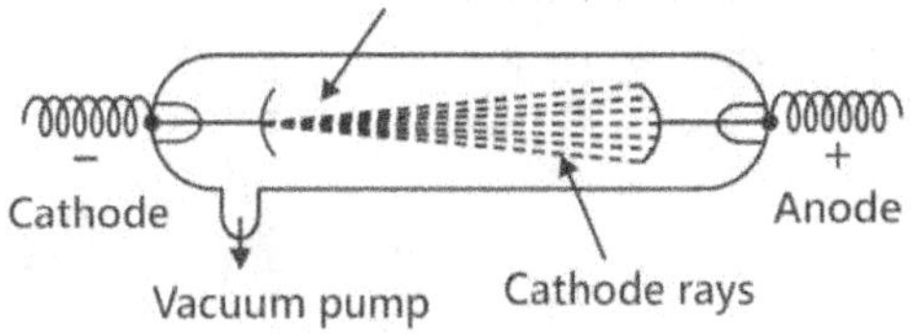

Figure 2.1: Cathode ray tube

Subsequently Thomson (1897) carried out the discharge through a vacuum tube which was filled with a gas at very low pressure (10^{-2} to 10^{-3} mm); he noticed the emission of invisible rays which produced fluorescence on the glass and influenced photographic plate. These rays were called cathode rays.

Important characteristics of cathode rays are as follows:

(a) Cathode rays travel in straight lines with high speed of the order of 10^7 ms^{-1}, producing shadows of the objects placed in their path.

(b) Cathode rays can pass through thin layers of matter.

(c) Cathode rays are emitted from the surface of the cathode; their direction is not affected by the position of the anode.

(d) Cathode rays have high kinetic energy and therefore:

  (i) They can exert mechanical pressure on the object on which they fall.

  (ii) They can produce heat when stopped by matter.

  (iii) They can ionize a gas.

(e) Cathode rays are deflected, both, by electric and magnetic fields.

(f) Cathode rays produce fluorescence when they fall on certain substances like ZnS. The color of fluorescence varies with the chemical nature of substance.

(g) Cathode rays can produce chemical changes and thus they affect photographic plates.

(h) Cathode rays produce X-rays, when they strike a metal target of high atomic number such as tungsten, which are highly penetrating. In 1897 J. J. Thomson determined the e/m value (charge/mass) of the electron by studying the deflection of cathode rays in electric and magnetic fields. The value of e/m has been found to be -1.7588×10^8 coulomb/g.

2.1.1 Millikan's Oil-Drop Experiment

In 1909, the first precise measurement of the charge on an electron was made by Robert A. Millikan using his oil drop experiment. The charge on the electron was calculated to be -1.6022×10^{-19} coulomb. An electron has the smallest charge known; so it was, designated as unit negative charge.

Mass of the electron: The mass of the electron is calculated from the values of e/m.

$$e.\ m = e = \frac{e}{e/m} = \frac{-1.6022\times10^{-19}}{-1.7588\times10^{8}} = 9.1096 \times 10^{-28}\ \text{g or } 9.1096 \times 10^{-31}\ \text{kg}$$

This is known as the rest mass of the electron; that is, the mass of the electron when it is moving with low speed. For calculating the mass of a moving electron, the following formula is used.

$$\text{Mass of moving electron} = \frac{\text{rest mass of electron}}{\sqrt{1-\left(\frac{\upsilon}{c}\right)^2}}$$

, where υ is the velocity of the electron and c is the velocity of light. When υ is equal to c, mass of the moving electron is infinity and when the velocity of the electron becomes greater than c, the mass of the electron becomes imaginary.

Note: An electron can, thus be defined as a subatomic particle which carries charge -1.60×10^{-19} coulomb, that is, one unit negative charge and has mass 9.1×10^{-19} g, that is, $\frac{1}{1837}$ the mass of the hydrogen atom (0.000549 amu).

2.2 Discovery of Protons – Positive Rays

(a) Goldstein (1886) repeated the discharge tube experiment but he used perforated cathode and noticed the emission of positive rays or canal rays.

Note: These rays do not originate from anode, and so it is wrong to call them anode rays.

(b) The specific charge (e/m) of canal rays particles varied with nature of gas and was found to be maximum if H_2 was used.

(c) The positive rays particles were thus, called positively charged gaseous atoms left after the removal of electron or ionized gaseous atoms. However, if H_2 gas is used in discharges, the positive rays particles are named as protons (usually represented as P).

(d) Thus, a subatomic particle, that is a fundamental constituent of all matter, is called a proton; it has a mass 1.673×10^{-27} kg and charge + 1.603×10^{-19} C.

NOMORECLASS CONCEPTS

- The specific charge (e/m) of proton is 9.58×10^7 C/kg.
- Mass of 1 mole proton= $N \times m_p = 6.023 \times 10^{23} \times 1.673 \times 10^{-27}$ kg = 1.0076×10^{-3} kg = 1.0076 g.
- The radius of proton = 1.53×10^{-13} cm.
- The volume of proton = $\frac{4}{3}\pi r^3 = \frac{4}{3} \times \frac{22}{7} \times (1.53 \times 10^{-13})^3$ = 1.50×10^{-38} cm^3.
- The charge on positive rays is usually +1 but it may have +2, +3 values.

The TV QUESTION

Q. What is the basic principle of a television picture tube or fluorescent light tubes?

Ans. The television picture tube is a cathode ray tube; the picture is produced due to fluorescence on the television screen coated with suitable material. Likewise florescent light tubes are also cathode ray tubes. These are coated inside with suitable materials which produce visible light when hit with cathode rays.

3. EARLIER MODELS OF ATOMS

3.1 Thomson's Model

After the discovery of protons and electrons, Thomson in 1898 proposed a watermelon model; the atom is considered as a sphere of positive charge with the electrons distributed within the sphere of radius of 10^{-10} m so as to give the most stable electrostatic agreement. In this model, the atom is visualized as a pudding or cake of positive charge with raisins (electrons) embedded in it. A major point of this model is that the mass of atom is considered to be spread uniformly over the atom.

3.2 Rutherford's Experiment

Rutherford conducted a series of experiments using α-particles. A beam of α-particles was directed against a thin foil of gold, platinum, silver, or copper. The foil was surrounded by a circular fluorescent zinc screen. Whenever an α-particle struck the screen, it produced a flash of light.

The following observations were made:

(a) Most of the α-particles went straight without suffering any deflection.

(b) Some of them were deflected through small angles.

(c) A very small number (about 1 in 20,000) did not pass through the foil at all but suffered large deflections or even rebound.

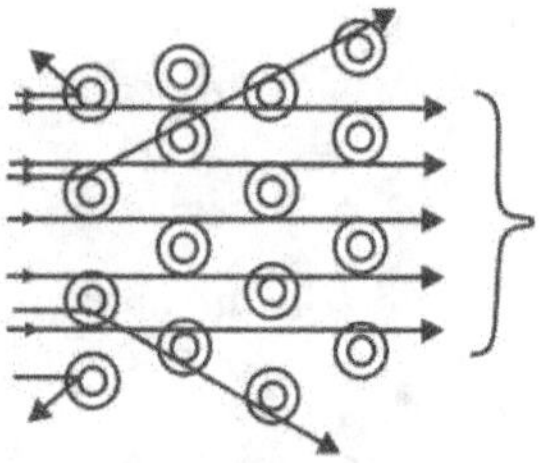

Figure 2.2: Gold leaf experiment

Following conclusions were drawn from these observations:

(a) Many of the particles went straight through the metal foil undeflected, indicating that there must be very large empty space within the atom.

(b) Some of the α-particles were deflected from their original paths through moderate angles, indicating that whole of the positive charge is concentrated in a space called nucleus. It is proposed to be present at the center of the atom.

(c) A very small number of the α-particles suffered strong deflections or even rebound on their path indicating that the nucleus is rigid and α-particles recoil due to direct collision with the positively charged heavy mass.

Note: Information of Rutherford's scattering equation can be memorized by the following relations:

(i) Kinetic energy of α-particles: $N = K_1/[(1/2)m\upsilon^2]^2$

(ii) Scattering angle 'θ': $N = K_2 / [(\sin^4(\theta / 2)]$

(iii) Nuclear charge: $N = K_3(Ze)^2$

Here, N = Number of α-particle striking the screen and K_1, K_2 and K_3 are the constants.

3.3 Moseley's Experiment

Moseley studied the X-ray spectra of 38 different elements, starting from aluminium to gold by measuring the frequency of principal lines of a particular series (the a-lines in the k-series) of the spectra. He observed that the frequency of the particular spectral line was related to the serial number of the element in the periodic table which he termed atomic number (Z). He presented the following relationship

$$\sqrt{v} = a(Z-b)$$

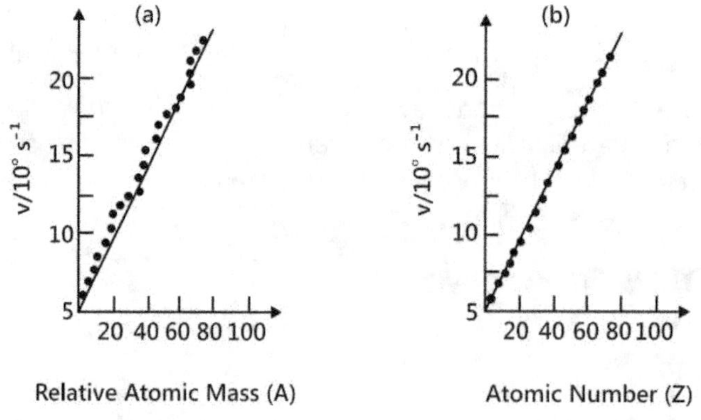

Figure 2.3: Moseley's plots

where ν = frequency of X-rays, Z = atomic number, a and b are constants. When the values of square roots of the frequency were plotted against atomic number of the element producing X-rays, a straight line was obtained. It was pointed out that the atomic number of an element is equal to the total positive charge of the nucleus.

NOMORECLASS CONCEPTS

Atomic number of the element
= Serial number of the element in periodic table
= Charge on the nucleus of the atom of the element
= Number of protons present in the nucleus of the atom of the element
= Number of extra nuclear electrons present in the atom of the element

3.4 Discovery of Neutron and Rutherford's Model

Chadwick bombarded beryllium with a stream of α-particles. He observed that the produced penetrations were not affected by electric and magnetic fields. These radiations consisted of neutral particles, which were called neutrons.

$$\underset{\text{Beryllium}}{{}^{9}_{4}Be} + \underset{\alpha\text{-particle}}{{}^{4}_{2}He} \rightarrow \underset{\text{Carbon}}{{}^{12}_{6}C} + \underset{\text{Neutron}}{{}^{1}_{0}n}$$

The mass of the neutron was determined to be 1.675×10^{-24} g, that is, nearly equal to the mass of proton.

Thus a neutron is a subatomic particle having a mass of 1.675×10^{-24} g, approximately 1 amu or nearly equal to the mass of proton or hydrogen atom and carrying no electrical charge. The e/m of neutron is zero.

Rutherford's Model of Atom: Rutherford proposed a model of the atom known as nuclear atomic model. As per this model,

(a) An atom consists of a positively charged heavy nucleus where all the protons and neutrons are present. The number of the positive charge on the nucleus is different for different atoms.

(b) The volume of the nucleus is very small and is only a very small fraction of the total volume of the atoms.

$$\frac{\text{Diameter of atom}}{\text{Diameter of the nucleus}} = \frac{10^{-8}\text{ cm}}{10^{-13}\text{ cm}} = 10^{5}$$

Thus, the diameter of an atom is 100,000 times the diameter of the nucleus.

The radius of a nucleus is proportional to the cube root of the mass number.

Radius of the nucleus = $1.33 \times 10^{-13} \times A^{1/3}$ cm where A is the mass number.

(c) There is an empty space around nucleus called extra nuclear part, where electrons are present. The number of electrons in an atom is always equal to the number of protons present in the nucleus. The volume of the atom is about 10^{15} times the volume of the nucleus.

$$\frac{\text{Volume of the atom}}{\text{Volume of the nucleus}} = \frac{(10^{-8})^3}{(10^{-13})^3} = \frac{10^{-24}}{10^{-39}} = 10^{15}$$

(d) The electrons revolve around the nucleus in closed orbits with high speeds. Centrifugal force is acting on the revolving electrons and is being counterbalanced by the force of attraction between electrons and the nucleus.

This model is similar to solar system; the nucleus representing the Sun and the electrons the Planets. The electrons are, therefore, generally referred to as planetary electrons.

Drawbacks of Rutherford's Model

(a) According to wave theory, when a charged particle moves under the influence of an attractive force, it loses energy continuously in the form of electromagnetic radiations. Thus, the electron which moves in an attractive field (created by protons present in the nucleus) will emit radiations.

As result the electron will lose energy at every turn and move closer and closer to the nucleus following a spiral path and finally fall into the nucleus, thereby making the atom unstable. But the atom is quite stable meaning the electrons do not fall into nucleus, thus this model does not explain the stability of the atom.

(b) If the electrons lose energy continuously, the observed spectra should be continuous. But the observed spectra consists of well-defined lines of definite frequency. Hence, in an atom, the loss of energy by the electrons is not continuous.

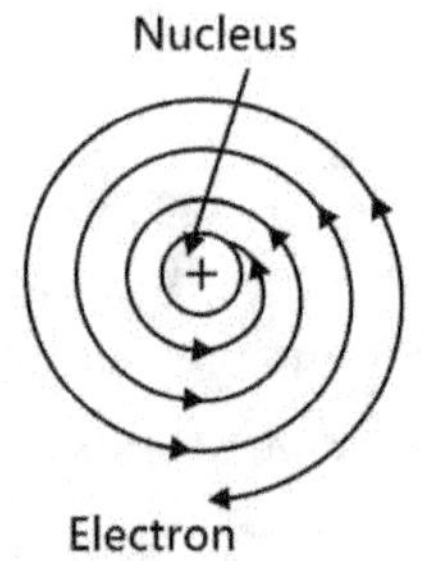

Figure 2.4: Spiral path of electron

3.5 Atomic Number and Mass Number

Atomic number of an element = Total number of protons present in the nucleus

= Total number of electrons present in the neutral atom

Mass number of an element = No. of protons + No. of neutrons

3.6 Isotopes, Isobars, Isotones and Isoelectronic

Such atoms of the same element having the same atomic number but different mass numbers are known as isotopes. Such atoms of different elements having different atomic numbers, but same mass number, e.g. ${}^{40}_{18}Ar$, ${}^{40}_{19}K$, ${}^{40}_{20}Ca$, are known as isobars. Such atoms of different elements containing the same number of neutrons, e.g. ${}^{14}_{6}C$, ${}^{15}_{7}N$, ${}^{16}_{8}O$, are called isotones. Each of these atoms contains eight neutrons. They differ in atomic number as well mass number. Such species (atoms or ions) containing the same number of electrons, e.g. O^{2-}, F^{-}, Na^{+}, Mg^{2+} Al^{3+}, Ne are called isoelectronic. Each of these contains 10 electrons each and so they are isoelectronic.

NOMORECLASS CONCEPTS

- **Retain in Memory:** Isotopes have same number of protons but different number of neutrons. Isobars have different number of protons as well as different number of neutrons. Isotones have different number of protons but same number of neutrons.
- **Retain in Memory:** In a neutral atom (${}^{A}_{Z}X$), No. of protons = No. of electrons = Atomic no.

  In a negative ion (X^{n-}), No. of electrons > No. of protons.

  No. of electrons = Z + n

  In a positive ion (M^{n+}), No. of electrons < No. of protons.

  No. of electrons = Z − n

  However, No. of protons is always = Z and No. of neutrons is always = A − Z.

Illustration 1: Find out the atomic number, mass number, number of protons, electrons and neutrons present in the element with the notation ${}^{238}_{92}U$. **(JEE MAIN)**

Sol: Number of protons = Number of electrons = Atomic number (Z)

Number of neutrons = Mass number − Atomic number

Atomic number (Z) = 92

Mass number (A) = 238

∴ Number of protons = 92 and Number of electrons = 92

Further, A − Z = 238 − 92 = 146

Illustration 2: The nuclear radius is of the order of 10^{-13} cm while atomic radius is of the order 10^{-8} cm. If the nucleus and the atoms are assumed to be spherical, what is the fraction of the atomic volume occupied by the nucleus? **(JEE MAIN)**

Sol: Considering the spherical conditions, the volume of a sphere = $4\pi r^3/3$ where r is the radius of the sphere.

∴ Volume of the nucleus = $4\pi r^3/3 = 4\pi\ (10^{-13})^3/3\ cm^3$

Similarly, volume of the atom = $4\pi r^3/3 = 4\pi\ (10^{-8})^3/3\ cm^3$

∴ Fraction of the volume of atom occupied by the nucleus = $\frac{4\pi(10^{-13})^3/3\ cm^3}{4\pi(10^{-8})^3/3\ cm^3} = 10^{-15}$

4. ELECTROMAGNETIC RADIATIONS

Electromagnetic Radiations (EMR) are energy radiations which do not need any medium for propagation, e.g. visible, ultraviolet, X-rays, etc. Following are the important characteristics of EMR:

(a) All electromagnetic radiations or waves travel with the velocity of light.

(b) These consist of electric and magnetic fields that oscillate in directions perpendicular to each other and perpendicular to the direction in which the wave is travelling.

4.1 Characteristics of Waves

A wave is always characterized by the following six characteristics:

(a) Wavelength: The distance between two nearest crests or nearest troughs is called the wavelength. It is denoted by λ (lambda) and is measured in units of centimetre (cm), angstrom (Å), micrometre, (μm) or nanometre (nm).

1 Å = 10^{-8} cm = 10^{-10} m

1 μm = 10^{-4} cm = 10^{-6} m

1 nm = 10^{-7}cm = 10^{-9} m

1 cm = 10^{8} Å = 104 μm = 10^{7} nm

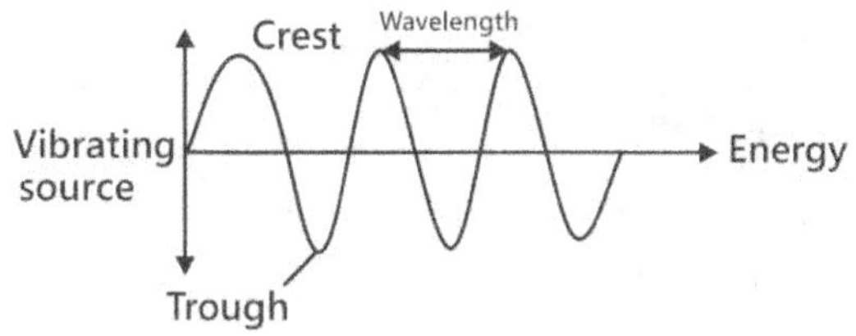

Figure 2.5: Representation of a wave and wavelength

(b) Frequency: It is defined as the number of waves passing through a point in one second. It is denoted by the symbol (nu) and is measured in terms of cycles (or waves) per second (cps) or hertz (Hz).

$\lambda\nu$ = Distance travelled in one second = Velocity = c or $\nu = \frac{c}{\lambda}$

(c) Velocity: It is defined as the distance covered in one second by the wave. It is denoted by the letter 'c'. All electromagnetic waves travel with the same velocity, that is, 3×10^{10} cm/sec. $\lambda\nu = 3 \times 10^{10}$. Thus, a wave of higher frequency has shorter wavelength while a wave of lower frequency has a longer wavelength.

(d) Wave number: This is the reciprocal of wavelength, i.e. the number of wavelength per centimeter. It is denoted by the symbol $\bar{\nu}$ (nu bar). $\bar{\nu} = \frac{1}{\lambda}$. It is expressed in cm^{-1} or m^{-1}.

(e) Amplitude: It is defined as the height of the crest or depth of the trough of a wave. It is denoted by the letter 'A'. It determines the intensity of the radiation.

(f) Time period: Time taken by one wave to complete a cycle or vibration is called time period. It is denoted by T. $T = \frac{1}{\nu}$

Unit: Second per cycle.

4.2 Electromagnetic Spectrum

The arrangement of various types of electromagnetic radiations in the order of their increasing or decreasing wavelengths or frequencies is known as electromagnetic spectrum.

$$\nu \equiv 3 \times 10^7 \text{ (cycle/sec)} \xrightarrow{\text{Frequency}} 3 \times 10^{21}$$

$$\lambda(\text{cm}) = 10^3 \xrightarrow{\text{Wavelength}} 10^{-11}$$

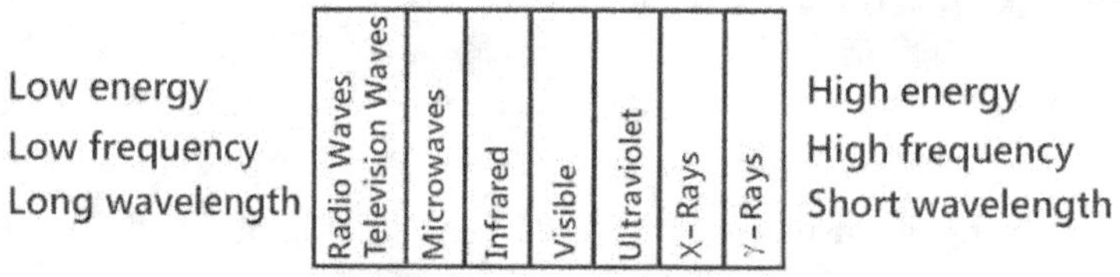

Figure 2.6: Electromagnetic spectrum

5. ATOMIC SPECTRA OF HYDROGEN

The impression produced on a screen when radiations of particular wavelengths are passed through a prism or diffraction grating is known as spectrum. It is broadly of two types: Emission spectra and Absorption spectra

Differences between emission spectrum and absorption spectrum

Emission spectrum	Absorptions spectrum
1. It gives bright lines (colored on the dark background.	1. It gives dark lines on the bright background.
2. Radiations from emitting source are analyzed by the spectroscope.	2. It is observed when the white light is passed through the substance and the transmitted radiations are analyzed by the spectroscope.
3. It may be continuous (if source emits white light) and may be discontinuous (if the source emits colored light).	3. These are always discontinuous.

5.1 Emission Spectra

It is obtained from the substances which emit light on excitation, that is, either by heating the substances on a film or by passing electric discharge through a thin filament of high melting point metal. Emission spectra are of two types:

(a) Continuous spectra: When white light is allowed to pass through a prism, the light gets resolved into several colours. This spectrum is a rainbow of colours, meaning violet merges into blue, blue into green and so on. This is a continuous spectrum.

(b) Discontinuous spectra: When gases or vapours of a chemical substance are heated in an electric arc or in a Bunsen flame, light is emitted. When a ray of this light is passed through a prism, a line spectrum is produced.

This spectrum consists of limited number of lines, each of which corresponds to a different wavelength of light. Each element has a unique line spectrum. Spectrum of hydrogen is an example of line emission spectrum or atomic emission spectrum. If an electric discharge is passed through hydrogen gas at low pressure, a bluish light is emitted. If a ray of this light is passed through a prism, a discontinuous line spectrum of many isolated sharp lines is obtained. The wavelengths of the different lines show that these lines are in the visible, ultraviolet and infrared regions. The lines observed in the hydrogen spectrum can be classified into six series.

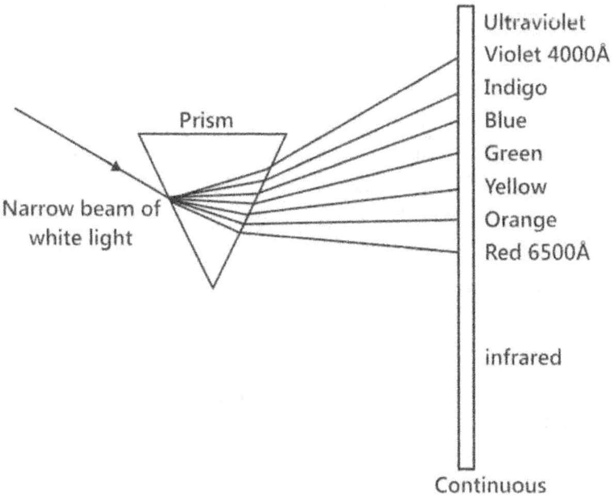

Figure 2.7: Continuous spectrum

Spectral series	Discovered by	Appearing in
Lyman series	Lyman	Ultraviolet region
Balmer series	Balmer	Visible region
Paschen series	Paschen	Infrared region
Brackett series	Brackett	Infrared region
Pfund series	Pfund	Infrared region
Humphrey series	Humphrey	Far-infrared region

5.2 Ritz Mathematical Formula

Ritz presented a mathematical formula to find the wavelengths of various hydrogen lines.

$\bar{v} = \frac{1}{\lambda} = \frac{v}{c} = \left(\frac{1}{n_1^2} - \frac{1}{n_2^2}\right)$ where, R is universal constant, known as Rydberg constant. Its value is 0.9678 cm^{-1}, n_1 and n_2 are integers (such that $n_2 > n_1$). For a given spectral series, n_1 remains constant while n_2 varies from line to line in the same series. The value of n_1 is 1, 2, 3, 4 and 5 for the Lyman, Balmer, Paschen, Brackett and Pfund series respectively. n_2 is greater than n_1 by at least 1.

Following are the values of n_1 and n_2 for various series:

Spectral series	Value of n_1	Value of n_2
Lyman series	1	2, 3, 4, 5,.....
Balmer series	2	3, 4, 5, 6,.....
Paschen series	3	4, 5, 6, 7,.....
Brackett series	4	5, 6, 7, 8,.....
Pfund series	5	6, 7, 8, 9,.....

Note: (i) Atoms give line spectra while molecules give band spectra.

(ii) Balmer, Paschen, Brackett, Pfund series are found in emission spectrum.

5.3 Absorption Spectra

When the radiations from a continuous source like a hot body (sunlight) containing the quanta of all wavelengths pass through a sample of hydrogen gas, then the wavelength missing in the emergent light gives dark lines on the bright background. This type of spectrum that contains lesser number of wavelengths in the emergent light than in incident light is called absorption spectrum.

Illustration 3: Which has a higher energy, a photon of violet light with wavelength 400 Å or a photon of red light with wavelength 7000Å? [$h = 6.62 \times 10^{-34}$ Js] **(JEE MAIN)**

Sol: Using photoelectric effect, $E = h\nu = h\dfrac{c}{\lambda}$

Given $h = 6.62 \times 10^{-34}$ Js, c = s, $c = 3\times 10^{8}\ ms^{-1}$

For a photon of violet light $\lambda = 4000$ Å $= 4000 \times 10^{-10}$ m; $E = 6.62 \times 10^{-34} \times \dfrac{3\times10^{8}}{4\times10^{-7}} = 4.96 \times 10^{-19}$

For a photon of red light, L = 7000 Å = 7000×10^{-10} m

$$E = 6.62 \times 10^{-34} \times \frac{3\times10^{8}}{7000\times10^{-10}} = 2.83\times10^{-19}\ J$$

Hence, a photon of violet light has higher energy than a photon of red light.

Illustration 4: Calculate the wavelength of the spectral line, when the electron in the hydrogen atom undergoes a transition from the energy level 4 to energy level 2. **(JEE MAIN)**

Sol: According to Rydberg equation $\dfrac{1}{\lambda} = R\left(\dfrac{1}{x^2} - \dfrac{1}{y^2}\right)$

$R = 109678\ cm^{-1}$; x = 2; y = 4; $\dfrac{1}{\lambda} = 109{,}678\left[\dfrac{1}{4} - \dfrac{1}{16}\right] = 109{,}678 \times \dfrac{3}{16}$

On solving, $\lambda = 486$ nm.

6. QUANTUM THEORY OF RADIATION

In 1905, Einstein pointed out that light can be considered to consist of a stream of particles, called photons. The energy of each photon of light depends on the frequency of the light, that is, $E = h\nu$. According to Einstein, energy is also related as $E = mc^2$ where m is the mass of photon. Thus, he pointed out that light has wave as well as particle characteristics (dual nature).

Though the wave theory successfully explains many properties of electromagnetic radiations such as reflection, refraction, diffraction, interference, polarization etc. it fails to explain some phenomena like blackbody radiation, photoelectric effect etc. A new theory which is known as quantum theory of radiation was presented by Max Planck in 1901, to explain the blackbody radiation and photoelectric effect. According to this theory, a hot body emits radiant energy not continuously but discontinuously in the form of small packets of energy called quantum (quanta in plural). The energy associated with each quantum of a given radiation is proportional to the frequency of the emitted radiation. $E \propto \nu$ or $E = h\nu$ where h is a constant known as Plank's constant. Its numerical value is 6.624×10^{-27} erg/sec. The energy emitted or absorbed by a body can be either one quantum or any whole number multiple of $h\nu$, that is, $2h\nu$, $3h\nu$, $4h\nu$,..., $nh\nu$ quanta of energy.

6.1 Photoelectric Effect

The emission of electrons from a metal surface when exposed to light radiations of appropriate wavelength is called photoelectric effect. The emitted electrons are called photoelectrons. Work function or threshold energy may be defined as the minimum amount of energy required to eject electrons from a metal surface. According to Einstein, maximum kinetic energy of the ejected electron = Absorbed energy − Work function $\frac{1}{2}mv_{max}^2 = h\nu - h\nu_0 = hc\left[\frac{1}{\lambda} - \frac{1}{\lambda_0}\right]$, where ν_0 and λ_0 are threshold frequency and threshold wavelength respectively.

Stopping potential: Stopping potential is the minimum potential at which the photoelectric current becomes zero. If ν_0 is the stopping potential, then $\nu_0 = h(\nu - \nu_0)$.

Laws of Photoelectric Effect

(a) Rate of emission of photoelectrons from a metal surface is directly proportional to the intensity of incident light.

(b) The maximum kinetic energy of photoelectrons is directly proportional to the frequency of incident radiation; also, it is independent of the intensity of light used.

(c) There is no time lag between incidence of light and emission of photoelectrons.

(d) For emission of photoelectrons, the frequency of incident light must be equal or greater than the threshold frequency.

NOMORECLASS CONCEPTS

If kinetic energy of the emitted photoelectrons is plotted against the frequency of the absorbed photons, a straight line of slope h is obtained as shown in Fig. 2.8 A. If kinetic energy of the photoelectrons is plotted against intensity of the incident radiation (keeping frequency constant), a horizontal line is obtained as shown in Fig. 2.8 B.

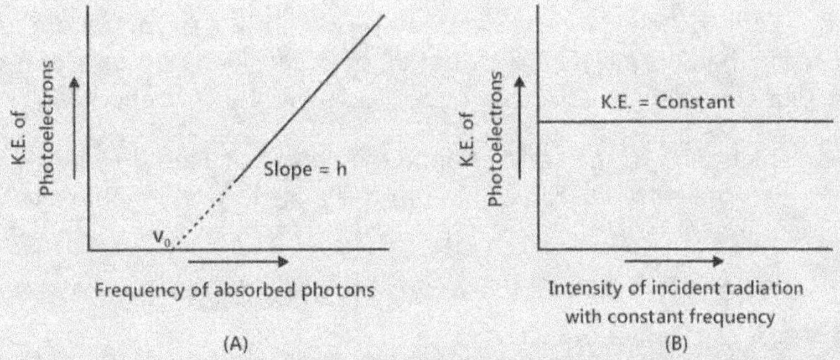

Figure 2.8: Plot of kinetic energy of photoelectrons emitted versus (A) frequency of absorbed photons and (B) intensity of the incident radiation with constant frequency

The Heated Iron!!!

Q. When an iron bar is heated, it first becomes red, then yellow and finally begins to glow with white light and then blue light. Why?

Ans. The change in colour is because of increase in the frequency of radiation emitted on heating as red colour lies in the lower frequency region while blue colour lies in the high frequency region.

Illustration 5: When electromagnetic radiation of wavelength 300 nm falls on the surface of sodium metal, electrons are emitted with a kinetic energy of 1.68×10^5 J mol^{-1}. What is the minimum energy needed to remove an electron from sodium? What is the maximum wavelength that will cause a photoelectron to be emitted?

(JEE MAIN)

Sol: The minimum energy needed to remove an electron is the threshold energy and the maximum wavelength would be obtained from the same equation. Thus, we use the following equations $E = h\nu = h\frac{c}{\lambda}$

For one electron, we multiply by the Avogadro's number, which gives us for one mole of photons, which is then subtracted from the kinetic energy and then divided by Avogadro's number again.

$E = E_0 + KE$

The maximum wavelength is found by the equation stated at the beginning.

Energy of a photon of radiation of wavelength 300 nm will be

$$E = h\nu = h\frac{c}{\lambda} = \frac{(6.626\times10^{-34}\,Js)(3.0\times10^{8}\,ms^{-1})}{(300\times10^{-9})m} = 6.626\times10^{-19}\,J$$

$\therefore$ Energy of 1 mole of photons = $(6.626 \times 10^{-19}\,J) \times (6.022 \times 10^{23}\,mol^{-1}) = 3.99 \times 10^{5}\,J\,mol^{-1}$

As $E = E_0 + KE$ of photoelectrons emitted.

$\therefore$ Minimum energy (E_0) required to remove 1 mole of electrons from sodium = E-KE

$= (3.99-1.68)\,10^{5}\,J\,mol^{-1} = 2.31\times10^{5}\,J\,mol^{-1}$

$\therefore$ Minimum energy required to remove one electron $= \dfrac{2.31\times10^{5}\,J\,mol^{-1}}{6.022\times10^{23}\,mol^{-1}} = 3.84\times10^{-19}\,J$

The wavelength corresponding to this energy can be calculated using the formula, $E = h\nu = h\dfrac{c}{\lambda}$

$$\therefore \lambda = \frac{hc}{E} = \frac{(6.626\times10^{-34}\,Js)(3.0\times10^{8}\,ms^{-1})}{3.84\times10^{-19}} = 5.17\times10^{-7}\,m = 517\times10^{-9}\,m = 517nm$$ which corresponds to green light.

Illustration 6: O_2 undergoes photochemical dissociation into one normal oxygen atom and one excited oxygen atom. 1.976 eV more energetic than normal. The dissociation of O_2 into two normal atoms of oxygen requires 498 kJ mol^{-1}. What is the maximum wavelength effective for photochemical dissociation of O_2? **(JEE MAIN)**

Sol: Frame the dissociation reaction of oxygen and arrange the data. Calculate the energy required for the simple dissociation and then add it to the extra energy quoted in the question. Using this energy, solve for maximum wavelength.

Given $O_2 \longrightarrow O_{Normal} + O_{Excited}$; H = ?

$O_2 \longrightarrow ON + ON;\ \Delta H = 498\ kJ/mol$

Energy required for simple dissociation of O_2 into two normal atoms $= \dfrac{498\times10^{3}}{6.023\times10^{23}} = 8.268\times10^{-19}\,J\ mol^{-1}$ Since, one atom has more energy (1.967 eV) in excited state when photochemical dissociation takes place, then energy required for photochemical dissociation of O_2 = $8.268 \times 10^{-19} + 1.967 \times 1.602 \times 10^{-19} = 114.19 \times 10^{-20}$ J

Now using $E = \dfrac{hc}{\lambda}$; $114.19 \times 10^{-20} = \dfrac{6.626\times10^{-34}\times3.0\times10^{8}}{\lambda}$; $\lambda = 1740.52 \times 10^{-10}$ m = 1740.52 Å

7. BOHR'S ATOMIC MODEL

Bohr proposed a quantum mechanical model of the atom, to overcome the objections of Rutherford's model and to explain the hydrogen spectrum. This model was based on the quantum theory of radiation and the classical laws of physics. The important postulates on which Bohr's model is based are the follows:

(a) The atom has a nucleus where all the protons and neutrons are present. The size of the nucleus is very small. It is present at the center of the atom.

(b) Negatively charged electrons are revolving around the nucleus in a similar way as the Planets are revolving around the Sun. The path of the electron is circular. The force of attraction between the nucleus and the electron is equal to centrifugal force of the moving electron.

Force of attraction toward nucleus = Centrifugal force

(c) Out of infinite number of possible circular orbits around the nucleus, the electron can revolve only on those orbits whose angular momentum is an integral multiple of $\frac{h}{2\pi}$, that is, $mvr = n\frac{h}{2\pi}$ where

m = Mass of the electron, v = Velocity of electron, r = Radius of the orbit and n = 1, 2, 3,.... number of the orbit. The angular momentum can have values such as, $\frac{h}{2\pi}, \frac{2h}{2\pi}, \frac{3h}{2\pi}$, etc., but it cannot have fractional value. Thus, the angular momentum is quantized. The specified or circular orbits (quantized) are called stationary orbits.

(d) When the electron remains in any one of the stationary orbits, it does not lose energy. Such a state is called ground of normals.

In the ground state potential energy of electron will be minimum, hence it will be the most stable state.

(e) Each stationary orbit is associated with a definite amount of energy. The greater is the distance of the orbit from the nucleus, more shall be the energy associated with it. These orbits are also called energy levels and are numbered as 1, 2, 3, 4, ...or K, L, M, N,... from nucleus outward, i.e. $E_1 < E_2 < E_3 < E_4$ .... $(E_2 - E_1) > (E_3 - E_2) > (E_4 - E_3)$.

(f) The emission or absorption of energy in the form of radiation can occur only when an electron jumps from one stationary orbit to another. $\Delta E = E_{high} - E_{low} = h\nu$; energy is absorbed when the electron jumps from an inner to an outer orbit and it is emitted when the electron moves from an outer to an inner orbit.

When the electron moves from an inner to an outer orbit by absorbing definite amount of energy, the new state of the electron is said to be excited state.

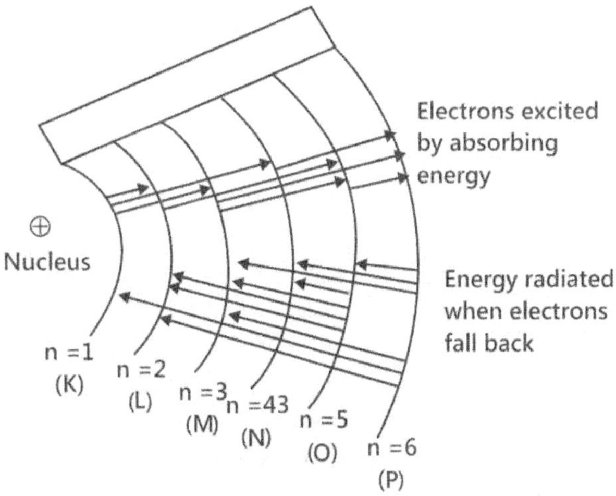

Figure 2.9: Excitation and De- excitation

7.1 Some Important Results of Bohr's Equation

Consider a H-atom in which an electron of mass 'm', charge 'e' revolving round a nucleus of charge lp (unit + ve) with a tangential velocity 'v' at a distance 'r' from the nucleus. The necessary centripetal force required to keep the electron moving is provided by the electrostatic attraction between the electron and the nucleus,

i.e. $\frac{e^2}{r^2} = \frac{mv^2}{r}$ ... (i)

Since $mvr = \frac{nh}{2\pi}$ ... (ii)

(a) The radius: Therefore, by Eqs. (1) and (2),

$$\frac{e^2}{r^2} = m\frac{n^2h^2}{m^2.r^2\,4\pi^2.r} \text{ or } r_n = \frac{n^2h^2}{4\pi^2me^2} = n^2\frac{h^2}{4\pi^2me^2}$$ ... (iii)

or $r_n = n^2 \times r_1$ ... (iv)

Also, $r_1 = \dfrac{h^2}{4\pi^2 me^2}$

For H-atom; $r_{1H} = \dfrac{(6.626\times10^{-27})^2}{4\times(3.14)^2\times(9.108\times10^{-28})\times(4.803\times10^{-10})^2}$; $r_{1H} = 0.529 \times 10^{-8}$ cm $= 0.529$ Å

For H like atoms; (He^+, Li^{2+}, Be^{3+},..., etc.). The charge on nucleus is +Z and therefore, $\dfrac{Ze^2}{r^2} = \dfrac{mv^2}{r}$

Similarly, we get

$$r_n = \frac{n^2h^2}{4\pi^2 mZe^2} \quad \text{... (v)}$$

$$\text{Or } r_{nH\text{-like atom}} = r_{nH}/Z \quad \text{... (vi)}$$

where Z is atomic number of atom

(b) Velocity: The velocity of an electron in nth orbit of H-atom can be derived by Eqs. (1) and (2), as

$$v_n = \frac{2\pi e^2}{nh} \quad \text{... (vii)}$$

$$= \frac{2\times3.14\times(4.803\times10^{-10})^2}{n\times6.626\times10^{-27}} = \frac{2.18\times10^8}{n}\text{cm sec}^{-1};\ \therefore \quad v_{1H} = 2.18\times10^8 \text{ cm sec}^{-1} \quad \text{... (viii)}$$

$$\text{Also, } v_n = \frac{v_1}{n} \text{ For H like atom; } \quad v_n = \frac{2\pi Ze^2}{nh} \quad \text{... (ix)}$$

$$\text{Or } v_{n_{H\text{ like atom}}} = v_{n_{Z\text{-atom}}} \times Z \quad \text{... (x)}$$

(c) Time required to complete one revolution by an electron round nucleus in an orbit: Let $2\pi r_n$ be the circumference of orbit and vn is velocity of electron in that orbit T for one revolution in an orbit $= \dfrac{2\pi r_n}{v_n}$ ... (xi)

(d) Number of revolutions per second made by an electron round the nucleus in an orbit:

$$\text{Number of revolutions} = \frac{v_n}{2\pi r_n} \quad \text{... (xii)}$$

(e) Energy: Total energy ET of an electron for H-atom in a shell can be given by,

$$E_T = \text{Potential energy} + \text{Kinetic energy};\ E_T = \left(-\frac{e^2}{r_n}\right)+\left(\frac{1}{2}\cdot\frac{e^2}{r_n}\right) = -\frac{e^2}{2r_n} \quad \text{... (xiii)}$$

$$\text{By Eqs. (3) and (13), } E_T = \frac{2\pi^2 me^4}{n^2h^2} \quad \text{... (xiv)}$$

$$= -\frac{2\times(3.14)^2\times9.108\times10^{-28}\times(4.803\times10^{-10})^4}{n^2\times(6.625\times10^{-27})^2}\text{er} = -\frac{21.77\times10^{-12}}{n^2}\text{erg} = -\frac{21.77\times10^{-19}}{n^2}\text{J (107 erg = 1 J)}$$

$$E_T = -\frac{13.6}{n^2}\text{eV } (1.602\times10^{-19}\text{ J} = 1\text{ eV});\ E_1H = -13.6\text{ eV} \quad \text{... (xv)}$$

For H like atom; $E_T = -\dfrac{2\pi^2 mZ^2e^4}{n^2h^2}$

$$E_T = \frac{-21.77\times10^{-12}}{n^2}\times Z^2 \text{ erg} \quad \text{... (xvi)}$$

Also, for H-atom: $E_n \propto -\dfrac{1}{n^2}$ and $E_n = \dfrac{E_1}{n_2}$; E_1 for H-atom $= -\dfrac{13.6}{1^2} = -13.6$ eV

E_2 for H-atom $= -\dfrac{13.6}{2^2} = 3.4$ eV; E_3 for H-atom $= -\dfrac{13.6}{3^2} = -1.51$ eV

E_∞ for H-atom $= -\dfrac{13.6}{\infty^2} = 0$ eV; E_1 for H like atom = E_1 for H × Z_2 = −13.6 × Z_2 eV ... (xvii)

(f) Frequency (ν), wavelength (λ) and wave number ($\bar{\nu}$) during electron transition:

Furthermore, $\Delta E = h\nu = E_{n_2} - E_{n_1}$ ... (xviii)

$$\Delta E = h\nu = \frac{hc}{\lambda} = \frac{2\pi^2 me^4}{h^2}\left[\frac{1}{n_1^2} - \frac{1}{n_2^2}\right] \quad \text{... (xix)}$$

$$\text{Or } \bar{\nu} = \frac{1}{\lambda} = \frac{2\pi^2 me^4}{ch^3}\left[\frac{1}{n_1^2} - \frac{1}{n_2^2}\right]\text{cm}^{-1} \quad \text{... (xx)}$$

where $\bar{\nu}$ is wave number, λ is wavelength of light involved during electronic transition.

$$\frac{1}{\lambda} = 1.096\times10^5\left[\frac{1}{n_1^2} - \frac{1}{n_2^2}\right]\text{cm}^{-1} \quad \text{... (xxi)}$$

7.2 Interpretation of Hydrogen Spectrum

Maximum number of lines produced when an electron jumps from n^{th} level to ground level is equal to $\dfrac{n(n-1)}{2}$.

For example, in the case of n = 4, the number of lines produced is six (4 → 3, 4 → 2, 4 → 1, 3 → 2, 3 → 1, 2 → 1). When an electron returns from n_2 to n_1 state, the number of lines in the spectrum will be equal to $\dfrac{(n_2 - n_1)(n_2 - n_1 + 1)}{2}$.

If the electron returns from energy level having energy E_2 to energy level having energy E_1, then the difference may be expressed in terms of energy of photon as: $E_2 - E_1 = \Delta E = h\nu$.

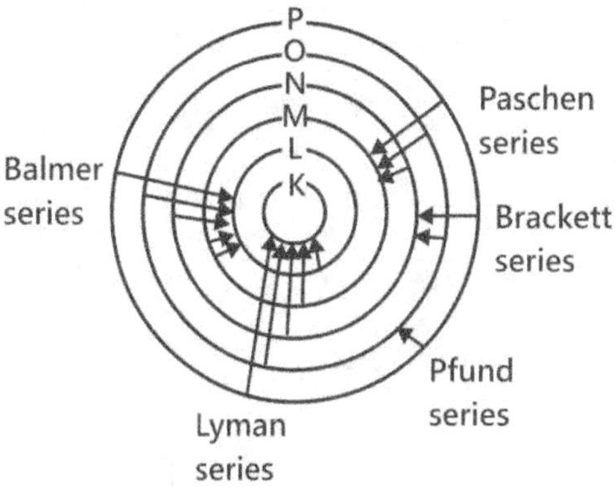

Figure 2.10: Interpretation of Hydrogen Spectrum

Or the frequency of the emitted radiation in given by $\nu = \dfrac{\Delta E}{h}$.

Since, ΔE can be only definite values depending on the definite energies of E_2 and E_1, ν will have only fixed values in an atom, or $\nu = \dfrac{c}{\lambda} = \dfrac{\Delta E}{h}$ or $\lambda = \dfrac{hc}{\Delta E}$.

Since, h and c are constants, ΔE corresponds to definite energy; thus, each transition from one energy level to another will produce a light of definite wavelength. This is actually observed as a line in the spectrum of hydrogen atom. Thus, the different spectral lines in the spectra of atoms correspond to different transitions of electrons from higher energy levels to lower energy levels.

7.3 Derivation of Rydberg Equation

Let an excited electron from n_2 shell come to the n_1 shell with the release of radiant energy. The wave number $\bar{v}$ of the corresponding spectral line may be calculated in the following manner:

$$\Delta E = E_2 - E_1 = (-)\frac{2\pi^2 mZ^2e^4}{n_2^2h^2} - (-)\frac{2\pi^2 mZ^2e^4}{n_1^2h^2};\ \frac{hc}{\lambda} = \frac{2\pi^2 mZ^2e^4}{h^2}\left(\frac{1}{n_1^2} - \frac{1}{n_2^2}\right)$$

Where $\Delta E = h\nu = \frac{hc}{\lambda};\ \therefore\ \bar{v} = \frac{1}{\lambda} = 2\pi^2 mZ^2e^4\left(\frac{1}{n_1^2} - \frac{1}{n_2^2}\right)$ Or $\bar{v} = RZ^2\left(\frac{1}{n_1^2} - \frac{1}{n_2^2}\right)$

where $R = \frac{2\pi^2 me^4}{ch^3}$ = Rydberg constant = 109,743 cm^{-1}

This value of R is in agreement with experimentally determined value 109,677.76 cm^{-1}. Rydberg equation for hydrogen may be given as, $\bar{v} = \frac{1}{\lambda} = R\left[\frac{1}{n_1^2} - \frac{1}{n_2^2}\right]$

7.4 Modification of Rydberg's Equation

According to the Rydberg's equation: $\bar{v}_{\text{wave number}} = \frac{2\pi^2 mZ^2e^4}{ch^3}\left[\frac{1}{n_1^2} - \frac{1}{n_2^2}\right]$

It can be considered that the electron and the nucleus revolve around their common center of mass. Therefore, instead of the mass of the electron, the reduced mass of the system was introduced and the equation becomes

$$\bar{v} = \frac{2\pi^2 \mu Z^2e^4}{2h^3}\left[\frac{1}{n_1^2} - \frac{1}{n_2^2}\right]$$

Reduced mass 'μ' can be calculated as, $\frac{1}{\mu} = \frac{1}{m} + \frac{1}{M}$

Where m = Mass of electron and M = Mass of nucleus. $\therefore\ \mu = \frac{mM}{m+M}$

Some Important Points to be Remembered

First Line of a Series

(a) First line of a series: It is the line of longest wavelength' or 'line of shortest energy'

$$n_2 = (n_1 + 1);\ \bar{v} = \frac{1}{\lambda_{\text{first}}} = R\left[\frac{1}{n_1^2} - \frac{1}{(n_1+1)^2}\right]$$

Similarly for second, third and fourth lines, $n_2 = n_1 + 2$; $n_2 = n_1 + 3$ and $n_2 = n_1 + 4$ respectively

$\therefore$ Rydberg's equation may be written as, $\bar{v} = \frac{1}{\lambda} = RZ^2\left[\frac{1}{n_1^2} - \frac{1}{(n_1+x)^2}\right]$ where x = Number of line in the spectrum.

e.g. x = 1, 2, 3, 4,... for first, second, third and fourth lines in the spectrum respectively.

Series Limit or Last Line of a Series

(b) Series limit or last line of a series: It is the line of shortest wavelength or line of highest energy.

For last line, $n_2 = \infty$; $\bar{\nu}_{last} = \frac{1}{\lambda_{Last}} = \frac{R}{n_1^2}$

Lyman limit = $\frac{R}{1^2}$; Balmer limit = $\frac{R}{2^2}$

Paschen limit = $\frac{R}{3^2}$; Brackett limit = $\frac{R}{4^2}$

Pfund limit = $\frac{R}{5^2}$; Humphrey limit = $\frac{R}{6^2}$

Intensity of Spectral Lines

(c) The intensities of spectral line in a particular series decrease with increase in the value of n_2, i.e. higher state e.g. **Lyman series** $(2 \to 1) > (3 \to 1) > (4 \to 1) > (5 \to 1)$,

$(n_2 \to n_1)$

Balmer series $(3 \to 2) > (4 \to 2) > (5 \to 2) > (6 \to 2)$

$(n_2 \to n_1)$ ⟶

Decreasing intensity of the spectral lines

7.5 Ionization Energy, Excitation Energy and Separation Energy

Excitation potential for $n_1 \to n_2 = \frac{E_{n_2} - E_{n_1}}{\text{Electronic charge}}$

Ionization potential for $n_1 \to \infty = \frac{E_{n_1}}{\text{Electronic charge}}$

The energy required to remove an electron from the ground state to form cation, i.e. to take the electron to infinity, is called ionization energy.

$IE = E_\infty - E_{ground}$; $IE = 0 - E_1(H) = 13.6 \text{ eV atom}^{-1} = 2.17 \times 10^{-18} \text{ J atom}^{-1}$

$$IE = \frac{Z^2}{n^2} \times 13.6 \text{ eV}; \frac{I_1}{I_2} = \frac{Z_1^2}{n_1^2} \times \frac{n_2^2}{Z_2^2}; (IE)Z = \frac{(IE)_H \times Z^2}{n^2}$$

If an electron is already present in the excited state, then the energy required to remove that electron is called separation energy. $E_{separation} = E_\infty - E_{excited}$

7.6 Limitations of Bohr's Model

(a) It does not explain the spectra of multi-electron atoms.

(b) By using a high resolving power spectroscope it is observed that a spectral line in the hydrogen spectrum is not a simple line but a collection of several lines which are very close to one another. This is known as fine spectrum. Bohr's theory does not explain the fine spectra of even the hydrogen atom.

(c) Spectral lines split into a group of inner lines under the influence of magnetic field (Zeeman effect) and electric field (Stark effect); but, Bohr's theory does not explain this.

(d) Bohr's theory is not in agreement with Heisenberg's uncertainty principle.

Illustration 7: Find out the number of waves made by a Bohr-electron in one complete revolution in its third Bohr orbit of H-atom. **(JEE MAIN)**

Sol: These 3 formulas need to be considered.

$$r_n = \frac{n^2h^2}{4\pi^2me^2}; v_n = \frac{2\pi e^2}{nh}; \text{ number of waves in one round } = \frac{2\pi r_3}{\lambda} = \frac{2\pi r_3}{h/mv_3}\left(\because \lambda = \frac{h}{mv}\right)$$

Number of waves made by an electron in Bohr's orbit is equal to number of orbits.

$$\therefore \text{ No. of waves in one round } = \frac{2\pi r_3 \times v_3 \times m}{h} = \frac{2\pi \times n^2h^2}{4\pi^2me^2} \times \frac{2\pi e^2}{nh} \times \frac{m}{h}; \ n = 3$$

Illustration 8: Calculate the shortest and longest wavelengths in hydrogen spectrum of Lyman series.

Or

calculate the wavelengths of the first line and the series limit for the Lyman series for hydrogen. (R_H = 109678 cm^{-1})

Sol: (a) For Lyman series, $n_1 = 1$

For shortest wavelength in Lyman series (i.e. series limit), the energy difference in two states showing transition should be maximum, i.e. $n = \infty$.

$$\text{So, } \frac{1}{\lambda} = R_H\left(\frac{1}{1^2} - \frac{1}{\infty^2}\right) = R_H; \lambda = \frac{1}{109678} = 9.117 \times 10^{-6} \text{ cm } = 911.7 \text{ Å}$$

(b) For longest wavelength in Lyman series (i.e. first line), the energy difference in two states showing transition should be minimum, i.e. $n_2 = 2$

$$\text{So, } \frac{1}{\lambda} = R_H\left[\frac{1}{1^2} - \frac{1}{2^2}\right] = \frac{3}{4}R_H \text{ or } \lambda = \frac{4}{3} \times \frac{1}{R_H} = \frac{4}{3 \times 109678} = 1215.7 \times 10^{-8} \text{cm } = 1215.7 \text{ Å}$$

Illustration 9: Calculate the ratio of the time required for an electron taking one round of second orbit of H atom and He^+ ion. **(JEE MAIN)**

Sol: Use the time formula $\frac{2\pi r_n}{v_n}$. This is the time required to complete one round in nth orbit where v is the velocity of electron.

For H-like atom, $T_{He^+} = \frac{2\pi r_1 \times n^2}{Z} \times \frac{n}{v_1 \times Z}$ needs to be used.

For second orbit of H-atom: When r_n and r_1 are radii of H-atom: v_n and v_1 are velocities of electron in nth and first orbit of H-atom. Time required to complete one round in nth orbit is $\frac{2\pi r_n}{v_n}$ where vn is velocity of electron in nth orbit of H-atom

Time required to complete one round in n^{th} orbit is $\frac{2\pi r_n}{v_n}$ where v_n is velocity of electron in n^{th} orbit and r_n is radius of nth orbit. $T_H = \frac{2\pi r_n}{v_n}$ $\quad (\because r_n = r_1 \times n_2 \text{ and } v_n = \frac{v_1}{n})$

$$= \frac{2\pi r_1 \times n^2 \times n}{v_1} \quad \left(\because v_1 = \frac{2\pi e^2}{h}\right) = \frac{2\pi r_1 \times n^3 \cdot h}{2\pi e^2} \quad T_H = \frac{r_1 n^3 h}{e^2} \qquad \text{...(i)}$$

For second orbit of He^+: $T_{He^+} = \frac{2\pi r_{nHe^+}}{v_{n\,He^+}}$; $[r_{nHe^+} = \frac{r_{1H} \times n^2}{Z}; v_{n\,He^+} = v_{nH} \times Z$ and $v_{nH} = \frac{v_1}{n}]$

$$T_{He^+} = \frac{2\pi r_1 \times n^2}{Z} \times \frac{n}{v_1 \times Z}; T_{He^+} = \frac{2\pi r_1 n^3 \times h}{Z^2 \times 2\pi e^2} = \frac{r_1 n^3 h}{e^2.Z^2}; \therefore \frac{T_{H^+}}{T_{He^+}} = Z^2 = 4 \qquad (Z = 2 \text{ for } He^+)$$

Illustration 10: Find out the quantum number 'n' corresponding the excited state of He^- ion if on de-excitation to the ground state that ion emits only two photons in succession with wavelength 1023.7 and 304 Å ($R_H = 1.097 \times 10^7\ m^{-1}$) **(JEE MAIN)**

Sol: For two successive transitions say from n to n_1 and n_1 to 1.

$$\Delta E = (E_n - E_{n_1})(E_{n_1} - E_1) = E_n - E_1 = \frac{hc.R_HZ^2}{n^2} + \frac{hcR_HZ^2}{1^2} = hcR_HZ^2\left[\frac{n^2-1}{n^2}\right] \quad \text{... (i)}$$

$$\text{Also given, } \Delta E_1 = \frac{hc}{\lambda_1} + \frac{hc}{\lambda_2} = hc\left[\frac{\lambda_1 + \lambda_2}{\lambda_1.\lambda_2}\right] \quad \text{...(ii)}$$

$$\therefore \text{By Eqs. (i) and (ii) } \frac{n^2-1}{n^2} = \left[\frac{\lambda_2 + \lambda_2}{\lambda_1\lambda_2}\right] \times \frac{1}{R_H.Z^2} = \frac{1}{1.097 \times 10^7 \times 4}\left[\frac{(1026.7 + 304) \times 10^{-10}}{1026.7 \times 304 \times 10^{-20}}\right] = 0.9716; \therefore n = 6$$

8. SOMMERFIELD'S EXTENSION OF BOHR'S THEORY

To explain the fine spectrum of hydrogen atom, Sommerfield in 1915 proposed that the moving electron might describe elliptical orbits in addition to circular orbits and the nucleus is situated at one of the foci. In a circular motion only the angle of revolution changes while the distance from the nucleus remains the same but in an elliptical motion both the angle of revolution and the distance of the electron from the nucleus change. The distance from the nucleus is termed as radius vector and the angle revolution is known as azimuthal angle. The tangential velocity of the electron at a particular instant can be resolved into two components: one along the radius vector called radial velocity and the other perpendicular to the radius vector called transverse or angular velocity. These two velocities give rise to radial momentum and angular or azimuthal momentum. Sommerfeld proposed that both the moments must be integral multiples of $\frac{h}{2\pi}$

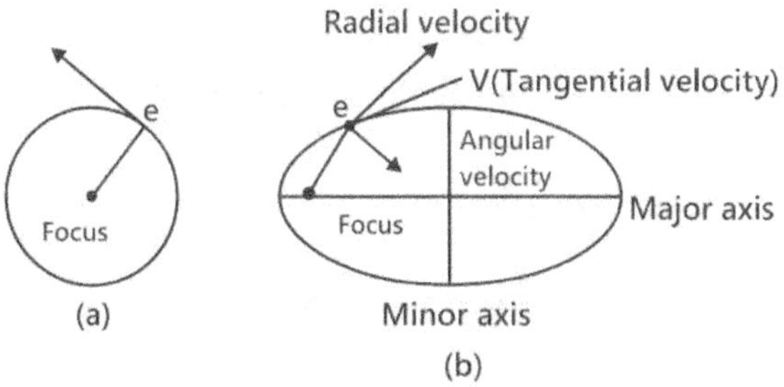

Figure 2.11: Sommerfield's extension

Radial momentum = $n_r \frac{h}{2\pi}$;

azimuthal momentum = $n_\phi \frac{h}{2\pi}$

n_r and n_ϕ are related to the main orbit 'n' as:

$n = n_r + n_\phi$ or

$$\frac{n}{n_\phi} = \frac{n_r + n_\phi}{n_\phi} = \frac{\text{Length of major axis}}{\text{Length of minor axis}}$$

(i) 'n_ϕ' cannot be zero because under this condition, ellipse shall take the shape of a straight line.

(ii) 'n_ϕ' cannot be more than 'n' because the minor axis is always smaller than major axis; when the major axis becomes equal to minor axis the ellipse takes the shape of a circle. Thus, n_2 can have all integral values up to 'n' but not zero. When the values are less than 'n', orbits are elliptical and when it becomes equal to 'n'. The orbit becomes circular in nature.

For n = 1, n_ϕ can have only one value, i.e. 1. Therefore, the first orbit is circular in nature.

For n = 2, n_ϕ can have two values 1 and 2, i.e. the second orbit has two sub orbits; one is elliptical and the other is circular in nature.

For n = 3, n_ϕ can have three values 1, 2 and 3, i.e. the third orbit has three sub orbits, two are elliptical and one is circular in nature. For n = 4, n_ϕ can have four values 1, 2, 3 and 4, i.e. fourth orbit has four suborbitssub orbits, theree are elliptical and fourth one is circular in nature (Fig.2.12). Sommerfield thus introduced the concept of sub-energy shells. In a main energy shell, the energies of sub-shells differ slightly from one another. Hence, the jumping of an electron from one energy shell to another energy shell will involve slightly different amount of energy as it will depend on sub-shell also. This explains to some extent the final spectrum of hydrogen atom. However, Sommerfield extension fails to explain the spectra of multi-electron atoms.

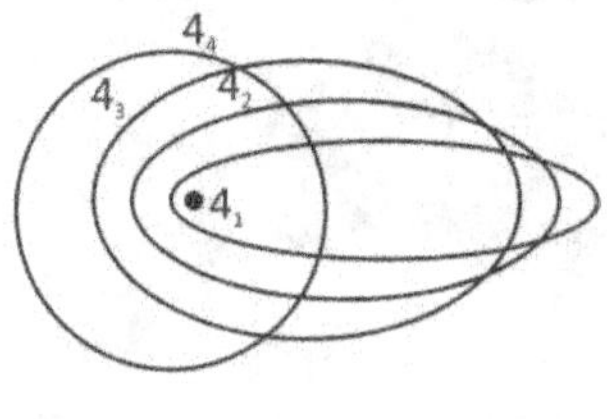

Figure 2.12: Elliptical path

9. PARTICLE AND WAVE NATURE OF ELECTRON

In 1924, **Louis de Broglie** proposed that an **electron**, like light behaves both as a material particle and as wave. This proposal gave birth to a new theory as **wave mechanical theory of matter**. According to this theory, the electrons, protons and even atoms, when in motion, possess wave properties.

De Broglie derived an expression for calculating the wavelengths of the wave associated with the electron. Using Planck's equation, $E = h\nu = h.\dfrac{c}{\lambda}$ ...(i)

On the basis of Einstein's mass–energy relationship the energy of a photon is $E = mc^2$ ...(ii)

where c is the velocity of the electron

Equating both the equations, we get $h\dfrac{c}{\lambda} = mc^2;\ \lambda = \dfrac{h}{mc} = \dfrac{h}{p}$

Momentum of the moving electron is inversely proportional to its wavelengths.

Let kinetic energy of the particle of mass 'm' is E. $E = \dfrac{1}{2}mv^2;\ 2Em = m^2v^2$

$\sqrt{2Em} = mv = p(\text{momentum});\ \lambda = \dfrac{h}{p} = \dfrac{h}{\sqrt{2Em}}$

9.1 Davisson and Germer Modification

Davisson and German made the following modification to de Broglie equation:

When a charged particle say an electron is accelerated with a potential of V; then the kinetic energy is given as:

$$\frac{1}{2}mv^2 = eV;\ m^2v^2 = 2eVm;\ mv^2 = \sqrt{2eVm} = p;\ \lambda = \frac{h}{\sqrt{2eVm}}$$

$\lambda = \dfrac{h}{\sqrt{2qVm}}$ for charged particles of charge q

De Broglie waves are not radiated into space, i.e. they are always associated with electron. Since the wavelength decreases if the value of mass (m) increases, in the case of heavier particles the wavelength is too small to be measured. De Broglie equation is applicable in the case of smaller particles like electron and has no significance for larger particles.

(a) de Broglie wavelengths associated with charged particles

(i) For electron: $\lambda = \frac{12.27}{\sqrt{V}}$ Å (ii)For proton: $\lambda = \frac{0.286}{\sqrt{V}}$ Å

(iii) For α- particles: $\lambda = \frac{0.101}{\sqrt{V}}$ Å where V = acceleration potential of these particles

(b) de Broglie wavelength associated with uncharged particles

(i) For neutrons: $\lambda = \frac{h}{\sqrt{2Em}} = \frac{6.62\times10^{-34}}{\sqrt{2\times1.67\times10^{-27}\times E}} = \frac{0.286}{\sqrt{E(eV)}}$ Å

(ii) For gas molecules: $\lambda = \frac{h}{m\times v_{rms}} = \frac{h}{\sqrt{3mkT}}$ where, k = Boltzmann constant

9.2 Bohr's Theory vs. de Broglie Equation

Bohr's Theory versus de Broglie equation: Bohr's theory postulates that angular momentum of an electron is an integral multiple of h/2π. This postulate can be derived with the help of de Broglie concept of wave nature of electron.

Consider an electron moving in a circular orbit around nucleus. The wave train would be associated with the circular orbit as shown in the figure. If the two ends of an electron wave meet to give a regular series of crests and troughs, the electron wave is said to be in phase, i.e. the circumference of Bohr orbit is equal to whole number multiple of the wavelength of the electron wave.

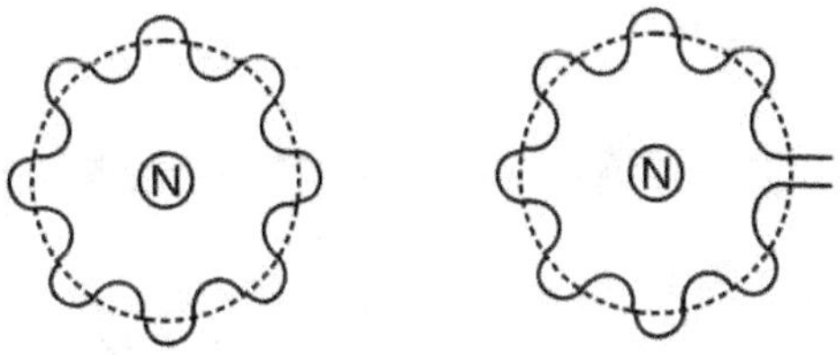

Figure 2.13: Wave in phase and out of phase

So, $2\pi r = m\lambda$ Or $\lambda = \frac{2\pi r}{n}$ ...(i)

From de Broglie equation, $\lambda = \frac{h}{mv}$ ...(ii)

Thus, $\frac{h}{mv} = \frac{2\pi r}{n}$ or $mvr = n.\frac{h}{2\pi}$ (v = velocity of electron and r = radii of the orbit)

i.e. Angular momentum = $n.\frac{h}{2\pi}$ ...(iii)

This proves that the concepts of de Broglie and Bohr are in perfect agreement with each other.

Illustration 11: With what velocity must an electron travel so that its momentum is equal to that of a photon of wavelength of λ = 5200 Å **(JEE MAIN)**

Sol: Momentum=mv and λ = h/mv. They are the momentum of an electron and photon respectively. Thus, by equating them would give the velocity.

$\because \lambda = h/mv \therefore$ Momentum of photon = $mv = h/\lambda = \frac{6.626\times10^{-34}}{5200\times10^{-10}}$ kg m sec^{-1}

Also, momentum of electron = mv = $9.108\times10^{-31}\times v$ kg m sec^{-1}

Since, both are same and thus equating the two $9.108 \times 10^{-31} \times v = \frac{6.626 \times 10^{-34}}{5200 \times 10^{-10}}$

$\therefore v = 1400 \text{ m sec}^{-1}$

Illustration 12: The wavelength of an electron moving in second orbit of H-atom is an integral multiple of its circumference. **(JEE MAIN)**

Calculate:

(a) Number of wave in the second orbit,

(b) Speed of electron in the second orbit.

(c) How much potential must be applied to an electron so that the electron becomes stationary at a point?

Sol: Circumference is $2\pi r$ and it is said that the wavelength is an integral multiple of the circumference. Thus, $n\lambda = 2\pi r$. Moreover, number of waves made by an electron in Bohr's orbit is equal to number of orbits.

(a) No. of waves (n) in the second orbit = 2

(b) Also $2\pi r_2 = 2 \times 3.14 \times 0.529 \times 10^{-8} \times 4 = 13.28 \times 10^{-8}$ cm

Since, $n\lambda = 2\pi r \therefore \lambda = \frac{13.28 \times 10^{-8}}{2} = 6.64 \times 10^{-8} \text{ cm} = 6.64 \times 10^{-10} \text{ m}$

$$\therefore v = \frac{h}{\lambda m} = \frac{6.626 \times 10^{-34}}{6.64 \times 10^{-10} \times 9.108 \times 10^{-31}} = 1.09 \times 10^6 \text{ m/s}$$

(c) Also $eV_0 = \frac{1}{2}mv^2 \therefore V_0 = \frac{9.108 \times 10^{-31} \times (1.09 \times 10^6)^2}{2 \times 1.602 \times 10^{-19}} = 3.38\text{V}$

10. HEISENBERG'S UNCERTAINTY PRINCIPLE

We see around us all moving particles, e.g. a car, a ball thrown in the air etc. move along definite paths. Hence their position and velocity can be measured accurately at any instant of time. Likewise is it possible to measure the position and velocity for the subatomic particle also?

Heisenberg, in 1927 gave a principle about the uncertainties in simultaneous measurement of position and momentum (mass × velocity) of small particles. This principle is due to the consequence of dual nature of matter.

This Principle States: 'It is impossible to measure simultaneously the position and momentum of a small microscopic moving particle with absolute accuracy or certainty', i.e. if an attempt is made to measure any one of these two quantities with higher accuracy, the other becomes less accurate. The product of the uncertainty in position (Δx) and the uncertainty in the momentum ($\Delta p = m.\Delta v$ where m is the mass of the particle and Δv is the uncertainty in velocity) is equal to or greater than $h/4\pi$ where h is the Planck's constant. Thus, the mathematical expression for the Heisenberg's uncertainty principle is readily written as $\Delta x.\Delta p \geq h/4\pi$

Explanation of Heisenberg's uncertainty principle: Let us attempt to measure both the position and momentum of an electron; to pinpoint the position of the electron we have to use light so that the photon of light strikes the electron and the reflected photon is seen in the microscope. As a result of the hitting, both the position and the velocity of the electron are disturbed. The accuracy with which the position of the particle can be measured depends upon the wavelength of the light used.

The uncertainty in position is $\pm \lambda$. The shorter the wavelength, the greater is the accuracy. But shorter wavelength means higher frequency and hence higher energy. This high energy photon on striking the electron changes its speed as well as direction. But this is not true for a moving macroscopic particle. Hence Heisenberg's uncertainty principle does not apply to macroscopic particles.

Illustration 13: Why cannot the electron exist inside the nucleus according to Heisenberg's uncertainty principle? **(JEE MAIN)**

Sol: Following the Heisenberg's Uncertainty principle, the location of the electron is justified. Use $\Delta x.\Delta p \geq h/4\pi$ or $\Delta x.\Delta p - h/4\pi$.

Diameter of the atomic nucleus is of the order of 10^{-15} m

The maximum uncertainty in the position of electron is 10^{-15}

Mass of electron = 9.1×10^{-31} kg.;

$\Delta x.\Delta p - h/4\pi$;

$\Delta x \times (m.\Delta v) - h/4\pi$;

$\Delta v = h/4\pi \times 1/\Delta x.\ M = 6.53 \times 10^{-34}/4x\ (22/7)x\ 1/10^{-15} \times 9.1 \times 10^{-31}$; $\Delta v = 5.80 \times 10^{10}\ ms^{-1}$

The value is much higher than the velocity of light and hence not possible.

Illustration 14: What is the uncertainty in the position of electron, if uncertainty in its velocity is 0.0058 m/s? **(JEE MAIN)**

Sol: $\Delta x \times \Delta v = h/4\pi m$; $\Delta x = 6.62 \times 10^{-34} / 4x\ 3.14 \times 9.1 \times 10^{-31} \times 0.0058$; $\Delta x = 0.01$ m

11. WAVE MECHANICAL MODEL OF ATOMS

Erwin Schrodinger in 1920 put forward this model by taking into account the de Broglie concept of dual nature of matter and Heisenberg's uncertainty principle. In this model, the discrete energy levels or orbits proposed by Bohr's model are replaced by mathematical function ψ (psi) which is related with probability of finding electrons around the nucleus. The wave equation for an electron wave propagating in 3-D space is: $\frac{\partial^2\psi}{\partial x^2}+\frac{\partial^2\psi}{\partial y^2}+\frac{\partial^2\psi}{\partial z^2}+\frac{8\pi^2 m}{h^2}(E-V)\psi = 0$

where y is the amplitude of the electron wave at point with coordinates x, y, z, E = total energy and V = potential energy of the electron; ψ is also called wave function and ψ^2 gives the probability of finding the electron at (x, y, z). The acceptable solutions of the above equation for the energy E are called Eigen values and the corresponding wave functions are called **Eigen functions**.

Every function is not an Eigen function. An acceptable solution for Schrodinger wave equation must satisfy the following conditions:

1. The function should be finite.
2. It should always bear a single value at a particular point in space.
3. It should be a continuous function. Schrodinger wave equation can be written as

$$\left(\frac{\partial^2}{\partial x^2}+\frac{\partial^2}{\partial y^2}+\frac{\partial^2}{\partial z^2}\right)\psi+\frac{8\pi^2 m}{h^2}(E-V)\psi = 0;\ \text{or}\ \nabla^2\psi+\frac{8\pi^2 m}{h^2}(E-V)\psi = 0$$

where $\nabla^2 = \frac{\partial^2}{\partial x^2}+\frac{\partial^2}{\partial y^2}+\frac{\partial^2}{\partial z^2}$ is called Laplacian operator This equation can be rewritten as $\nabla^2\psi = -\frac{8\pi^2 m}{h}(E-V)\psi$;

or $\left(-\frac{h^2}{8\pi^2 m}\nabla^2+V\right)\psi = E\psi$; or $H\psi = E\psi$, where $H = -\frac{h^2}{8\pi^2 m}\nabla^2+V$, is called Hamiltonian operator

In this operator, the first term represents kinetic energy operator (T) and the second term represents potential energy operator (V).

Significance of ψ: It represents the amplitude of an electron wave. It can be positive or negative. It has no physical value.

Significance of ψ^2: It is a probability function. It determines the probability of finding an electron within a smaller region of space around nucleus. The space in which there is maximum probability of finding an electron is termed as orbital.

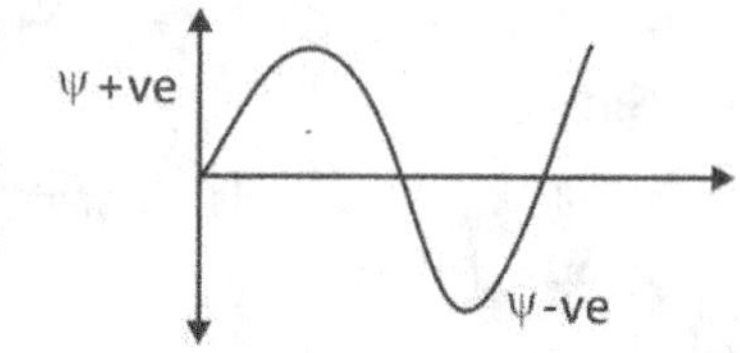

Figure 2.14: Significance of ψ

12. QUANTUM NUMBERS

An atom contains large number of shells and sub-shells. These are distinguished from one another on the basis of their size, shape and orientation (direction) in space. The parameters are given in terms of different numbers called **quantum numbers**.

Quantum numbers may be defined as a set of four numbers with the help of which we can get complete information about all the electrons in an atom. It tells us the address of the electron i.e. location, energy, the type of orbital occupied and orientation of that orbital.

12.1 Principal Quantum Number

(a) This is denoted by n, an integer.

(b) The values of n are from 1 to n. n = 1 K shell; n = 2 L shell

n = 3 M shell; n = 4 N shell

(c) 'n' represents the major energy shell to which an electron belongs.

(d) The values of 'n' signify the size and energy level of major energy shells.

(e) As the value of 'n' increases, the energy of the electron increases and thus, the electron is less tightly held with nucleus.

(f) Angular momentum can be calculated using principal quantum number: $mvr = nh/2\pi$

12.2 Azimuthal Quantum Number

This is denoted by l.

(a) The values of l are from 0 to (n − 1)

(b) l = 0, s-sub-shell, spherical (The representation is independent of the value of n)

l = 1, p-sub-shell, dumbbell

l = 2, d-sub-shell, double dumbbell or like leaf

(c) The letters s, p, d, f designate old spectral terms.

Sharp (s), principal (p), diffuse (d), fundamental (f)

(d) For a given value of n, total values of 'l' are n.

(e) The values of l signify the shape and energy level of sub-shells in a major energy shell.

(f) The angular momentum of an electron in an orbital is given by $nh/2\pi$.

(g) The energy level for sub-shells of a shell shows the order: $s < p < d < f$

12.3 Magnetic Quantum Number

(a) Denoted by m_l, an integer.

(b) Zeeman effect: Zeeman studied the fine spectrum of H using a spectroscope of high resolving power as well as putting the source under the influence of magnetic field. He noticed that the spectral line splits up to more than one component

(c) Each frequency of radiation emitted by the atom in the presence of magnetic field splits up into components if the angular momentum of the electron along the magnetic field are restricted to the value, $m_l = \frac{h}{2\pi}$

(d) The values of m_l lie from $\pm l$ through zero.

(e) The positive values of magnetic quantum number m_l represent the angular momentum component of the orbital in the direction of the applied magnetic field whereas the negative values of m_l account for the angular momentum component of orbital in the opposite direction of applied magnetic field.

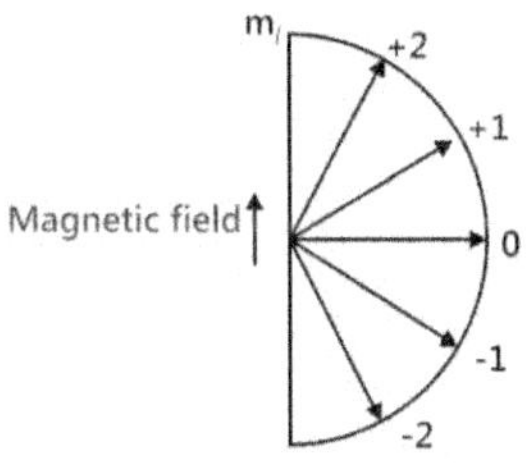

Figure 2.15: Magnetic quantum number

(f) Total values of m_l for a given value of $n = n^2$.

(g) Total values of m_l for a given value of $l = (2l + 1)$

(h) The values of m_l signify the possible numbers of orientations of a sub-shell.

(i) In the absence of magnetic field, the three p-orbitals are equivalent in energy and are said to be threefold degenerate, i.e. sub sub-shell (orbitals) having same energy level are known as **degenerate orbitals**.

12.4 Spin Quantum Number

(a) Wave mechanical treatment required no more than three quantum number n, l and m. The existence of multiple, i.e. doublet structure led to the introduction of a spin quantum number m_s.

(b) The values of ms are $+\frac{1}{2}$ and $-\frac{1}{2}$. This is due to the fact of the doublet structures of spectral lines which can be explained by proposing only two directions of spin of electron along its own axis.

(c) The values of ms signify the direction of rotation or spin of an electron in its axis during its motion.

(d) Spin angular momentum is given by $\frac{h}{2\pi}\sqrt{m_s(m_s+1)} = \frac{h}{2\pi}\sqrt{\frac{1}{2}\left(\frac{1}{2}+1\right)} = \frac{\sqrt{3}\,h}{4\ \pi} = \frac{\sqrt{3}}{2}h.$

(e) The spin may be clockwise $\left(+\frac{1}{2}\text{or}-\frac{1}{2}\right)$ or anticlockwise $\left(-\frac{1}{2}\text{or}+\frac{1}{2}\right)$

(f) Spin multiplicity of an atom = $\sqrt{s(s+1)}$.

12.5 Shapes of Orbitals

The electron cloud represents the shape of the orbital. It is not uniform but it is dense where the probability for finding the electron is maximum.

(a) s-orbitals do not vary with angles, i.e. they do not have directional dependence. Thus, all s-orbitals are called spherically symmetrical. Their size increases with increases in the value of n. 1s-orbital has no nodal plane (the plane at which zero electron density is noticed). 2s-orbital has one nodal plane; 3s-orbital has two nodal planes. Thus it is evident that the number of nodal planes increases with increasing value of principal quantum number.

(b) All orbitals with $l \neq 0$ have angular dependence. Therefore, p and d and other higher angular momentum orbitals are not spherically symmetrical. p-orbitals consist of two lobes to form dumbbell shaped structure. The three p-orbitals along *x*, *y*, *z*-axes named as p_x, p_y, p_z orbitals are perpendicular to each other. All the three p-orbitals of a sub-shell have the same size and shape but differ from each other in orientation. The subscripts *x*, *y* and *z* indicate the axis along which the orbitals are oriented and possess maximum electron density. Also, the orbitals of a sub-shell having same energy are referred as **degenerate orbitals**.

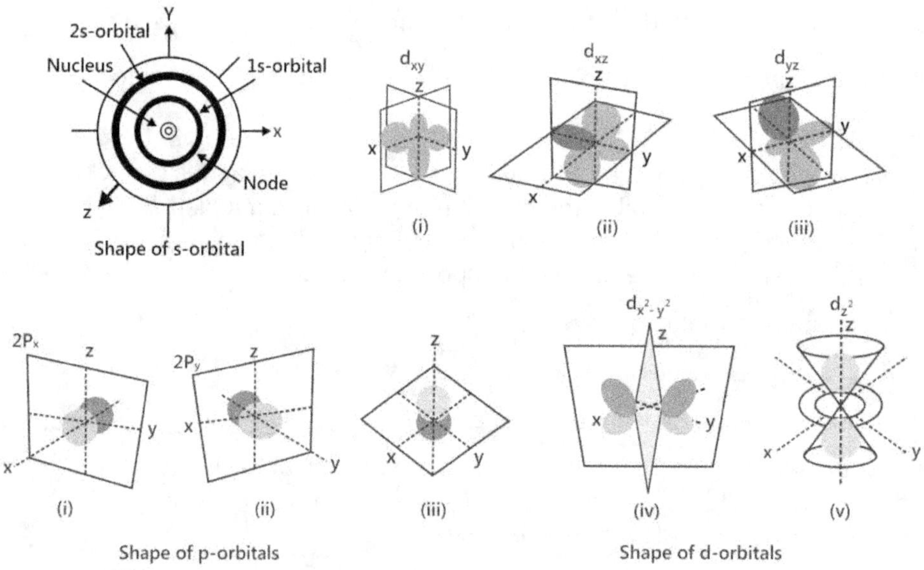

Figure 2.16: Shapes of orbitals

Illustration 15: Given below are the sets of quantum numbers for given orbitals. Name these orbitals. **(JEE MAIN)**

(a)	(b)	(c)	(d)
$n = 2$	$n = 4$	$n = 3$	$n = 4$
$l = 1$	$l = 2$	$l = 1$	$l = 2$
$m_l = 2$	$m_l = 0$	$m_l = \pm 1$	$m_l = \pm 2$

Sol: (a) $\because$ n = 2 and l = 1 $\therefore$ 2p Since, $m_l = -1$ $\therefore$ $2p_y$ or $2p_x$ similarly for others

(b) $4d_{z^2}$ (c) $3p_x$ or $3p_y$ (d) $4d_{x^2-y^2}$ or $4d_{xy}$

Illustration 16: For n = 6 suggest **(JEE MAIN)**

(a) Total number of electrons that it can have

(b) Total number of sub-shells which can exist

(c) Total number of sub sub-shells (orbitals) which can exist.

Sol: Having the knowledge of quantum numbers and the existing combinations, the following can be predicted.

(a) 72 (b) 6 (c) 36

13. SCHRODINGER WAVE EQUATION FOR H-ATOM

Since the electron in the hydrogen atom seems to be a spherically symmetric potential spherical polar coordinates are used to solve the equation. The potential energy is simply that of a point charge, $U(r) = -e^2/4\pi\varepsilon_0 r$. When the value of potential energy is substituted in the equation and is solved in polar coordinate form, the total energy is found to be $E = -me^4/8\varepsilon^2 n^2 h^2$, which is the same as given by the Bohr's model. The solutions of Schrodinger wave equation is obtained by separating the variables so that the wave function is represented by the product – $\psi(r,\theta,\phi) = \underbrace{R(r)}_{\text{Radial part}}\ \underbrace{\Theta(\theta)\Phi(\phi)}_{\text{Angular part}}$. Since the function R depends only on r_1, it describes the distribution of the electron as function of r from the nucleus. These two functions, Θ and Φ depend upon two quantum numbers, n and l and taken together give the angular distribution of the electron. The radial part of the wave function for some orbitals may be given as,

n	l	Rn	
1s	1	0	$2\left(\frac{Z}{a_0}\right)^{3/2} e^{-Z/a_0}$;
2s	2	0	$\frac{1}{\sqrt{3}}\left(\frac{Zr}{2a_0}\right)^{3/2} e^{-Zr/a_0}$

where Z = atomic number, a_0 = radius of first Bohr orbit of hydrogen

14. PLOTS OF GRAPHS

14.1 Plots of Graphs of R (Radial Wave Function) vs. r (Radius of Atom)

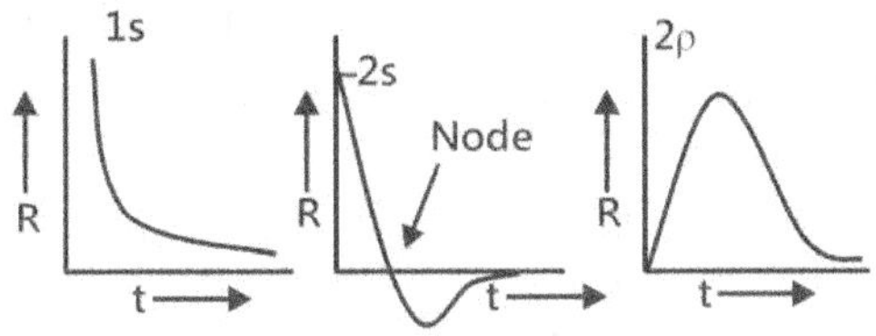

Note: The point where the wave function changes its sign is called node. The number of radial nodes can be determined by the formula: (n – l - 1).

14.2 Plots of Graphs of R^2 (Radial Probability Density) vs. r (Radius of Atom)

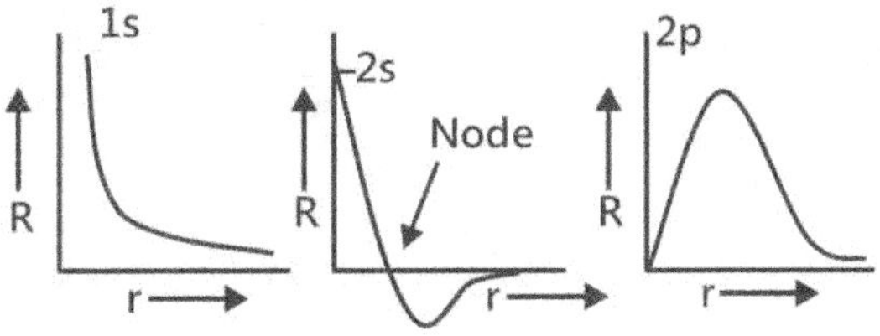

(a) For 1s and 2s orbitals, the function is not defined at r = 0 (asymptotic behaviour) as the probability of finding electron inside nucleus is not known but there can be a probability of finding electron just outside the nucleus as s orbital lies very closer to nucleus.

(b) For 2p orbital, at r = 0, probability is zero.

14.3 Plots of Graphs of R^2r (Radial Probability Distribution) vs. r (Radius of Atom)

In order to visualize the electron cloud, consider the space around nucleus divided into a large number of small concentric spherical shells of radius dr. The volume of such a shell can be –

$dv = 4\pi/3(r+dr)^3\ 4\pi/3r^3$ So, $dv = 4\pi r^2 dr$

This volume is called radial volume and the probability of finding an electron within this shell is called radial probability distribution function.

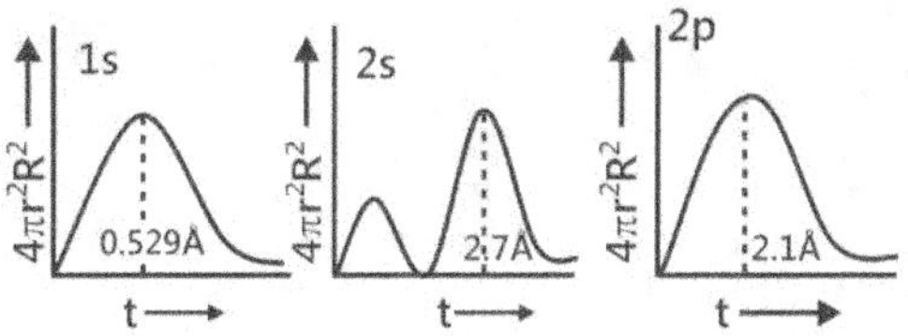

R.P.F. = (Volume of spherical shell) × probability density

= $(4\pi r^2\ dr) \times R$

Radial probability distribution = $4\pi r^2 dr\ R^2$

NOMORECLASS CONCEPTS

- For 1s orbital, radial probability increases with increase in distance from the nucleus, reaches a maximum and then decreases. The maxima are the maximum probability of finding an electron which is also called 'radius of maximum probability' and is also same as Bohr's radius.
- For 2s, the graph has two maxima. In between these two maxima, the curve passes through a zero value indicating that there is zero probability of finding the electron at that distance. This point is a nodal point which can be a radial/spherical node.

14.4 Angular Node/Nodal Plane

The probability of finding an electron in nucleus is zero; so it is called a nodal point. Any plane passing through that point where the probability of finding an electron is zero is called a nodal plane.

NOMORECLASS CONCEPTS

- s orbital doesn't have any nodal plane.
- p orbital has one nodal plane. Nodal plane for p_x is YZ plane, for p_y is XZ plane and for p_z is XY plane.
- d orbital has two nodal planes. Nodal planes for d_{xy} are XZ and YZ, for d_{xz} are YZ and XY, for d_{yz} are XY and XZ, for $d_{x^2-y^2}$ are the lines inclined at 45° with X and Y axes.
- So, number of angular nodes = l

Important Points: Key Take-Away

Type of information	Principal Quantam No. (n)	Azimuthal Quantam No. (l)	Magnetic Quantam m_l	Spin Quantam No. m_s
1. Whys is it required?	To explain the main lines of a spectrum	To explain the line structure of the line spectrum	To explain the splitting of lines in a magnetic field	To explain the magnetic properties of substances
2. What does it tell?	(i) Main shell in which the electron resides	(i) No. of sub-shells present in any main shell	No. of orbitals present in any sub-shell or the number of orientiation of each sub-shell	Direction of electron spin, i.e. clockwise or anti clockwise
	(ii) Approx distance of the electron from the nucleus	(ii) Relative energies of the sub-shell		
	(iii) Energy of the shell	Shapes of orbitals		
	(iv) Max. no. of electrons present in the sheell ($2n_1^2$)			
3. What are the symbols?	n	l	m or m_l	s or m_s
4. What are the values?	1, 2, 3, 4 etc. i.e. any integer	For a particular value of n, l = 0 to n - 1	For a particular value of i.e.= -1 to +1 including zero	For a particular value of m,
5. Othe designations?	K, L, M, N, etc.	l = 0, s-sub-shell; l = 1, p--sub-shell; l = 2, d-sub-shell; l = 3, f-sub-shell	For p-sub-shell m_l = -1, 0, +1, designated as p_x, p_y and p_z	Two arrows pointing in opposite directions, i.e. ↑ and ↓

NOMORECLASS CONCEPTS

- Number of radial nodes = $(n - l - 1)$
- Number of angular nodes = l
- Total number of nodes = $(n - l)$
- Number of nodal planes = l

15. PAULI'S EXCLUSION PRINCIPLE

The principle states that no two electrons in an atom can have the same set of all the numbers. In other words, no orbital can have more than two electrons.

Conclusion:

(a) The maximum capacity of a main energy shell is equal to $2n^2$ electrons.

(b) The maximum capacity of a sub-shell is equal to $2(2l + 1)$ electrons.

(c) Number of sub-shells in a main energy shell is equal to the value of n.

(d) Number of orbitals in a main energy shell is equal to n^2.

(e) One orbital cannot have more than two electrons. If two electrons are present, their spins should be in opposite directions.

16. AUFBAU PRINCIPLE

The word 'aufbau' originates from the German word 'Aufbauen' which means 'to build'. This gives us a sequence in which various sub-shells are filled up depending on the relative order of the energy of the sub-shells. The sub-shell of the lowest energy is filled up first, then the next sub-shell of higher energy starts filling. The sequence in which various sub-shells are filled is the following:

1s, 2s, 2p, 3s, 4s, 3d, 5s, 4d, 5s, 4d, 5p, 6s, 4d, 5d, 6p, 7s, 5f, 6d, 7p

Using (n + l) Value:

The sequence in which various sub-shells are filled up can also be determined with the help of (n + l) value. When two or more sub-shells have same (n + l) value, the sub-shell with the lowest value of 'n' is filled up first.

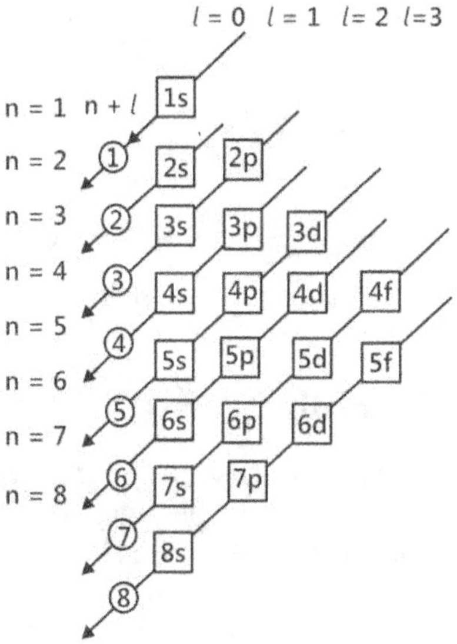

Figure 2.17: Aufbau rule

Sub-shell	n	l	$(n+l)$	
1s	1	0	0	
2s	2	0	2	
2p	3	0	3	Lowest value of n
3s	3	0	3	
3p	3	1	4	Lowest value of n
4s	4	0	4	
3d	3	2	5	Lowest value of n
4p	4	1	5	
5s	5	0	5	
4d	4	2	6	Lowest value of n
5p	5	1	6	
6s	6	0	6	
4f	4	3	7	Lowest value of n
5d	5	2	7	
6p	6	1	7	
7s	7	0	7	
5f	5	3	8	Lowest value of n
6d	6	2	8	
7p	7	1	8	

The principal quantum number solely determines the energy of the electron in a hydrogen atom and other single electron species like He^+, Li^{2+} and Be^{3+}. The energy of orbitals in hydrogen and hydrogen like species increases as follows: 1s < 2s = 2p < 3s = 3p = 3d < 4s = 4p = 4d = 4f < .....

Exceptions to Aufbau Principle: In some instances it is noted that actual electronic arrangement is slightly different from the arrangement expected by Aufbau principle. A simple reason behind this is that half-filled and full-filled sub-shells have got extra stability.

Cr_{24} → $1s^2, 2s^2\ 2p^6, 3s^2\ 3p^6\ 3d^4, 4s^2$ (wrong)

→ $1s^2, 2s^2\ 2p^6, 3s^2\ 3p^6\ 3d^5, 4s^1$ (right)

Cr_{29} → $1s^2, 2s^2\ 2p^6, 3s^2\ 3p^6\ 3d^9, 4s^2$ (wrong)

→ $1s^2, 2s^2\ 2p^6, 3s^2\ 3p^6\ 3d^{10}, 4s^1$ (right)

Similarly the following elements have slightly different configurations than expected:

$Nb_{41} \longrightarrow [Kr]4d^4\ 5s^1$

$Mo_{42} \longrightarrow [Kr]4d^5\ 5s^1$

$Ru_{44} \longrightarrow [Kr]4d^7\ 5s^1$

$Rh_{45} \longrightarrow [Kr]4d^8\ 5s^1$

$Pd_{46} \longrightarrow [Kr]4d^{10}\ 5s^0$

$Ag_{47} \longrightarrow [Kr]4d^{10}\ 5s^1$

$Pt_{78} \longrightarrow [Xe]4f^{14}\ 5d^9\ 6s^1$

$Au_{79} \longrightarrow [Xe]4f^{14}\ 5d^{10}\ 6s^1$

$La_{57} \longrightarrow [Kr]4d^{10}\ 5s^2\ 5p^6\ 5d^1\ 6s^2$

$Ce_{58} \longrightarrow [Kr]\ 4d^{10}\ 5s^2\ 5p^6\ 5d^0\ 6s^2$

$Gd_{64} \longrightarrow [Kr]\ 4d^{10}\ 4f^7\ 5s^2\ 5p^6\ 5d^1\ 6s^2$

17. HUND'S RULE OF MAXIMUM MULTIPLICITY (ORBITAL DIAGRAMS)

It states that electrons are distributed among the orbitals of sub-shell in such a way as to give the maximum number of unpaired electrons with parallel spins. This means that the orbitals available in a sub-shell are first filled singly before they begin to pair i.e. the pairing of electrons occurs with the introduction of the second electron in the s-orbital, the fourth electron in the p-orbitals, the sixth electron in the d-orbitals and the eighth electron in the f-orbitals. The rule is based on the fact that electrons have the same charge and repel each other and hence try to keep further apart from each other as much as possible. The electrons thus occupy different orbitals of the sub-shell as to minimize the inter-electronic repulsion and increase the stability of the atom. Orbitals tend to become half-filled or completely filled since such an arrangement will be more stable on account of symmetry.

The orbital diagram for nitrogen, oxygen, fluorine and neon are as follows:

Nitrogen	(7)	[He]	↑↓ ↑ ↓ ↑ (2p)
Oxygen	(8)	[He]	↑↓ ↑↓ ↑ ↑ (2p)
Fluorine	(9)	[He]	↑↓ ↑↓ ↑↓ ↑ (2p)
Neon	(10)	[He]	↑↓ ↑↓ ↑↓ ↑↓ (2p)

The orbital diagrams of elements from atomic number 21 to 30 can be represented on similar lines as below:

	3d					4s
Sc[Ar]$3d^1$ $4d^2$	↑	–	–	–	–	↑↓
Ti [Ar]$3d^2$ $4s^2$	↑	↑	–	–	–	↑↓
V[Ar]$3d^3$ $4s^2$	↑	↑	↑	–	–	↑↓
Cr[Ar]$3d^5$ $4s^1$	↑	↑	↑	↑	↑	↑
Mn[Ar]$3d^5$ $4s^2$	↑	↑	↑	↑	↑	↑↓

	3d					4s
Fe[Ar]$3d^6$ $4s^2$	↑↓	↑	↑	↑	↑	↑↓
Co[Ar]$3d^7$ $4s^2$	↑↓	↑↓	↑	↑	↑	↑↓
Ni[Ar]$3d^8$ $4s^2$	↑↓	↑↓	↑↓	↑	↑	↑↓
Cu[Ar]$3d^{10}$ $4s^1$	↑↓	↑↓	↑↓	↑↓	↑↓	↑
Zn[Ar]$3d^{10}$ $4s^2$	↑↓	↑↓	↑↓	↑↓	↑↓	↑↓

All those atoms which consist of at least one of the orbitals singly occupied behave as paramagnetic materials because these are weakly attracted to a magnetic field, while all those atoms in which all the orbitals are doubly occupied behave as diamagnetic materials because they have no attraction for magnetic field. However, these are slightly repelled by magnetic field due to induction.

Magnetic moment may be calculated as, 1 BM (Bohr magneton) = where n = No. of unpaired electrons

Illustration 17: Predict total spin for each configuration. **(JEE MAIN)**

(a) $1s^2, 2s^2, 2p^5$ (b) $1s^2, 2s^2, 2p^3$ (c) $1s^2, 2s^2\ 2p^6, 3s^2\ 3p^6\ 3d^5, s^2$ (d) $1s^2, 2s^2$

Sol: Use Total spin of an electron = $\left(\pm\frac{1}{2}\right)$ × no. of unpaired electrons

$\therefore$ (a) Total spin = $1\times\left(\pm\frac{1}{2}\right)=\pm\frac{1}{2}$ (b) Total spin = $3\times\left(\pm\frac{1}{2}\right)=\pm(3/2)$

(c) Total spin = $5\times\left(\pm\frac{1}{2}\right)=\pm(5/2)$ (d) Total spin = $0\times\left(\pm\frac{1}{2}\right)=0$

Illustration 18: Write the electronic configurations of the elements with the following atomic numbers: 3, 8, 14, 17, 21, 38, 57. Also mention the groups of the periodic table to which they belong. **(JEE MAIN)**

Sol:

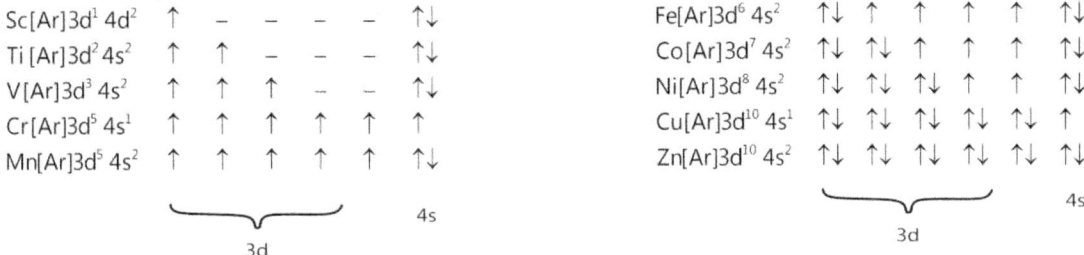

Atomic No.	Electronic configuration	Group of periodic Table
3	$1s^2\ 2s^1$	1
8	$1s^2, 2s^2, 2p_{x^2}, 2p_{y^1}, 2p_{z^1}$	16
14	$1s^2, 2s^2, 2p^6, 3s^2, 3p_{x^1}, 3p_{y^1}$	14
17	$1s^2, 2s^2, 2p^6, 3s^2, 3p_{x^2}, 3p_{y^2}, 3p_{z^2}$	17
21	$1s^2, 2s^2, 2p^6, 3s^2, 3p^6, 3d^1, 4s^2$	3

Atomic No.	Electronic configuration	Group of periodic Table
22	$1s^2, 2s^2, 2p^6, 3s^2, 3p^6, 3d^{10}, 4s^2, 4p^6, 5s^2$	2
57	$1s^2, 2s^2, 2p^6, 3s^2, 3p^6, 3d^{10}, 4s^2, 4p^6, 4d^{10}, 5s^2, 5p^6, 5d^1, 6s^2$	2

Note: Here one electron enters the 5d orbital before filling up of the 4f begins.

Rules for finding the group number

(i) If the last shell contains one , two electrons, then the group number is 1, 2 respectively.

(ii) If the last shell contains more than two electrons, then the group number is the total number of electrons in the last shell plus 10.

(iii) If electrons are present in (n – 1)d orbital in addition to those in the ns orbital, then the group number is equal to the total number of electrons present in the (n – 1)d orbital and ns orbital.

Illustration 19: Give the electronic configuration of the following ions:

(i) Cu^{2+} (ii) Cr^{3+} (iii) Fe^{2+} and Fe^{3+} (iv) H^- (v) S^{2-} **(JEE MAIN)**

Sol: During the formation of cations, electrons are lost while in the formation of anions, electrons are added to the valence shell. The number of electrons added or lost is equal to the numerical value of the charge present on the ion.

Following this general concept, we can write the electronic configurations of all the ions given in the equation.

(i) $Cu^{2+} = {}_{29}Cu - 2e^- = 1s^2\ 2s^2\ 3s^2\ 3p^6\ 3d^{10}\ 4s^1 - 2e^- = 1s^2\ 2s^2\ 2p^6\ 3s^2\ 3p^6\ 3d^9$

(ii) $Cr^{3+} = {}_{24}Cr - 3e^- = 1s^2\ 2s^2\ 2p^6\ 3s^2\ 3p^6\ 3d^5\ 4s^1 - 3e^- = 1s^2\ 2s^2\ 2p^6\ 3s^2\ 3p^6\ 3d^3$

$Fe^{3+} = {}_{26}Fe - 3e^- = 1s^2\ 2s^2\ 2p^6\ 3s^2\ 3p^6\ 3d^5 = 1s^2\ 2s^2\ 2p^6\ 3s^2\ 3p_{x^2}\ 3p_{y^2}\ 3p_{z^2}\ 3d^5$

(iii) $H^- = {}_1H + 1e^- = 1s^1 + 1e^1 = 1s^2$

(iv) $S^{2-} = {}_{16}S + 2e^- = 1s^2\ 2s^2\ 2p^6\ 3s^2\ 3p_{x^2}\ 3p_{y^1}\ 3p_{z^1} + 2e^- = 1s^2\ 2s^2\ 2p^6\ 3s^2\ 3p_{x^2}\ 3p_{y^2}\ 3p_{z^2}$

POINTS TO REMEMBER

Representation of a chemical symbol	$X_Z^{A......A=Z+n}$	X = Element, A = Mass number = No. of protons (p) + No. of neutrons(n) Z = Atomic number = No. of protons
Millikan's oil drop experiment	$m = \frac{m_0}{[1-(v/c)^2]^{1/2}}$	m = Moving mass of an electron m_0 = Rest mass of an electron
Reduced mass (μ) $\frac{1}{\mu} = \frac{1}{M} + \frac{1}{m} = \frac{mM}{m+M}$	Radius of the nucleus (R_n) $R_n = R_1(A)^{1/3}$, $R_1 = 1.33 \times 10^{-13}$ cm A = Mass number	
Moseley's experiment $\sqrt{v} = a(Z - b)$ where v = Frequency of X-rays, Z = atomic number, a and b are constants,	Ritz mathematical formula $\frac{1}{\lambda_{vac}} = \bar{v} = R\left(\frac{1}{n_1^2} - \frac{1}{n_2^2}\right)$ Where $\bar{v}$ = Wave number, R = Rydberg's constant, n_1 and n_2 are integers (such that $n_2 > n_1$) integers or energy levels.	Maximum kinetic energy of the ejected electron Max. KE = Absorbed energy – Work function $\frac{1}{2}mv_{max}^2 = hv - hv_0 = hc\left[\frac{1}{\lambda} - \frac{1}{\lambda_0}\right]$

Bohr's model inputs Radius of an atom = $r_a = \frac{n^2}{Z} \times \frac{h^2}{a\pi^2 e^2 m}$ Velocity of an electron = $v = \frac{z}{n} \times \frac{2\pi e^2}{h}$ Energy of an electron= $E_n = \frac{E_1}{n^2} z^2 - \frac{2\pi}{n} = -z^2$	De Broglie's hypothesis $\lambda = \frac{h}{p} = \frac{h}{\sqrt{2Em}}$ where λ = wavelength, p = Momentum, h = Planck's constant, E = Kinetic energy	Heisenberg's uncertainty principle $\Delta x \, \Delta p \geq \frac{h}{4\pi}$ Δx = Change in position, Δp = Change in momentum
Orbital angular momentum $L = \frac{h}{2\pi}\sqrt{l(l+1)}$	Spin angular momentum $S = \frac{h}{2\pi}\sqrt{S(S+1)}$	Magnetic moment (μ) = $\mu\sqrt{n(n+2)}$ B.M. n = Number of unpaired e^-
Some important points to be remembered	Total no. of e^- in an energy level = $2n^2$ Total no. of e^- in a sublevel = $2(2l + 1)$ Maximum no. of e^- in an orbital =2 Total no. of orbitals in a sublevel = $(2l + 1)$ No. of sub-shells in main energy shell = n No. of orbitals in a main energy shell = n^2 l = 0 1 2 3 4 s p d f g	

Solved Examples

JEE Main/Boards

Example 1: When a certain metal was irradiated with light of frequency 3.2×10^{16} Hz, the photoelectrons emitted had twice the energy as did photoelectrons emitted when the same metal was irradiated with light of frequency 2.0×10^{16} Hz. Calculate v_0 for the metal.

Sol: Threshold frequency needs to be calculated. The incident frequency has been given and a condition of photons emission has been given.

Applying photoelectric equation,

$KE = hv - hv_0$

It can be also written as $(v - v_0) = \frac{KE}{h}$; Given $KE_2 = 2KE_1$

$$v_2 - v_0 = \frac{KE_2}{h} \quad \text{... (i)}$$

$$\text{and } v_1 - v_0 = \frac{KE_1}{h} \quad \text{...(ii)}$$

Dividing equation (i) by equation (ii)

$$\frac{v_2 - v_0}{v_1 - v_0} = \frac{KE_2}{KE_1} = \frac{2KE_1}{KE_1} = 2; \text{ or } v_2 - v_0 = 2v_1 - 2v_0$$

or $v_0 = 2v_1 - v_2 = 2(2.0 \times 10^{16}) - (3.2 \times 10^{16})$

$= 8.0 \times 10^{15}$ Hz

Example 2: An electron moves in an electron field with a kinetic energy of 2.5 eV. What is the associated de Broglie wavelength?

Sol: The de-Broglie equation is $\lambda = \frac{h}{p} = \frac{h}{mv}$. Modifying the equation according to the given data and taking the help of kinetic energy, we solve it.

$$\text{Kinetic energy} = \frac{1}{2}mv^2 \left(v = \frac{h}{m\lambda} \right)$$

$$= \frac{1}{2}m\left(\frac{h}{m\lambda}\right)^2 = \frac{1}{2}\frac{h^2}{m\lambda^2} \text{ or } \lambda^2 = \frac{1}{2}\frac{h^2}{m \times KE}$$

$$\lambda = \frac{h}{\sqrt{2m \times KE}} \begin{pmatrix} m = 9.108 \times 10^{-28} \text{ g} \\ h = 6.626 \times 10^{-27} \text{ erg.sec} \\ 1ev = 1.602 \times 10^{-12} \text{ erg} \end{pmatrix}$$

$$= \frac{6.626 \times 10^{-27}}{\sqrt{2 \times 9.108 \times 10^{-28} \times 2.5 \times 1.602 \times 10^{-12}}}$$

$= 7.7 \times 10^{-8}$ cm

Example 3: The minimum energy required to overcome the attractive force between an electron and the surface of Ag metal is 5.52×10^{-19} J. What will be the maximum kinetic energy of electron ejected out from Ag which is being exposed to UV light λ = 362 Å?

Sol: The photoelectric equation gives the minimum energy and the maximum kinetic energy.

Energy of the photon absorbed

$$= \frac{h.c}{\lambda} = \frac{6.625 \times 10^{-27} \times 3 \times 10^{10}}{360 \times 10^{-8}} = 5.52 \times 10^{-11} \text{ erg}$$

$= 5.52 \times 10^{-18}$ J

E(photon) = work function + KE

KE = $5.52 \times 10^{-8} - 7.52 \times 10^{-19} = 47.68 \times 10^{-19}$ J

Example 4: How many orbits, orbitals and electrons are there in an atom having atomic mass 24 and atomic number 12?

Sol: The atomic number gives the no. of electrons which when arranged according to the Aufbau's principle gives the orbits and the orbitals of the given atom.

Atomic Number = No. of protons =

No. of electrons = 12

Electronic configuration =2, 8, 2

No. of orbits = (K, L and M)

No. of orbitals on which electrons are present

= (one 1s + one 2s + three 2p + one 3s)

Example 5: Consider the hydrogen atom to be a proton embedded in a cavity of radius ae (Bohr's radius) whose charge is neutralized by the addition of an electron to the cavity in vacuum, infinitely slowly?

(a) Estimate the average of total energy of an electron in its ground state in a hydrogen atom as the work done in the above neutralization process. Also, if the magnitude of the average kinetic energy is half the magnitude of the average potential energy, find the average potential energy.

(b) Also derive the wavelength of the electron when it is a0 from the proton. How do we compare with the wavelength of an electron in the ground state Bohr's orbit?

Sol: As explained below in the respective steps.

Work obtained in the neutralization process is given by:

$$W = -\int_{e}^{da} F.da = -\int_{e}^{da} \frac{1}{4\pi\varepsilon_0} \frac{(-)e^2}{o_n^2} - da_0$$

$$W = -\frac{e^2}{4\pi\varepsilon_0.a_0}$$

(a) This work is to be called potential energy. However in doing so, one should note that this energy is simply lost during the process of attraction between proton and electron. As reported in the problem at this condition, the electron simply possesses potential energy.

$$\text{Thus, TE = PE + KE − PE} = -\frac{e^2}{4\pi\varepsilon_0.a_0} \quad ...(i)$$

Now in order, the electron to be captured by the proton to form a ground state hydrogen atom, it should also attain kinetic energy $e^2/(8\pi r_2 a_2)$ (as it is half of the potential energy given in question). Thus, the energy of the electron if it attains the ground state in H-atom.

$$TE = PE + KE = -\frac{e^2}{4\pi\varepsilon_0 a_0} + \frac{e^2}{8\pi\varepsilon_0 a_0}$$

$$TE = -\frac{e^2}{8\pi\varepsilon_0 a_0}$$

(b) The wavelength of electron when it is simply at a distance a0 from the proton can be given as:

$$\lambda = \frac{h}{mv} = \frac{h}{p} \text{ Also, KE} = \frac{1}{2}mv^2 = \frac{p^2}{m}; (\because p = mv)$$

$$\text{Thus, } \lambda = \frac{h}{\sqrt{2m(KE)}}$$

Since, KE = 0 at this situation, thus l = ∞

Also, when electron is at a distance a_0 in

$$\text{Bohr's orbit of H-atom. } \lambda = \frac{h}{\sqrt{2m(KE)}} = \frac{h}{\sqrt{\frac{2me^2}{2a_0.4\pi\varepsilon_0}}}$$

Example 6: The velocity of an electron in a certain Bohr orbit of H-atom bears the ratio 1:275 to the velocity of light.

(a) What is the quantum no. of orbit?

(b) Calculate the wave numbers of radiation when electron jumps from (n + 1) state to ground state.

Sol: Use the Rydberg's equation to get the wave number and the application of the given statement to get the velocity of the electron. This in turn can be used to find the quantum no. by $u_n = \frac{2\pi e^2}{nh}$.

Given velocity of electron in a certain Bohr orbit of H-atom = (1/275) × velocity of light

= (1/275) × 3 × 10^8 = 1.09 × 10^8 cm sec^{-1}.

(a) Since, $u_n = \frac{2\pi e^2}{nh}$;

$$\therefore 1.09 \times 10^8 = \frac{2.\times 3.14 \times (4.803\times 10^{-10})^2}{6.626\times 10^{-27}\times n}$$

∴ n = 2.006 = 2 (an integer value)

(b) Thus, during the jump of electron from (n + 1), i.e. 3rd shell to ground state

$$\bar{v} = \frac{1}{\lambda} = R_H\left[\frac{1}{1^2} - \frac{1}{3^2}\right] = 109678\left[\frac{1}{1} = \frac{1}{9}\right]$$

= 9.75 × 10^4 cm^{-1}

Example 7: The series limit of Balmer series of H spectrum occurs at 3664 Å.

(a) ionization energy of H-atom

(b) wavelength of the photon that would remove the electron in the ground state of the H-atom

Sol: Since it is the Balmer series, n_1 = 2. Using the given λ and n_2 = ∞, calculate ionization energy and wavelength by $\Delta E = E_\infty - E_2$

$= -E_2 - \frac{hc}{\lambda}$ and the usual energy formula respectively.

Given series is Balmer series, i.e.

λ = 3664 Å and thus, n_1 = 2; n_2 = ∞

(a) ∵ E/photon of series limit = $\Delta E = E_\infty - E_2$

$$= -E_2 - \frac{hc}{\lambda} = \frac{6.626\times 10^{-34}\times 3\times 10^8}{3664\times 10^{-10}}$$

Or E_2 = −5.42 × 10^{-19} J

∵ $E_1 = E_2 \times n_2$ = −4 × 5.42 × 10^{-19} J

= −21.68 × 10^{-19} J

(b) Now for removal of electron from the first orbit

$$\frac{hc}{\lambda} = 21.68\times 10^{-19};\ 21.68\times 10^{-19}$$

$$= \frac{6.626\times 10^{-34}\times 3\times 10^8}{\lambda};\ \lambda = 916\ Å$$

Example 8: How many elements would be there in the second period of the periodic table if the spin quantum number m could have the values $-\frac{1}{2}$, 0, $\frac{1}{2}$?

Sol: For second period n = 2, hence

l	m_l	m_s
0	0	$+\frac{1}{2}$, 0, $-\frac{1}{2}$
1	−1	$+\frac{1}{2}$, 0, $-\frac{1}{2}$
	0	$+\frac{1}{2}$, 0, $-\frac{1}{2}$
	+1	$+\frac{1}{2}$, 0, $-\frac{1}{2}$

Hence, total number of electrons = 12

(= total values of spin quantum number)

Example 9: The uncertainty in momentum of a particle is 3.31 × 10^{-2} kg m sec^{-1}. Calculate uncertainty in its position.

Sol: Use Heisenberg's uncertainty principle to determine the above.

$$\Delta x = \frac{h}{4\pi} - \Delta p$$

Δp = 3.31 × 10^{-2} kg m sec^{-1}

Since, $\Delta p.\ \Delta x = \frac{h}{4\pi}\ \therefore \Delta X$

$$= \frac{h}{4\pi.\Delta p} = \frac{6.626\times 10^{-34}}{4\times 3.14\times 3.31\times 10^{-2}}$$

= 1.6 × 10^{-33} m

JEE Advanced/Boards

Example 1: The wave function (ψ) of 2s-orbital is given by:

$$\psi_{2s} = \frac{1}{\sqrt{32\pi}}\left[\frac{1}{a_0}\right]^{3/2}\left[2 - \frac{r}{a_0}\right]e^{t/2a_0}$$

At r = r_0, radial node is formed. Calculate r_0 in terms of a_0.

Sol: $\psi_{2s} = \frac{1}{\sqrt{32\pi}}\left[\frac{1}{a_0}\right]^{3/2}\left[2-\frac{r}{a_0}\right]e^{t/2a_0}$

For radial node at r = r_0, $\psi_{2s}^2 = 0$. This is possible only when $\left[2-\frac{r_0}{a_0}\right] = 0$; ∴ $r_0 = 2a_0$

Example 2: 2.4 mole of H_2 sample was taken. In one experiment, 60% of the sample exposed to continuous radiation of frequency 4.47 × 10^{15} Hz, of which all the electrons are removed from the atom. In another experiment, remaining sample was irradiated with light of wavelength 600 Å, when all the electrons are removed from the atom. In another experiment, remaining sample was irradiated with light of wavelength 600Å , when all the electrons are removed from the surface. Calculate the ratio of maximum velocity of the ejected electron in each case. Assume that ejected electrons do not interact with any photon (Ionization potential of H = 13.6 eV).

Sol: Calculate 60% of the sample exposure from the given data and apply the photoelectric equation.

Moles of H_2 exposed to radiation of 4.47 × 10^{15} Hz

$-\frac{60}{100}\times 2.4 = 1.44$

Moles of atoms obtained by 60% sample

= 1.44 × 2 = 2.88

No. of atoms obtained = 2.88 × 6.023 × 10^{23} = 1.73 × 10^{14}

∴ No. of electron ejected = 1.73 × 10^{24}

(Each H-atom has one electron)

Applying photoelectric effect,

$h\nu$ = KE +IE; $h\nu$ = KE + 13.6 × 1.6 × 10^{-19}

KE = [6.626 × 10^{-14} × 4.47 × 10^{15}] − [13.6 × 1.6 × 10^{-19}] = 7.86 × 10^{-19} (∵ = 4.47 × 10^{15} Hz);

∵ KE = $mv_1^2/2$

$$v_1 = \sqrt{\frac{1\times 7.86\times 10^{-39}}{9.1\times 10^{-31}}} = 1.3\times 10^{6}\ \text{m/s}$$

Applying photoelectric effect

$h\nu$ = KE + 13.6 × 1.6 × 10^{-19}

$$KE = \left[\frac{6.626\times 10^{-34}\times 3\times 10^{8}}{600\times 10^{-10}}\right] - [13.6\times 1.6\times 10^{-19}]$$

$$= 1.137\times 10^{-18}\ \text{J}\left[\because \nu = \frac{c}{\lambda} = \frac{3\times 10^{8}}{600\times 10^{-10}}\right]$$

$$v_2 = \sqrt{\frac{2\times 1.137\times 10^{-18}}{9.1\times 10^{-31}}};\ v_2 = 1.58\times 10^{9}\ \text{m/s}$$

$$\frac{v_1}{v_2} = \frac{1.3\times 10^{6}}{1.56\times 10^{6}} = 0.83;\ \frac{v_2}{v_1} = 1.22$$

Example 3: Calculate the following:

(i) Velocity of electron in first Bohr orbit of H-atom (r = a_0)

(ii) De Broglie wavelength of electron in first Bohr orbit of H-atom

(iii) Orbit angular momentum of 2p-orbitals in term of $\frac{h}{2\pi}$ unit. $-\sqrt{l(l+1)}\times\frac{h}{2\pi}$

$= \sqrt{2}\times\frac{h}{2\pi} = \sqrt{2}h$

Sol: Using one of the Bohr's postulates, apply the centripetal force equation. Secondly, solve the De-Broglie equation for the wavelength.

(i) $mvr = \frac{nh}{2\pi}$; ∴ $v = \frac{nh}{2\pi mr}$

$$= \frac{1\times 6.626\times 10^{-34}}{2\times 3.14\times 9.108\times 10^{-31}\times 0.529\times 10^{-10}}$$

(ii) $\lambda = \frac{h}{mv} = \frac{6.626\times 10^{-34}}{9.108\times 10^{-31}\times 2.19\times 10^{6}} = 3.32\times 10^{-10}\text{m}$

Example 4: A gas of identical H-like atom has some atoms in the lowest (ground) energy level A and some atoms in a particular upper (excited) energy level B and there are not atoms in any other energy level. The atoms of the gas make transition to a higher energy level by absorbing monochromatic light of photon energy 2.7 eV. Subsequently, the atoms emit radiation of only six different photon energies. Some of the emitted photons have energy 2.7 eV. Some have more and some have less than 2.7 eV.

(i) Find the principal quantum number of initially excited level B.

(ii) Find the ionization energy for the gas atoms.

(iii) Find the maximum and the minimum energies of the emitted photons.

Sol: The electrons being present in *l* shell and another shell n_1. These are excited to higher level n_2 by absorbing 2.7 eV and on de-excitation emits six l and thus excited state n_2 comes to be 4 [6 = ΣE_n = $\Sigma(n_2 - 1)$ ∴ n_2 = 4]

Now, $E_1 = -\frac{R_H ch}{1^2}$;

$E_n = -\frac{R_H ch}{n_1^2}$;

$E_4 = -\frac{R_H ch}{4^2}$

Since, de-excitation leads to different l having photon. Energy < 2.7 eV and thus absorption of 2.7 eV energy causing excitation to fourth shell then reemitting photon of > 2.7 eV is possible only when $n_1 = 2$ (the de-excitation from fourth shell occurs in first, second and third shells).

$E_4 - E_2 = 2.7$ eV; $E_4 - E_3 < 2.7$ eV

$E_4 - E_1 > 2.7$ eV

$\therefore E_n = E_2 = \frac{R_H \times c \times h}{2^2} = \frac{E_1}{2^2}$ since $n_1 = 2$

Also, $E_4 - E_2 - 2.7$ eV; $\therefore \left[\frac{-E_1}{4^2} + \frac{-E_1}{2^2}\right] = 2.7$ eV

$\therefore E_1 = -14.4$ eV; IE = 14.4 eV

$$E_{max.} = E_4 - E_1 = -\frac{E_1}{4^2} + E_1; -\frac{14.4}{16} + 14.4 = 13.5 \text{ eV};$$

$$E_{min.} = E_4 - E_3 = \left[-\frac{E_1}{4^2} + \frac{E_1}{3^2}\right] = 0.7 \text{ eV}$$

Note: It is $_1H^2$ atom

Example 5: Two hydrogen atoms collide head on and end up with zero kinetic energy. Each atom then emits a photon of wavelength 131.6 nm. Which transition leads to this wavelength? How fast were the hydrogen atoms travelling before collision?

[$R_H = 1.097 \times 10^7$ m^{-1} and $m_H = 1.67 \times 10^{-27}$ kg]

Sol: With the given data calculate the transition levels using the Rydberg equation and solve for velocity by equating kinetic energy and the energy of a photon.

Wavelength emitted in UV region and thus

$n_1 = 1$; For H-atom $\therefore \frac{1}{\lambda} = R_H\left[\frac{1}{1^2} - \frac{1}{n^2}\right]$

$$\frac{1}{121.6 \times 10^{-9}} = 1.097 \times 10^7\left[\frac{1}{1^2} - \frac{1}{n^2}\right] \quad \therefore n = 2$$

Also the energy released is due to collision and all the kinetic energy is released in the form of photon. Thus,

$$\frac{1}{2}mv^2 = \frac{hc}{\lambda}$$

Or $\frac{1}{2} \times 1.67 \times 10^{-27} \times v^2 = \frac{6.626 \times 10^{-34} \times 3 \times 10^8}{121.6 \times 10^{-3}}$

$\therefore v = 4.43 \times 10^4$ m sec^{-1}

Example 6: Let a light of wavelength λ and intensity T strike a metal surface to emit x electrons per second. Average energy of each electron is 'y' unit, What will happen to x and y when (a) λ is halved (b) intensity I is double?

Sol: (a) Rate of emission of electron is independent of wavelength. Hence, 'x' will be unaffected. Kinetic energy of photoelectron = absorbed energy – Threshold energy $y = \frac{hc}{\lambda} - w_0$

When I is halved, average energy will increases but it will not become double.

(b) Rate of emission of electron per second 'x' will become double when intensity I is double. Average energy of ejected electron, i.e. 'y' will be unaffected by increases in the intensity of light.

Example 7: The IP_1 of H is 13.6 eV. It is exposed to electromagnetic waves of 1028 Å and gives out induced radiation. Find the wavelength of these induced radiations:

Sol: From energy of H-atom, solve for the level. Thus, calculate the consecutive wavelength.

E_1 of H atom = −13.6 eV

Energy given to H atom

$$-\frac{6.625 \times 10^{-34} \times 3.0 \times 10^8}{1028 \times 10^{-10}} = 1.933 \times 10^{-18} \text{ J} = 12.07 \text{ eV}$$

$\therefore$ Energy of H atom after excitation

= −13.6 + 12.07 = −1.53 eV

$\therefore E_n = \frac{E_1}{n^2}; \therefore n^2 = \frac{-13.6}{-1.53} = 9; \therefore n = 3$

Thus, electron in H atom is excited to third shell.

$\therefore$ I induced $\lambda_1 = \frac{hc}{E_3 - E_1}$

$\therefore E_1 = -13.6$ eV; $E_3 = -1.53$ eV

$$\therefore \lambda_1 = \frac{6.625 \times 10^{-34} \times 3.0 \times 10^8}{(-1.53 + 13.6 \times 1.602) \times 10^{19}} = 1028 \times 10^{10} \text{ m}$$

$\therefore \lambda = 1028Å;$

$$\therefore \text{II induced } \lambda_2 = \frac{hc}{(E_2 - E_1)}$$

$$E_1 = -13.6 \text{ eV}; E_2 = -\frac{13.6}{4}\text{eV}$$

$$\therefore \lambda_2 = \frac{6.625\times10^{-14}\times3.0\times10^8}{\left(-\frac{13.6}{4}+13.6\right)\times1.602\times10^{-19}}$$

$= 1216 \times 10^{-16}$ m $= 1216$ Å

$$\therefore \text{III induced } \lambda_3 = \frac{hc}{(E_2 - E_1)}$$

$$\therefore \lambda_2 = -\frac{6.625\times10^{-34}\times3.0\times10^8}{\left(-\frac{13.6}{9}+\frac{13.6}{4}\right)\times1.602\times10^{-19}}$$

$= 6568 \times 10^{-10}$m $= 6568$ Å

Example 8: How many elements would be in the third period of the periodic table if the spin quantum number m_2 could have the value $-\frac{1}{2}$, 0 and $+\frac{1}{2}$?

Sol: Apply the data to the formulas of the quantum numbers.

n= 1, l = 0, m=0 $\qquad m_s = -1/2, 0, +1/2$

l=1; m = −1, 0, +1 $\quad \begin{cases} m_s = -\frac{1}{2}, 0, +\frac{1}{2} \\ m_s = -\frac{1}{2}, 0, +\frac{1}{2} \\ m_s = -\frac{1}{2}, 0, +\frac{1}{2} \end{cases}$

l=2 ; m =−2, −1, 0, 0, +1, +2 $\left\{ m_f = -\frac{1}{2}, 0, +\frac{1}{2} \right.$ for each value of magnetic quantum no.$\}$

Number of elements = 3s(3e)

3p(9e)

3d(15e)

$\therefore$ 27 elements will be there in third period of periodic table.

Example 9: Consider the following two electronic transition possibilities in a hydrogen atom as pictured below:

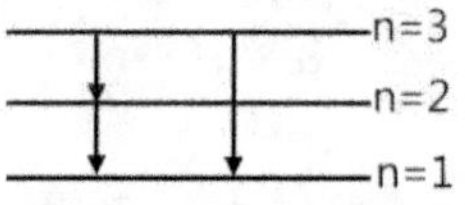

(a) The electron drops from third Bohr orbit to second Bohr orbit followed with the next transition from second to first Bohr orbit.

(b) The electron drops from third Bohr orbit to first Bohr orbit directly. Show that the sum of energies for the transitions n = 3 to n = 2 and n = 2 to n = 1 is equal to the energy of transition for n =3 to n = 1.

Sol: Applying $\Delta E = R_H\left[\frac{1}{n_1^2} - \frac{1}{n_2^2}\right]$

For n = 3 to n = 2

$$\Delta E_{3\to2} = R_H\left[\frac{1}{2^2} - \frac{1}{3^2}\right] = R_H \times \frac{5}{36} \qquad \text{... (i)}$$

For n = 2 to n = 1

$$\Delta E_{2\to1} = R_H\left[\frac{1}{1^2} - \frac{1}{2^2}\right] = R_H \times \frac{3}{4} \qquad \text{... (ii)}$$

For n = 3 to n = 1;

$$\Delta E_{3\to1} = R_H\left[\frac{1}{1^2} - \frac{1}{3^2}\right] = R_H \times \frac{8}{9} \qquad \text{... (iii)}$$

Adding equation (i) and (ii)

$$\left(\frac{5}{36}+\frac{3}{4}\right) = R_H\left(\frac{5+27}{36}\right) = R_H \times \frac{8}{9}$$

Thus, $\Delta E_{3\to1}\ \Delta E_{3\to2} + \Delta E_{2\to1}$

Example 10: Consider the hydrogen atom to be a proton embedded in a cavity of radius a_0 (Bohr radius) whose charge is neutralized by the addition of an electron to the cavity in vacuum infinitely slowly. Estimate the average total energy of an electron in its ground state in a hydrogen atom as the work done in the above neutralization process. Also, if the magnitude of average KE is half the magnitude of average potential energy, find the average potential energy.

Sol: Coulombic force of attraction = Centrifugal force

$$\frac{1}{4\pi\epsilon_0}\frac{Ze\times e}{a_0^2} = \frac{m\upsilon^2}{a_0}$$

Where, υ = velocity of electron

a_0 = distance between electron and nucleus

$$\frac{1}{4\pi\epsilon_0}\frac{Ze^2}{a_0} = m\upsilon^2$$

$$KE = \frac{1}{2}m\upsilon^2 = \frac{1}{4\pi\varepsilon_0}\frac{Ze^2}{2a_0}$$

$$PE = -2 \times KE = -2\times\frac{1}{4\pi\varepsilon_0}\times\frac{Ze^2}{2a_0} = -\frac{1}{4\pi\varepsilon_0}\frac{Ze^2}{a_0}$$

Exercise 1

Q.1 In an oil drop experiment, the following charges (in arbitrary units) were found on a series of oil droplets: 4.5×10^{-18}, 3.0×10^{-18}, 6.0×10^{-18}, 7.5×10^{-18}, 9.0×10^{-18}. What is the charge on electron (in the same unit)?

Q.2 Arrange electron (e), proton (p), neutron (n) and α-particle (a) in the increasing order of their specific charges.

Q.3 With what velocity should an α-particle travel towards the nucleus of a copper atom so as to arrive at a distance 10^{-13} m from the nucleus of the copper atom? Atomic number of copper is 29. Mass of alpha particle is 4 amu.

Q.4 The wavelength of the Kα line for an element of atomic number 57 is λ. What is wavelength of the Kα line for the element of atomic number 29.

Q.5 The wavelength of an electromagnetic radiation is 600 nm. What is its frequency?

Q.6 Calculate the energy per quanta of an electron magnetic radiation of wavelength 6626Å.

Q.7 Calculate the Rydberg constant R_H if He^+ ions are known to have the wave length difference between first (of the longest wavelength) lines of Balmer and Lyman series equal to 133.7 nm.

Q.8 Suppose 10^{-17} J of light energy is needed by the interior of the human eye to see an object. How many photons of green light (λ = 550 nm) are needed to generate this minimum amount of energy?

Q.9 The bond energy of H–H bond is 104 kcal/mol. What is the largest wavelength of electromagnetic radiation needed to dissociate H_2 molecules? Assume that one photon may dissociate only one molecule.

Q.10 The work function of potassium is 2.25 eV. Would photoelectron emit when red light of 660 nm falls on potassium surface? If yes, what would be the maximum kinetic energy of electron liberated?

Q.11 The threshold frequency of a metal is 1.8×10^{14} Hz. Calculate the maximum kinetic energy of photoelectron liberated when the metal is irradiated by an electromagnetic radiation of wavelength 4000Å.

Q.12 A particle of mass 'm' moves along a circular orbit in a centro-symmetrical potential field $U(r) = Kr^2/2$. Using the Bohr quantization condition, find the permissible orbital radii of that particle.

Q.13 Light of wave length 2000Å falls on an aluminium surface (work function of aluminium 4.2 eV). Calculate (a) The kinetic energy of the fastest and slowest emitted photo electrons (b) Stopping potential. (c) Cut-off wavelength for Aluminium.

Q.14 Calculate the speed of electron revolving in the 3rd orbit of hydrogen atom.

Q.15 An electron is revolving at a distance of 4.761 Å from the hydrogen nucleus. Determine its speed.

Q.16 Calculate the ratio of time period of electron in the 2nd orbit of H-atom to that in the 3rd orbit of He^+ ion.

Q.17 Calculate the angular frequency of an electron occupying the second Bohr orbit of He^+ ion.

Q.18 Calculate the first four energy levels for electron in hydrogen atom.

Q.19 The dissociation energy of H_2 is 103.2 k cal $mole^{-1}$. Suppose H_2 molecules are irradiated with wavelength, λ = 2537Å. Assume that one photon is absorbed by and dissociated one molecule of H_2. How much of the photon energy is converted into kinetic energy of the dissociated atoms.

Q.20 Calculate the kinetic, potential and total energy of electron in the 3rd orbit of He^+ ion.

Q.21 Calculate the excitation energy of Li^{2+} ion in the ground state.

Q.22 Calculate the binding energy of electron in the ground state of He^+ ion.

Q.23 Electromagnetic radiation of wavelength 24 nm is just sufficient to ionize sodium atom. Calculate the ionization energy of sodium atom.

Q.24 A beam of electron accelerated with 4.64 V is passed through a tube containing mercury vapours. As a result of absorption, electronic changes occurred with mercury atoms and light was emitted if the full energy of single electron was converted into light, what was the wave number of emitted light?

Q.25 A proton and an electron, both at rest initially, combine to form a hydrogen atom in the ground state. A single photon is emitted in this process. What is its wavelength?

Q.26 Calculate the frequency of the radiation absorbed in the transition n = 2 to n = 4 in hydrogen atom.

Q.27 When electromagnetic radiations of wavelength λ nm fall on hydrogen atoms, electron excite from the ground state to a particular upper energy state. Subsequently, the atoms emit the radiations of six different wavelengths. Calculate the value of λ.

Q.28 The wavelength of H line in the Balmer series of hydrogen spectrum is 660 nm. What is the wavelength of H-line of this series

Q.29 Calculate the momentum of a photon of wavelength 10 Å.

Q.30 Electrons which have absorbed 10.20 eV and 12.09 eV in hydrogen atom can cause radiations to be emitted when they come back to ground state. Calculate in each come back to ground state. Calculate in each case the principal quantum no. of the orbit to which electron goes and the wavelength of the radiations emitted if it drops back to ground state.

Exercise 2

Single Correct Choice Type

Q.1 Which of the following does not characterize x-rays?

(A) The radiation can ionize gases

(B) It causes ZnS to show fluorescence

(C) Deflected by electric and magnetic fields

(D) Have wavelength shorter than ultraviolet rays.

Q.2 Which of the following is false regarding Bohr's model

(A) It introduces the idea of stationary states

(B) It explains the line spectrum of hydrogen

(C) It gives the probability of the electron near the nucleus

(D) It predicts that the angular momentum of electron in H-atom = $nh/2\pi$.

Q.3 The energy of an orbit in a hydrogen atom is given by the relation $E = \frac{\text{Constant}}{n^2}(\text{kJ mol}^{-1})$. Which of the following properties represents the constant in the above relation

(A) Electron affinity (B) Ionization energy

(C) Electro negativity (D) Bond energy

Q.4 The ratio of the energy of a photon of 2000Å wavelength radiation to that of 4000Å radiation is

(A) 1/4 (B) 4 (C) ½ (D) 2

Q.5 The energy of electron is maximum at

(A) Nucleus

(B) Ground state

(C) First excited state

(D) Infinite distance from the nucleus

Q.6 Which quantum number is not related with Schrödinger equation?

(A) Principal (B) Azimuthal

(C) Magnetic (D) Spin

Q.7 The shortest wavelength of He atom in Balmer series is X, then longest wavelength in the Paschen series of Li^{+2} is

(A) $\frac{36x}{5}$ (B) $\frac{16x}{7}$ (C) $\frac{36x}{5}$ (D) $\frac{5x}{9}$

Q.8 An electron in a hydrogen atom in its ground state absorbs energy equal to the ionization energy of Li^{+2}. The wavelength of the emitted electron is:

(A) 3.32×10^{-10} m (B) 1.17 Å

(C) 2.32×10^{-9} nm (D) 3.33 pm

Q.9 Given ΔH for the process $Li(g) \rightarrow Li^{+3}(g) + 3e^-$ is 19800 kJ/mole and IE_1 for Li is 520 then IE_2 and IE_3 of Li+ are respectively (approx. value)

(A) 11775, 7505 (B) 19280, 520

(C) 11775, 19280 (D) Data insufficient

Q.10 The ratio of difference in wavelengths of 1st and 2nd lines of Lyman series in H-like atom to difference in wavelength for 2nd and 3rd lines of same series is:

(A) 2.5: 1 (B) 3.5: 1 (C) 4.5: 1 (D) 5.5: 1

Q.11 The ratio of the radii of the first three Bohr orbit is

(A) 1: 0.5: 033 (B) 1: 2: 3

(C) 1: 4: 9 (D) 1: 8: 27

Q.12 Which combination of quantum number n, l, m, s for the electron in an atom does not provide a permissible solution of the wave equation?

(A) 3, 2, −2, +1/2 (B) 3, 3, 1, − ½

(C) 3, 2, 1, +1/2 (D) 3, 1, 1, −1/2

Q.13 The ratio of the energy of a photon of 2,000 Å wavelength radiation to that of 4,000 Å radiation is

(A) 1/4 (B) 4 (C) 1/2 (D) 2

Q.14 The orbital angular momentum of an electron in 2s orbital is

(A) $+\frac{1}{2}\left(\frac{h}{2\pi}\right)$ (B) Zero (C) $h/2\pi$ (D) $\sqrt{2} \times \frac{h}{2\pi}$

Q.15 If n and l are respectively the principal and azimuthal quantum number, then the expression for calculating the total number of electrons in any energy level is

(A) $\sum_{i=0}^{\ell=n-\ell} 2(2\ell+1)$ (B) $\sum_{i=0}^{\ell=n-\ell} 2(2\ell+1)$

(C) $\sum_{i=0}^{\ell=n+\ell} 2(2\ell+1)$ (D) $\sum_{i=0}^{\ell=n-\ell} 2(2\ell \times 1)$

Q.16 The wavelength of a tennis ball of mass 6.0×10^{-2} kg moving at a speed of about 140 miles per hour is ($h = 6.63 \times 10^{-27}$ erg s)

(A) 1.8×10^{-30} cm (B) 1.8×10^{-32} cm

(C) 1.8×10^{-34} cm (D) None of these

Q.17 If radius of second stationary orbit (in Bohr's atom) is R. Then radius of third orbit will be

(A) R/3 (B) 9R (C) R/9 (D) 2.25 R

Q.18 The ratio of wave length of photon corresponding to the α-line of Lyman series in H-atom and β-line of Balmer series in He^+ is

(A) 1: 1 (B) 1: 2 (C) 1: 4 (D) 3: 16

Q.19 The value of $(n_2 + n_1)$ and for He^+ ion in atomic spectrum are 4 and 8 respectively. The wavelength of emitted photon when electron jump from n_2 to n_1 is

(A) $\frac{32}{9}R_H$ (B) $\frac{9}{32}R_H$ (C) $\frac{9}{32R_H}$ (D) $\frac{32}{9R_H}$

Q.20 Number of possible spectral lines which may be emitted in bracket series in H atom if electron present in 9th excited level returns to group level, are

(A) 21 (B) 6 (C) 14 (D) 7

Q.21 The first use of quantum theory to explain the structure of atom was made by:

(A) Heisenberg (B) Bohr

(C) Planck (D) Einstein

Q.22 The wavelength associated with a golf weighing 200 g and moving at a speed of 5 m/h of the order

(A) 10^{-10} m (B) 10^{-20} m

(C) 10^{-30} m (D) 10^{-40} m

Q.23 The longest wavelength of He^+ in Paschen series is "m", then shortest wavelength of Be^{+3} in Paschen series is (in terms of m):

(A) $\frac{5}{36}m$ (B) $\frac{64}{7}m$ (C) $\frac{53}{8}m$ (D) $\frac{7}{64}m$

Q.24 Consider the following nuclear reactions involving X and Y.

$$X \rightarrow Y + {}_2^4He;\ Y \rightarrow {}_8O^{18} + {}_1H^1$$

If both neutrons as well as protons in both the sides are conserved in nuclear reaction then moles of neutrons in 4.6 gm of X

(A) 2.4 NA (B) 2.4

(C) 4.6 (D) 0.2 NA

Q.25 Electromagnetic radiations having λ = 310Å are subjected to a metal sheet having work function = 12.8 eV. What will be the velocity of photoelectrons with maximum Kinetic energy

(A) 0, no emission will occur (B) 2.15×10^6 m/s

(C) $2.18\sqrt{2}\times10^6 m/s$ (D) 8.72 × 106 m/s

Q.26 Assuming Heisenberg Uncertainty Principle to be true what could be the minimum uncertainty in de-Broglie wavelength of a moving electron accelerated by Potential Difference of 6V whose uncertainty in position is 7/22 n.m.

(A) 6.25Å (B) 6Å

(C) 0.625 Å (D) 0.3125 Å

Q.27 Correct statement(s) regarding 3P, orbital is/are

(A) Angular part of wave function is independent of angles θ and ϕ)

(B) No. of maxima when a curve is plotted between $4\pi r^2R^2(r)$ vs. r are '2'

(C) 'rz' plane acts as nodal plane

(D) Magnetic quantum number must be '–1'.

Q.28 Choose the incorrect statement(s):

(A) Increasing order of wavelength is Micro waves > Radio waves > IR waves > visible waves > UV waves

(B) The order of Bohr radius is (r_n: where n is orbit number for a given atom) $r_1 < r_2 < r_3 < r_4$

(C) The order of total energy is (E_n: where n is orbit number for a given atom)

(D) The order of velocity of electron in H, He^+, Li^+, Be^{3+} species in second Bohr orbit is

$Be^{1+} > Li^{+2} > He^+ > H$

Q.29 Which is/are correct statement.

(A) The difference in angular momentum associated with the electron present in consecutive orbits of H-atom is $(n-1)\dfrac{h}{2\pi}$

(B) Energy difference between energy levels will be changed if, PE. At infinity assigned value of other than zero.

(C) Frequency of spectral line in a H-atom is in the order of $(2 \rightarrow 1) < (3 \rightarrow 1) < (4 \rightarrow 1)$

Previous Years' Questions

Q.1 Who discovered neutron **(1982)**

(A) James Chadwick (B) William Crooks

(C) J.J. Thomson (D) Rutherford

Q.2 The radius of an atom is of the order of **(1985)**

(A) 10^{-10} cm (B) 10^{-13} cm

(C) 10^{-15} cm (D) 10^{-8} cm

Q.3 Which one of the following constitutes a group of the isoelectronic species **(2008)**

(A) NO^+, C_2^{2-}, CN^-, N_2 (B) CN^-, N_2, O_2^{2-}, C_2^{2-}

(C) N_2, O_2^-, NO^+, CO (D) C_2^{2-}, O^-_2, CO, NO

Q.4 Which one of the following groupings represents a collection of isoelectronic species **(2003)**

(A) Na^+, Ca^{2+}, Mg^{2+} (B) $N3^-$, F^-, Na^+

(C) Be, A^{3+}, Cl^- (D) Ca^{2+}, Cs^+, Br

Q.5 The radius of which of the following orbit is same as that of the first Bohr's orbit of hydrogen atom **(2004)**

(A) He^+ (n = 2) (B) Li^{2+} (n = 2)

(C) Li^{2+} (n = 3) (D) Be^{3+} (n = 2)

Q.6 The wavelength of the radiation emitted, when a hydrogen atom electron falls from infinity to stationary state 1, would be (Rydberg constant = 1.097×10^7 m^{-1}) **(2004)**

(A) 406 nm (B) 192 nm

(C) 91 nm (D) 9.1×10^{-8} nm

Q.7 The ionization enthalpy of hydrogen atom is 1.313×10^6 J mol^{-1}. The energy required to excite the electron in the atom from n = 1 to n = 2 is **(2008)**

(A) 6.56×10^5 J mol^{-1} (B) 7.56×10^5 J mol^{-1}

(C) 9.84×10^{-5} J mol^{-1} (D) 8.51×10^5 J mol^{-1}

Q.8 A gas absorbs a photon of 355 nm and emits at two wavelengths. If one of the emissions is at 680 nm, the other is at **(2011)**

(A) 1035 nm (B) 325 nm

(C) 743 nm (D) 518 nm

Q.9 Calculate the wavelength (in nanometer) associated with a proton moving at 1.0×10^3 ms^{-1} (Mass of proton = 1.67×10^{-27} kg and h = 6.63×10^{-34} Js): ***(2009)***

(A) 0.032 nm (B) 0.40 nm

(C) 2.5 nm (D) 14.0 nm

Q.10 In an atom, an electron is moving with a speed of 600 m/s with an accuracy of 0.005%. Certainty with which the position of the electron can be located is (h = 6.6×10^{-34} kg m^2 s^{-1}, mass of electron m = 9.1×10^{-31} kg) ***(2009)***

(A) 1.52×10^{-4} m (B) 5.10×10^{-3} m

(C) 1.92×10^{-3} m (D) 3.84

Q.11 Which of the following sets of quantum number represents the highest energy of an atom? ***(2007)***

(A) n = 3, l = 1, m = 1, s = ±1/2

(B) n = 3, l = 2, m=1, s = −1/2

(C) n = 4, l = 0, m = 0, s = ±1/2

(D) n = 3, l = 0, m = 0, s = ±1/2

Q.12 In a multi-electron atom, which of the following orbital's described by the three quantum numbers will have the same energy in the absence of magnetic and electric fields

(1) n = 1, l = 0, m = 0 (2) n = 2, l = 0, m = 0

(3) n = 2, l = 1, m= 1 (4) n = 3, l = 2, m = 0

(5) n = 3, l = 2, m = 1 ***(2005)***

(A) (1) and (2) (B) (2) and (3)

(C) (3) and (4) (D) (4) and (5)

Q.13 The electronic configuration of an element is $1s^2$ $2s^2$ $3p^6$ $3s^2$ $3p^6$ $3d^5$ $4s^1$. This represent its ***(2000)***

(A) Excited sate (B) Ground state

(C) Cationic form (D) Anionic form

Q.14 Which of the following sets of quantum number is correct for an electron in 4f orbital? ***(2004)***

(A) n = 4, l = 3, m = +1, s = +1/2

(B) n = 4, l = 4, m = −4, s = −1/2

(C) n = 4, l = 3, m = +4, s = +1/2

(D) n = 3, l = 2, m = −2, s = +1/2

Q.15 The electrons identified by quantum numbers n and λ ***(2012)***

(1) n = 4, λ = 1 (2) n = 4, λ = 0

(3) n = 3, λ = 2 (4) n = 3, λ = 1

Can be placed in order of increasing energy as

(A) (3) < (4) < (2) < (1) (B) (4) < (2) < (3) < (1)

(C) (2) < (4) < (1) < (3) (D) (1) < (3) < (2) < (4)

Q.16 In an atom, an electron is moving with a speed of 600m/s with an accuracy of 0.005%. Certainity with which the position of the electron can be located is $\left(h = 6.6 \times 10^{-34} \text{ kg m}^2 \text{ s}^{-1} \text{ mass of electron, } e_m = 9.1 \times 10^{-31} \text{ kg}\right)$ ***(2009)***

(A) 1.52×10^{-4} m (B) 5.10×10^{-3} m

(C) 1.92×10^{-3} m (D) 3.84×10^{-3} m

Q.17 Calculate the wavelength (in nanometer) associated with a proton moving at 1.0×10^3 ms^{-1} (Mass of proton = 1.67×10^{-27} kg and h = 6.63×10^{-34} Js): ***(2009)***

Q.18 The energy required to break one mole of Cl–Cl bonds in Cl_2 is 242 kJ mol^{-1}. The longest wavelength of light capable of breaking a single Cl – Cl bond is $\left(c = 3 \times 10^8 \text{ ms}^{-1} \text{ and } N_A = 6.02 \times 10^{23} \text{ mol}^{-1}\right)$ ***(2010)***

(A) 594 nm (B) 640 nm

(C) 700 nm (D) 494 nm

Q.19 Ionisation energy of He^+ is 1.96×10^{-18} J $atom^{-1}$. The energy of the first stationary state (n = 1) of Li^{2+} is ***(2010)***

(A) 4.41×10^{-16} J $atom^{-1}$

(B) -4.41×10^{-17} J $atom^{-1}$

(C) -2.2×10^{-15} J $atom^{-1}$

(D) 8.82×10^{-17} J $atom^{-1}$

Q.20 Which of the following is the energy of a possible excited state of hydrogen? ***(2015)***

(A) +13.6 eV (B) −6.8 eV

(C) −3.4 eV (D) +6.8 eV

Exercise 1

Q.1 What is the relationship between the eV and the wavelength in meter of the energetically equivalent photon?

Q.2 What electronic transition in the He^+ would emit the radiation of the same wavelength as that of the first Lyman transition of hydrogen (i.e., for an electron jumping from n = 2 to n = 1)? Neglect the reduced – mass effect. Also calculate second ionization potential of He and first Bohr orbit for He^+. (e = 1.65×10^{-19} coulomb, m = 9.1×10^{-31} kg, h = 6.626×10^{-34} J, sec c = 2.997×10^8 meter/sec and $e_0 = 8.854 \times 10^{-12}$ columb^2/newton metre^2)

Q.3 What acceleration potential is needed to produce an electron beam with an effective wavelength of 0.090 Å?

Q.4 In view of the uncertainty principle explain that the motion of an electron cannot be described in terms of orbit as proposed by Bohr.

Q.5 With what velocity should an α-particle travel towards the nucleus of a Cu-atom so as to arrive at distance 10^{-13} meter from the nucleus of the Cu-atom? (Cu = 29, e = 1.6×10^{-19}C $a_0 = 8.85 \times 10^{-12}$ J^{-1} C^{-1} C^2m^{-1} $m_0 = 9.1 \times 10^{-31}$ kg)

Q.6 Calculate the velocity of an electron in the third orbit of the hydrogen atom. Also calculate the number of revolutions per second made by this electron around the nucleus.

Q.7 According to Bohr theory, the electronic energy of hydrogen atom in the nth Bohr atom is given by $E = \frac{-21.76 \times 10^{-19}}{n^2}$ J. Calculate the longest wavelength of light that will be needed to remove an electron from the third Bohr orbit of the He^+ ion.

(h = 6.626×10^{-34} J sec, c = 3×10^8 m sec^{-1})

Q.8 Why the concept of orbit has been replaced by probability picture?

Q.9 What is meant by atomic orbital? Explain the concept of orbital in terms of probability density?

Q.10 Explain hydrogen spectrum.

Q.11 The λ of H_α line of Balmer series of 6500 Å. What is the l of H_β line of Balmer series.

Q.12 Calculate the frequency of the spectral line emitted when the electron in n = 3 in H atom de excites to ground state R_H = 109737 cm^{-1}.

Q.13 Estimate the difference in energy between 1st and 2nd Bohr orbit for a H atom. At what minimum atomic no., a transition from n = 2 to n = 1 energy level would result in the emission of X-ray with $\lambda = 3.0 \times 10^8$ m. Which hydrogen atom like species does this atomic no. corresponds to?

Q.14 What does the shape of an atomic orbital represent?

Q.15 How many nodes and spherical nodes are there in p_x orbital?

Q.16 Why is the electronic configuration $1s^2\ 2s^2\ 2p^2\ 2p^6\ 3s^2\ 3p^6\ 4s^2\ 3d^4$ not correct for chromium? What is the its correct configuration (Atomic number of Cr is 24). Give proper explanation.

Q.17 The photo electric emission requires a threshold frequency v_0 for a certain metal λ_1 = 2200Å and λ_2 = 1900Å produce electrons with maximum kinetic energy KE_1 and KE_2. If $KE_2 = 2KE_1$ calculate v_0 and corresponding value.

Q.18 A near ultraviolet photon of 300 nm is absorbed by a gas and then re-emitted as two photons? One photon is red with wavelength 760 nm. What would be the wavelength of the second photon?

Q.19 Show that the wavelength of a 150 rubber ball moving with a velocity 50 m sec^{-1} is short enough to be observed.

Q.20 When a certain metal was irradiated with light frequency 1.6×10^{16} Hz, the photo electrons emitted had twice the kinetic energy as did photoelectrons emitted when the same metal was irradiated with light of frequency 1.0×10^{16} Hz. Calculate v_0 (threshold frequency) for metal.

Q.21 Magnetic moment of X^{st} ion of 3d series is B.M. What is atomic number of X^{34}?

Q.22 Iodine molecule dissociates into atoms after absorbing light of 4500 Å if one quantum of radiation is absorbed by each molecule. Calculate the kinetic energy of iodine?

Q.23 Energy required to stop the ejection of electron from Cu plate is 0.24 eV. Calculate the work function when radiations of λ = 253.7 nm strikes the plate.

Q.24 Calculate the energy emitted when electrons of 1.0 g atom of hydrogen undergo transition giving the spectral lines lowest energy in the visible region of its atomic spectra

$R_H = 1.1 \times 10^3\ m^{-1}$, $C = 3 \times 10^8$ m sec^{-1} and $h = 6.62 \times 10^{-34}$ Js.

Q.25 The characteristics X-rays wavelength for the lines of the K_α series in elements X and Y are 9.87 Å and 2.29 Å respectively. If Moseley's equation $v = 4.9 \times 10^7 (Z - 0.75)$ is followed, what are atomic numbers of X and Y

Q.26 In the Balmer series of atomic spectra of hydrogen there is a line corresponding to wavelength 4744 Å. Calculate the number of higher orbits from which the electron drops to generate other line [$R \times C = 3.289 \times 10^{13}$]

Q.27 Assuming a spherical second and third Bohr orbits of the hydrogen atom is -5.42×10^{-12} ergs and -2.41×10^{-11} ergs respectively. Calculate the wavelength of the emitted radiation when the electron drops from third to second orbit.

Q.28 Assuming a spherical shape for the F nucleus, calculate the radius and the nuclear density of F nucleus of mass number 19.

Q.29 What conclusion may be drawn from the following results of? If a 10×10^{-1} kg body is traveling along the x-axis at 1 meter/sec within 0.01 meter/sec. Calculate the theoretical uncertainty in its position.

Q.30 What conclusion may be drawn from the following of? If an electron is traveling at 100 meter/sec. within 1 meter/sec. Calculate the theoretical uncertainty in its position.

[$h = 6.63 \times 10^{-34}$ J s, mass of electron = 9.109×10^{-31} kg]

Exercise 2

Single Correct Choice Type

Q.1 Bohr's concept of an orbit in an atom contradicts

(A) de Broglie's equation (B) Pauli's principle

(C) Uncertainty principle (D) Hund's rule

Q.2 It is a data sufficiency problem in which it is to be decided on the basis of given statements whether given question can be answered or not. No matter whether the answer is yes or no.

Question: Is the orbital of hydrogen atom $3p_x$?

Statement-I: The radial function of the orbital is

$$R(r) = \frac{1}{9\sqrt{6}a_0^{3/2}}(4-\sigma)\sigma e^{-\sigma/2}, \sigma = \frac{r}{2}$$

Statement-II: The orbital has 1 radial node and 0 angular mode.

(A) Statement-I alone is sufficient.

(B) Statement-II alone is sufficient

(C) Both together is sufficient.

(D) Neither is sufficient

Comprehension Type

Paragraph 1: The only electron in the hydrogen atom resides under ordinary conditions on the first orbit. When energy is supplied, the electron moves to higher energy orbit depending on the amount of energy absorbed. When this electron returns to any of lower orbits, it emits energy. Lyman series is formed when the electron returns to the lowest orbit while Balmer series is formed when the electron returns to second orbit. Similarly, Paschen, Brackett and Pfund series are formed when electron returns to the third, fourth and fifth orbits from higher energy orbits respectively.

Maximum number of lines produced when an electron jumps from nth level to ground level is equal to $\frac{n(n-1)}{2}$. For example, in the case of n = 4, number of lines produced is 6.

(4 → 3, 4 → 2, 4 → 1, 3 → 2, 3 → 1, 2 → 1). When an electron returns from n_2 to n_1 state, the number of lines in the spectrum will be equal to $\frac{(n_2 - n_1)(n_2 - n_1 + 1)}{2}$

If the electron comes back from energy level having energy E_2 to energy level having energy E_1, then the difference may be expressed in terms of energy of photon as:

$$E_2 - E_1 = \Delta E, \lambda = \frac{hc}{\Delta E}$$

Since h and c are constants, DE corresponds to definite energy, thus each transition from one energy level to another will produce a light of definite wavelength. This is actually observed as line in the spectrum of hydrogen atom. Wave number of line is given by the formula

$\bar{v} = R\left(\frac{1}{n_1^2} - \frac{1}{n_2^2}\right)$ Where R is a Rydberg's constant

($R = 1.1 \times 10^7\ m^{-1}$)

Q.3 The energy photon emitted corresponding to transition n = 3 to n = 1 is [$h = 6\ 0 \times 10^{-34}$ J –sec]

(A) 1.76×10^{-18} J (B) 1.98×10^{-18} J

(C) 1.76×10^{-17} J (D) None of these

Q.4 In a collection of H-atom, electrons make transition from 5th excited state to 2nd excited state then maximum number of different types of photons observed are

(A) 3 (B) 4 (C) 6 (D) 15

Q.5 The difference in the wavelength of the 1st line of Lyman series and 2nd line of Balmer series in a hydrogen atom is

(A) $\frac{9}{2R}$ (B) $\frac{4}{R}$

(C) $\frac{88}{15R}$ (D) None of these

Q.6 The wave number of electromagnetic radiation emitted during the transition of electron in between two levels of Li^{2+} ion whose principal quantum number sum is 4 and difference is 2 is

(A) 3.5 R (B) 4R

(C) 8R (D) $\frac{8}{9}R$

Paragraph 2: In the Rutherford's experiment, α-particles were bombarded towards the copper atoms so as to arrive at a distance of 10^{-13} meter from the nucleus of copper and then getting either deflected or traversing back. The α-particles did not move further closer.

Q.7 The velocity of the α-particles must be

(A) 8.32×10^8 cm/sec (B) 6.32×10^8 cm/sec

(C) 6.32×10^8 m/sec (D) 6.32×10^8 km/sec

Q.8 Which of the following metals can be used instead of gold in α-scattering experiment

(A) Pt (B) Na (C) K (D) Cs

Q.9 From the Rutherford's α-particle scattering, it can be concluded that

(A) $N \propto \sin^4 \frac{\theta}{2}$ (B) $N \propto \frac{1}{\sin^4 \theta}$

(C) $N \propto \frac{1}{\sin^4 \theta/2}$ (D) $N = \sin\frac{\theta}{2}$

Q.10 Rutherford observed that

(A) 50% of the α-particles got deflected

(B) 99% of the α-particles got deflected

(C) 99% of the α-particles went straight without suffering any deflection.

(D) Nucleus is negatively charged.

Match the Columns

Q.11 Match the entries in column I with entries in column II

Column I	Column II
(A) Electron moving in 2nd orbit in He^+ ion electron is	(p) Radius of orbit in which moving is 0.529Å
(B) Electron moving in 3rd orbit in H-atom	(q) Total energy of electron is (–)13.6 × 9eV
(C) Electron moving in 1st orbit in Li^{+2} ion	(r) Velocity of electron is $\frac{2.188 \times 10^8}{3}$ m/sec
(D) Electron moving in 2nd orbit is Be^{+3} ion	(s) De-broglie wavelength of Electron is $\sqrt{\frac{150}{13.6}}$ Å

Q.12 Column I and column II contain data on Schrodinger Wave–Mechanical model, where symbols have their usual meanings. Match the columns.

Column I	Column II
(A) Ψ_r 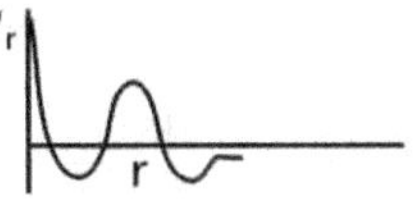	(p) 4s
(B) $\Psi^2_r 4\pi r^2$ 	(q) $5p_x$
(C) $\psi(\theta, \phi)$ = K (independent of θ and ϕ)	(r) 3s
(D) at least one angular node is present	(s) 6d

True/False Type

Q.13 Statement-I: Emitted radiations will fall in the visible range when an electron jumps from higher level to n = 2 in Li^{+2} ion.

Statement-II: Balmer series radiations belong to visible range in all H-like atoms.

(A) Statement-I true; Statement-II is true; Statement-II is the correct explanation of Statement-I.

(B) Statement-I true; Statement-II is true; Statement-II is not the correct explanation of Statement-I.

(C) Statement-I is true; Statement-II is false.

(D) Statement-I is false; Statement-II is true.

Previous Years' Questions

Q.1 The increasing order (lowest first) for the values of e/m (charges/mass) for electron (e), proton (p) ***(1984)***

(A) e, p, n α (B) n, p, e, α

(C) n, p, α, e (D) n, α, p, e

Q.2 Which of the following does not characterize X-rays ***(2000)***

(A) The radiation can ionize gases

(B) It causes ZnS to fluoresce

(C) Deflected by electric and magnetic fields

(D) Have wavelengths shorter than ultraviolet rays

Q.3 The number of nodal planes in a px orbital is ***(2001)***

(A) One (B) Two (C) Three (D) Zero

Q.4 If the nitrogen atom had electronic configuration $1s^1$ it would have energy lower than that of the normal ground state configuration $1s^2\ 2s^2\ 2p^3$, because the electrons would be closer to the nucleus, yet $1s^7$ is not observed because it violates ***(2002)***

(A) Heisenberg uncertainty principle

(B) Hund's rule

(C) Pauli exclusion principle

(D) Bohr postulate of stationary orbits

Q.5 Which hydrogen like species will have same radius as that of Bohr orbit of hydrogen atom? ***(2003)***

(A) n = 2, Li^{2+} (B) n = 2, Be^{3+}

(C) n = 2, He^{+} (D) n = 3, Li^{2+}

Q.6 The number of radial nodes in 3s and 2p respectively are ***(2005)***

(A) 2 and 0 (B) 0 and 2

(C) 1 and 2 (D) 2 and 1

Q.7 When alpha particles are sent through a thin metal foil, most of them go straight through the foil because ***(1984)***

(A) Alpha particles are much heavier than electrons

(B) Alpha particles are positively charged

(C) Most part of the atom is empty space

(D) Alpha particles move with high velocity

Q.8 The ground state electronic configuration of nitrogen atom can be represented by ***(2004)***

(A) [↑↓][↑↓][↑][↑][↑] (B) [↑↓][↑↓][↑][↓][↑]

(C) [↑↓][↑↓][↑][↓][↓] (D) [↑↓][↑↓][↓][↓][↓]

Read the following questions and answer as per the direction given below:

(A) Statement-I true; Statement-II is true; Statement-II is the correct explanation of Statement-I.

(B) Statement-I true; Statement-II is true; Statement-II is not the correct explanation of Statement-I.

(C) Statement-I is true; Statement-II is false.

(D) Statement-I is false; Statement-II is true.

Q.9 Statement-I: The first ionization energy of Be is greater than that of B.

Statement-II: 2p orbital is lower in energy than 2s. ***(2000)***

Paragraph 1: The hydrogen-like species Li^{2+} is in a spherically symmetric state S_1 with one radial node. Upon absorbing light the ion undergoes transition to a state S_2 has one radial node and its energy is equal to the ground state energy of the hydrogen atom. ***(2010)***

Q.10.1 The state S_1 is

(A) 1s (B) 2s (C) 2p (D) 3s

Q.10.2 Energy of the state S_1 in units of the hydrogen atom ground state energy is

(A) 0.75 (B) 1.50 (C) 2.25 (D) 4.50

Q.10.3 The orbital angular momentum quantum number of the state S_2 is

(A) 0 (B) 1 (C) 2 (D) 3

Q.11 According to Bohr's theory En = total energy

K_n = Kinetic energy K_n = Kinetic energy

V_n = Potential energy R_n = Radius of nth orbit

Match the following: ***(2006)***

Column I	Column II
(A) V_n/K_n =	(p) V_n/K_n =
(B) If radius of nth orbit $\propto E^x_n$, x=	(q) −1
(C) Angular momentum is lowest orbital	(r) −2
(D) $\frac{1}{r^n} \propto Z^y, y = ?$	(s) 1

Q.12 Match the entries in Column I with the correctly related quantum number (s) in column II. ***(2008)***

Column I	Column II
(A) Orbital angular momentum of the electron in a hydrogen-like atomic orbital	(p) Principal quantum number
(B) A hydrogen-like one-electron wave function obeying Pauli's principle	(q) Azimuthal quantum number
(C) Shape, size and orientation of hydrogen-like atomic orbital	(r) Magnetic quantum number
(D) Probability density of electron at the nucleus in Hydrogen-like atom	(s) Electron spin quantum number

Q.13 The maximum number of electrons that can have principal quantum number, n = 3 and spin quantum number, $m_s = -\frac{1}{2}$ is ***(2011)***

Q.14 The work function (Φ) of some metals is listed below. The number of metals which will show photoelectric effect when light of 300 nm wavelength falls on the metal is ***(2011)***

Metal	Li	Na	K	Mg	Cu	Ag	Fe	Pt	W
Ö(eV)	2.4	2.3	2.2	3.7	4.8	4.3	4.7	6.3	4.75

Q.15 (a) The Schrodinger wave equation for hydrogen atom is: $\psi_{2s} = \frac{1}{4(2\pi)^{1/2}}\left(\frac{1}{a_0}\right)^{3/2}\left(2-\frac{r}{a_0}\right)e^{-r/2a_0}$

Where, a_0 is Bohr's radius. Let the radial node in 2s be at r_0. Then find r in terms of a_0.

(b) A baseball having mass 100 g moves with velocity 100 m/s. Find out the value of wavelength of baseball. ***(2004)***

Q.16 (a) Calculate velocity of electron in first Bohr orbit of hydrogen atom (Given, r = a_0)

(b) Find de-Broglie wavelength of the electron in first Bohr orbit

(c) Find the orbital angular momentum of 2p-orbital in terms of h/2π units. ***(2005)***

Q.17 The atomic masses of He and Ne are 4 and 20 a.m.u., respectively. The value of the de Broglie wavelength of He gas at −73°C is "M" times that of the de Broglie wavelength of Ne at 727°C. M is ***(2013)***

Q.18 Not considering the electronic spin, the degeneracy of the second excited state(n = 3) of H atom is 9, while the degeneracy of the second excited state of H^- is ***(2015)***

Q.19 The kinetic energy of an electron in the second Bohr orbit of a hydrogen atom is [is Bohr radius] ***(2015)***

(A) $\frac{h^2}{4\pi^2 ma_0^2}$ (B) $\frac{h^2}{16\pi^2 ma_0^2}$

(C) $\frac{h^2}{32\pi^2 ma_0^2}$ (D) $\frac{h^2}{64\pi^2 ma_0^2}$

Important Questions

JEE Main/Boards

Exercise 1

Q.1	Q.4	Q.9
Q.13	Q.16	Q.27

Exercise 2

Q.5	Q.7	Q.18
Q.19	Q.28	

Previous Years' Questions

Q.7	Q.15

JEE Advanced/Boards

Exercise 1

Q.1	Q.7	Q.17
Q.19		

Exercise 2

Q.4	Q.6	Q.9
Q.12		

Previous Years' Questions

Q.8	Q.10.3	Q. 11

Answer Key

JEE Main/Boards

Exercise 1

Q.1 [1.5×10^{-18} unit.]

Q.2 [$n < \alpha < p < e$]

Q.3 [6.35×10^6 m/s]

Q.4 [$\lambda_1 = 4\lambda$]

Q.5 [$5 \times 10^{14}\ s^{-1}$]

Q.5 [3×10^{-19}]

Q.7 $1.095 \times 10^5\ cm^{-1}$

Q.8 [≈ 28]

Q.9 [2741 Å]

Q.10 [No photoelectron will emit]

Q.11 [3.78×10^{-19} J]

Q.12 [$r = \left(\dfrac{n^2h^2}{4\pi^2mk}\right)^{1/4}$]

Q.13 (A) 2.0 eV; (B) 2V; (C) 2970Å

Q.14 7.293×10^5 m/s

Q.15 7.293×10^5 m/s

Q.16 32/27

Q.17 $2.09 \times 10^{16}\ s^{-1}$

Q.18 $E_a = E_1/n^2$

Q.19 6×10^{-20} J

Q.20 −6.044eV; + 6.044 eV; −12.088 eV.

Q.21 91.8 eV

Q.22 54.4eV

Q.23 494.73 kJ/mol

Q.24 3.75×10^4 cm^{-1}

Q.25 912.37 Å

Q.26 1.01×10^{15} Hz

Q.27 97.86 nm

Q.28 488.89 nm

Q.29 6.626×10^{-25} kg ms^{-1}

Q.30 n = 2 and 3; λ = 1216 Å; λ = 1020Å

Exercise 2

Single Correct Chioce Type

Q.1 C	**Q.2** C	**Q.3** B	**Q.4** D	**Q.5** D	**Q.6** D	**Q.7** B
Q.8 B	**Q.9** A	**Q.10** B	**Q.11** C	**Q.12** B	**Q.13** D	**Q.14** B
Q.15 A	**Q.16** D	**Q.17** D	**Q.18** A	**Q.19** C	**Q.20** A	**Q.21** B
Q.22 C	**Q.23** D	**Q.24** B	**Q.25** C	**Q.26** B	**Q.27** B	**Q.28** A
Q.29 C						

Previous Years' Questions

Q.1 A	**Q.2** D	**Q.3** A	**Q.4** B	**Q.5** D	**Q.6** C	**Q.7** C
Q.8 C	**Q.9** B	**Q.10** C	**Q.11** B	**Q.12** D	**Q.13** B	**Q.14** A
Q.15 B	**Q.16** C	**Q.17** B	**Q.18** A	**Q.19** B	**Q.20** C	

JEE Advanced/Boards

Exercise 1

Q.1 1eV = 1.24×10^{-6} meter

Q.2 $n_1 = 2$, $n_2 = 4$, I.P. = 8.67×10^{-18} J,

r = 2.64×10^{-11} volts

Q.3 1.86×10^3 volts

Q.5 6.34×10^6 m/sec

Q.6 2.4×10^{14} sec^{-1}

Q.7 2.055×10^{-7} m/s

Q.11 4814.8 Å

Q.12 2.92×10^{15}

Q.13 Z = 2 He^+

Q.17 $v_0 = 1.148 \times 10^{15}$ sec^{-1}; λ_0 = 2614.6 Å

Q.18 495 nm

Q.19 8.84×10^{-35} m

Q.20 $v_0 = 4\times10^{15}$ Hz

Q.21 26

Q.22 2.2×10^{-20} J

Q.23 4.65 eV

Q.24 182.5 kJ

Q.25 $Z_x = 12$, $Z_y = 24$

Q.26 5

Q.27 $\lambda = 6.6 \times 10^3$ Å

Q.28 Volume = 2×10^{-37} cm^3 Density of molecules = 1.44×10^{14} gm/cm^3

Q.29 $\Delta n = 3 \times 10^{-30}$ m

Q.30 $\Delta n = 3 \times 10^{-5}$ m

This shows uncertainty in positions for larger particles is less whereas the uncertainty in position for smaller particles is larger. Hence for macro particles uncertainty in positions is not significant

Exercise 2

Single Correct Chioce Type

Q.1 C **Q.2** B

Comprehension Type

Paragraph 1: **Q.3** A B **Q.4** C **Q.5** B **Q.6** C

Paragraph 2: **Q.7** B **Q.8** A **Q.9** C **Q.10** C

Match the Columns

Q.11 $A \rightarrow s; B \rightarrow r; C \rightarrow q; D \rightarrow p$

Q.12 $A \rightarrow p; B \rightarrow p, q, s; C \rightarrow p, r; D \rightarrow q, s$

Assertion Reasoning Type

Q.13 D

Previous Years' Questions

Q.1 D **Q.2** C **Q.3** A **Q.4** C **Q.5** B **Q.6** A **Q.7** A, C

Q.8 A, D **Q.9** C **Q.10.1** B **Q.10.2** C **Q.10.3** B

Q.11 $A \rightarrow r; B \rightarrow q; C \rightarrow p; D \rightarrow s$

Q.12 $A \rightarrow q; B \rightarrow p, q, r, s; C \rightarrow p, q, r; D \rightarrow p, q, r$

Q.13 9 **Q.14** 4

Q.15 (A) Ro = 2ro, (B) 6.625×10^{-25} Å

Q.16 (A) $2.18 \times 10^{6}\ ms^{-1}$ (B) 3.3 Å (C) $\sqrt{2}\left(\frac{h}{2\pi}\right)$

Q.17 5 **Q.18** 3 **Q.19** C

Solutions

JEE Main/Boards

Exercise 1

Sol 1: In an oil drop experiment, the obtained charges will be integral multiples of charge of electron.

Let charge of electron be u

u = GCD (4.5×10^{-18}, 3×10^{-18}, 6×10^{-18}, 7.5×10^{-18}, 9×10^{-18}) and we know that

GCD (4.5, 3.6, 7.5, 9) = 1.5

$\therefore$ Charge of electron = u = 1.5×10^{-18} C

Sol 2: Let charge of electron be 'e_0' and mass of proton and neutron = m_0

$$SC_e = \frac{|e_0|}{\frac{m_0}{1837}} = 1837\frac{|e_0|}{m_0}$$

$$SC_p = \frac{|e_0|}{m_0}$$

$$SC_n = \frac{0}{m_0} = 0$$

$$SC_\infty = \frac{|2e_0|}{4m_0} = \frac{1}{2}\frac{|e_0|}{m_0}$$

[since α particle = He^{2+}]

$\therefore SC_n < SC_\alpha < SC_p < SC_e$

i.e. $[n < \alpha < p < e]$.

Sol 3: By conservation of energy

$\Delta PE = \Delta KE$

$$\Rightarrow \frac{1}{4\pi\epsilon_0} \times \frac{q_1 q_2}{r} = \frac{1}{2}mv^2$$

q_1 = charge of copper nucleus

q_2 = charge of α-particle

$$\Rightarrow \frac{9\times10^9 \times 29 \times 1.6\times10^{-19} \times 2 \times 1.6 \times 10^{-19}}{10^{-13}}$$

$$= \frac{1}{2} \times 4 \times 1.67 \times 10^{-27} \times v^2$$

$$v^2 = \frac{133.6}{3.3} \times 10^{12} \Rightarrow v = 6.34 \times 10^6 \text{ m/s}$$

Sol 4: By Moseley's law

$E = (10.2 \text{ eV}) (2 - 1)^2$

$$\lambda \propto \frac{1}{E} \Rightarrow \lambda \propto \frac{1}{(Z-1)^2}$$

$$\frac{\lambda_{29}}{\lambda_{57}} = \frac{(57-1)^2}{(29-1)^2} = \frac{56^2}{28^2} = 4$$

$$\Rightarrow \lambda_{29} = 4\lambda_{57} \Rightarrow \lambda' = 4\lambda.$$

Sol 5: Frequency, $\nu = \frac{c}{\lambda} = \frac{3\times10^8}{600\times10^{-9}}$

$= 5 \times 10^{14}$ /sec

Sol 6: $E = \frac{hc}{\lambda} = \frac{6.66\times10^{-34}\times3\times10^8}{6626\times10^{-10}} = 3 \times 10^{-19}$ J

Sol 7: First line of Lyman is He^{2+}

$$\frac{1}{\lambda} = R_H \times Z^2 \left[\frac{1}{1^2} - \frac{1}{2^2}\right] = R_H \times 4\left(\frac{3}{4}\right)$$

$$\frac{1}{\lambda} = 3R_H \qquad \lambda = \frac{1}{3R_H}$$

$$\therefore \Delta\lambda = \frac{1}{R_H}\left(\frac{9}{5} - \frac{1}{3}\right) = \frac{22}{15R_H} = 133.7 \times 10^{-17} \text{ cm}$$

$\therefore R_H = 1.085 \times 10^5 \text{ cm}^{-1}$

First line of Balmer is He^{2+}

$$\frac{1}{\lambda} = R_H \times Z^2 \left[\frac{1}{2^2} - \frac{1}{3^2}\right] = R_H \times Z^2\left(\frac{5}{36}\right)$$

$$\frac{1}{\lambda} = \frac{5R_H}{9}$$

$$\lambda = \frac{9}{5R_H}$$

$\therefore R_H = 1.095 \times 10^5 \text{ cm}^{-1}$

Sol 8: Energy of each photon $= E_0 = \frac{hC}{\lambda}$

$$= \frac{6.6\times10^{-34}\times3\times10^8}{550\times10^{-9}} = 3.6 \times 10^{-19} \text{ J}$$

Number of photon required $= \frac{E}{E_0} = \frac{10^{-17}}{3.6\times10^{-19}}$

≈ 27.7

So we need 28 ph otons.

Sol 9: Energy required of H_2 molecule is

$$E_0 = \frac{104\times10^3\times4.2\text{J}}{6\times10^{2.3}} = 72.8 \times 10^{-20}$$

$= 7.28 \times 10^{-19}$ J

$$E_0 = \frac{hc}{\lambda} \Rightarrow \lambda = \frac{hc}{E_0} = \frac{6.6\times10^{-34}\times3\times10^8}{7.28\times10^{-19}}$$

$= 2.74 \times 10^{-7}$ m = 2740 Å

Sol 10: Energy of photon $= \frac{hc}{\lambda}$

$$= \frac{6.6\times10^{-34}\times3\times10^8}{660\times10^{-9}} = 3 \times 10^{-19} \text{ J}$$

Work function $= 2.25 \times 1.6 \times 10^{-19}$ J

$\approx 3.6 \times 10^{-19}$ J

As work function is greater than energy of photon, no photoelectron will be emitted.

Sol 11: Energy of photon $= \frac{hc}{\lambda}$

$$= \frac{6.6\times10^{-34}\times3\times10^8}{4\times10^3\times10^{-10}} = 4.95 \times 10^{-19} \text{ J}$$

Work function $= h\nu_0$

$= 6.6 \times 10^{-34} \times 1.8 \times 10^{14} = 1.18 \times 10^{-19}$ J

and $KE_{max} = E_{photon}$ − work function

$= 3.78 \times 10^{-19}$ J

Sol 12: According to Bohr,

$mvr = \frac{nh}{2\pi}$ and for equilibrium of electron.

$$\frac{mv^2}{r} = F$$

$$F = \frac{\partial U}{\partial r} = \frac{d\left(\frac{1}{2}kr^2\right)}{dr} = kr$$

$$\frac{mv^2}{r} = kr \qquad \text{...(i)}$$

and from first equation

$$mvr = \frac{nh}{2\pi}$$

$$\Rightarrow m^2v^2r^2 = \frac{n^2h^2}{4\pi^2} \qquad \text{...(ii)}$$

If we equate v_2 in both the equations

$$\Rightarrow \frac{kr^2}{m} = \frac{n^2h^2}{4\pi^2m^2r^2} \Rightarrow r^4 = \frac{n^2h^2}{4\pi^2mk}$$

$$\Rightarrow r = \left(\frac{n^2h^2}{k4\pi^2m}\right)^{1/4}$$

Sol 13: Energy of photon $= \frac{hc}{\lambda}$

$$= \frac{6.6\times10^{-34}\times3\times10^{8}}{2\times10^{3}\times10^{-10}}$$

$\approx 9.9 \times 10^{-19}$ J

$$\approx \frac{9.9\times10^{-19}}{1.6\times10^{-19}} \text{ eV} = 6.2 \text{ eV}$$

(a) For fastest moving electron

$KE = \frac{hc}{\lambda} - E_0 = 6.2 \text{ eV} - 4.2 \text{ eV} = 2.0 \text{ eV}$

for slowest moving electron

KE = 0

(b) Stopping potential $= \frac{KE_{max}}{e} = \frac{2eV}{e} = 2V$

(c) Cut-off wavelength $= \lambda_0 = \frac{hc}{E_0}$

$$= \frac{6.6\times10^{-34}\times3\times10^{8}}{4.2\times1.6\times10^{-19}}$$

$\approx 2.97 \times 10^{-7}$ m = 2970 Å

Sol 14: Speed of an electron in H-atom in nth orbit is

$$v_n = \frac{2\pi ke^2}{h}\times\frac{Z}{n} = 2.18 \times 10^6 \times \frac{1}{n} \text{ and } n = 3$$

$\therefore$ v = 7.29×10^5 m/s

Sol 15: Radius of nth orbit in H is

$r_n = 0.529 \times n^2$ Å

$\Rightarrow 0.529\ n^2 = 4.761 \Rightarrow n^2 = 9 \Rightarrow n = 3$

$$v^3 = \frac{2.18\times10^{6}}{3} = 7.29 \times 10^5 \text{ m/s}$$

Sol 16: Time period of electron $= \frac{2\pi r}{v}$

$$\therefore t \propto \frac{n^2/Z}{Z/n} \Rightarrow t \propto \frac{n^3}{Z^2}$$

$$\therefore \frac{T_1}{T_2} = \frac{\frac{2^3}{1^2}}{\frac{3^3}{2^2}} = \frac{2^5}{3^3} = \frac{32}{27}$$

Sol 17: Angular frequency of an electron in an orbit is

$$f = \frac{2\pi}{T} = \frac{V}{2\pi r} \times 2\pi = \frac{2.18\times10^{6}\times\frac{2}{2}}{2\times3.14\times0.529\times\frac{2^2}{2}\times10^{-10}} \times$$

2×3.14

2.08 × 10^{16} /sec

Sol 18: Energy of nth orbit in H atom is

$$E = \frac{-2\pi^2k^2me^4}{n^2h^2} = \frac{-13.6}{n^2} \text{ eV}$$

$E_1 = -13.6 \text{ eV}E_2 = -3.4$ eV

$E_3 = -1.51 \text{ eV}E_4 = -0.85$ eV

Sol 19: Energy of photon $= \frac{hC}{\lambda}$

$$= \frac{6.6\times10^{-34}\times3\times10^{8}}{2.537\times10^{-7}} = 7.8 \times 10^{-19} \text{ J}$$

Energy needed to dissociate 1 molecule of H_2

$$= \frac{103.2\times10^{3}\times4.2}{6\times10^{23}} = 7.22 \times 10^{-19} \text{ J}$$

$\therefore$ KE = (7.8 − 7.2) × 10^{-19} J = 6 × 10^{-20} J

Sol 20: TE of 3rd orbit of He+ ion

$= 13.6 \times \frac{Z^2}{n^2}$ eV and Z = 2, n = 3

$\therefore$ TE = $-13.6 \times \frac{4}{9} = -6.044$ eV

PE = 2TE = −12.088 eV

KE = −TE = 6.044 eV

Sol 21: Excitation energy = Energy of 2nd orbit – Energy of 1st orbit

$E_{ex} = \frac{-13.6\times Z^2}{2^2} - \left(\frac{-13.6\times Z^2}{1^2}\right)$ and Z = 3

$\Rightarrow E_{ex} = 13.6 \times 9\left(\frac{3}{4}\right) = 91.8$ eV

Sol 22: BE of electron in ground of He^{2+} ion

= − total energy of electron in that orbit.

$\therefore$ BE = −TE

$= -\left(-13.6\times\frac{Z^2}{n^2}\right) = 13.6 \times 4 = 54.4$ eV

Sol 23: IE of Na = Energy of photon × N

$= \frac{hC}{\lambda} = \frac{6.6\times10^{-34}\times3\times10^{8}}{242\times10^{-9}}\times N$

$= 8.18 \times 10^{-19} \times 6 \times 10^{23}$

$= 4.9 \times 10^{3}$ J/mol = 4.9 kJ/mole

Sol 24: E_0 = Energy of electron = 4.64 eV

$= 4.64 \times 1.6 \times 10^{-19}$ J

Energy of photon = $hc\bar{\nu}$

$\Rightarrow \bar{\nu} = \frac{E}{hc} \Rightarrow \bar{\nu} = \frac{4.64\times1.6\times10^{-19}}{6.6\times10^{-34}\times3\times10^{8}}$

$= 3.74 \times 10^{4}$ /cm

Sol 25: Change in energy = $TE_H = -13.6$ eV

$\Rightarrow 13.6\text{ eV} = \frac{hc}{\lambda}$

$\Rightarrow \lambda = \frac{hc}{13.6\text{eV}}$

$= \frac{6.6\times10^{-34}\times3\times10^{8}}{13.6\times1.6\times10^{-19}} = 9.12 \times 10^{-8}$ m

≈ 912 Å

Sol 26: Energy absorbed in the transition

$= E_4 - E_2$

$= \frac{-13.6}{4^2} - \left(\frac{13.6}{2^2}\right) = 13.6\left[\frac{1}{4}-\frac{1}{16}\right]$

$= \frac{5\times13.6}{16} = 4.2$ eV

Energy of photon = hν

$\Rightarrow \nu = \frac{E}{h}$

$= \frac{4.2\times1.6\times10^{-19}}{6.6\times10^{-34}} = 1.01 \times 10^{15}$ Hz

Sol 27: Assume that the electron excites to orbit no. 'n'.

No. of subsequent emissions

$= \frac{(\Delta H)(\Delta n+1)}{2} = \frac{(n-1)(n-1+1)}{2}$

Given no. of subsequent emission = 6

$\Rightarrow \frac{n(n-1)}{2} = 6 \Rightarrow n = 4$

$\Rightarrow \frac{hc}{\lambda} = 13.6\left[\frac{1}{1^2}-\frac{1}{4^2}\right]$

$\Rightarrow \lambda = \frac{13.6\times\frac{15}{16}\times1.6\times10^{-19}}{6.6\times10^{-34}\times3\times10^{8}}$

$= 9.78 \times 10^{-8}$ m = 97.8 nm.

Sol 28: $\lambda_{H_\alpha} = \frac{hc}{E} = \frac{hc}{13.6\left[\frac{1}{2^2}-\frac{1}{3^2}\right]}$

$\lambda_{H_\beta} = \frac{hc}{13.6\left[\frac{1}{2^2}-\frac{1}{4^2}\right]}$

$\Rightarrow \frac{\lambda_{H_\beta}}{\lambda_{H_\alpha}} = \frac{\left[\frac{1}{4}-\frac{1}{9}\right]}{\left[\frac{1}{4}-\frac{1}{16}\right]}$

$$\Rightarrow \lambda_{H_\beta} = 660 \times \frac{\frac{5}{36}}{\frac{3}{16}} = 488.8 \text{ Å}$$

Sol 29: Momentum of a photon = P = $\frac{h}{\lambda}$

$$= \frac{6.6\times10^{-34}}{10\times10^{-10}} = 6.6 \times 10^{-25} \text{ kg m/s}$$

Sol 30: Energy of 2nd orbit of H = $\frac{-13.6}{2^2} = -3.4$ eV

given transitions have energies greater than 3.4 eV.

So $n_1 = 1$.

$$10.2 = 13.6\left[\frac{1}{n_1^2} - \frac{1}{n_2^2}\right]$$

$$\Rightarrow \frac{3}{4} = \frac{1}{1^2} - \frac{1}{n_2^2}$$

$$\Rightarrow \frac{1}{n_2^2} = \frac{1}{4} \Rightarrow n_2 = 2$$

$$\lambda = \frac{hC}{E} \text{ or } \frac{912\text{ Å}}{\left[\frac{1}{n_1^2} - \frac{1}{n_2^2}\right]}$$

$$= \frac{912\text{ Å}}{\frac{1}{1^2} - \frac{1}{2^2}} = 912 \times \frac{4}{3} = 1216 \text{ Å}$$

For 12.09 eV

$$12.09 = 13.6\left[\frac{1}{n_1^2} - \frac{1}{n_2^2}\right]$$

$$\frac{1}{1^2} - \frac{1}{n_2^2} = \frac{8}{9}$$

$$\Rightarrow n_2 = 3$$

$$\therefore \lambda = \frac{912\text{ Å}}{\left[\frac{1}{1^2} - \frac{1}{3^2}\right]} = \frac{912\times9}{8} = 1020 \text{ Å}$$

Exercise 2

Single Correct Choice Type

Sol 1: (C) (A) As x-rays are high energetic photons, it can ionise gases.

(B) ZnS shows fluorescence in x-rays.

(C) As x-rays are photons, they are neither deflected by electric nor magnetic fields

(D) $E_{rays} > E_{uv\text{-}rays}$, so $\lambda_{x\text{-}rays} < \lambda_{uv\text{-}rays}$

Sol 2: (C) Bohr's model doesn't say anything about probability of finding an electron near nucleus. It gives discrete orbitals as locus for finding electrons.

Sol 3: (B) If we put x = 1 in E = $\frac{\text{constant}}{n^2}$ kJ/mole.

Constant is the negative of energy an electron first orbit.

So, it also the ionisation energy of H-atom.

Sol 4: (D) E = $\frac{hc}{\lambda}$

$$\text{So } \frac{E_1}{E_2} = \frac{hc/\lambda_1}{hc/\lambda_2} = \frac{\lambda_2}{\lambda_1} = \frac{4000}{2000} = 2$$

Sol 5: (D) Let's assume that the electron is at a distance 'r' from nucleus.

$$E_{electron} = \frac{-k.Z_e.e}{r}$$

as the energy is negative, the energy is maximum when $r \to \infty$, $E \to 0$.

Sol 6: (D) Schrodinger's equation depends on radius, shape and orbital orientation. So, it depends on n, ℓ, m.

But it doesn't depend on spin of the electron.

∴ So Schrondinger's equation is not related spin quantum number.

Sol 7: (B) $\frac{1}{x} = R_H \times Z^2\left[\frac{1}{2^2} - \frac{1}{n^2}\right]$

Z = 2

and shortest wavelength comes when $n \to \infty$

$$\Rightarrow \frac{1}{x} = R_H \times 4\left[\frac{1}{4}\right] = R_H$$

$\therefore$ Let x' be longest wavelength is Paschen of Li^{2+}.

Longest wavelength: $3 \rightarrow 4$

$$\frac{1}{x'} = R_H \times Z^2 \left[\frac{1}{3^2} - \frac{1}{4^2}\right]$$

$$\Rightarrow \frac{1}{x'} = R_H \times 9 \left[\frac{1}{9} - \frac{1}{16}\right] = \frac{7R_H}{16}$$

$$\frac{1}{x'} = \frac{7}{x16} \Rightarrow x' = \frac{16x}{7}$$

Sol 8: (B) $\frac{1}{\lambda} = R_H \times Z^2 \left[\frac{1}{n_1^2} - \frac{1}{n_2^2}\right]$

$$\Rightarrow \lambda = \frac{912\,Å}{9\left[\frac{1}{1^2} - \frac{1}{\infty^2}\right]}$$

$\therefore$ E = 13.6 × 9 eV

Ionisation energy of H = 13.6 eV

KE of e^- = 13.6 × 8 eV

$$\lambda = \frac{h}{mv} = \frac{h}{\sqrt{2mkE}}$$

$$= \frac{6.6 \times 10^{-34}}{\sqrt{2 \times 9.1 \times 10^{-31} \times 13.6 \times 8 \times 1.6 \times 10^{-19}}}$$

$$= \frac{6.6 \times 10^{-34}}{6 \times 10^{-24}}; \approx 1.17\,Å$$

Sol 9: (A) IE_3 of Li is $Li^{2+} \xrightarrow{\Delta E} Li^{+3}$

$\Delta E = 13.6 \times Z^2$ eV per atom per mole.

$$IE_3 = \frac{13.6 \times 9 \times 1.6 \times 10^{-19} \times 6 \times 10^{23}}{1 \times 10^3} \text{ kJ/mole}$$

= 11775 kJ/mole

IE_2 = 19800 − 11775 − 520 = 7505 kJ/mole

Sol 10: (B) $\Delta_1 = \frac{912}{\left[\frac{1}{1^2} - \frac{1}{2^2}\right]} - \frac{912}{\left[\frac{1}{1^2} - \frac{1}{3^2}\right]}$

1st line $1 \rightarrow 2$

2nd line $1 \rightarrow 3$

3rd line $1 \rightarrow 4$

$$= 912\left(\frac{4}{3} - \frac{9}{8}\right) = 912\left(\frac{5}{24}\right)$$

$$\Delta_2 = \frac{912}{\left[\frac{1}{1^2} - \frac{1}{3^2}\right]} - \frac{912}{\left[\frac{1}{1^2} - \frac{1}{4^2}\right]}$$

$$= 912\left[\frac{9}{8} - \frac{16}{15}\right] = 912\left(\frac{7}{120}\right)$$

$$\Delta_1 : \Delta_2 = \frac{5}{24} \times \frac{120}{7} = \frac{25}{7}$$

≈ 3.5

Sol 11: (C) $r_n = 0.529 \times n^2\,Å \Rightarrow r_n \propto n^2$

$\therefore r_1 : r_2 : r_3 = 1 : 2^2 : 3^2$

Sol 12: (B) For a permissible solution of n, ℓ, m, s

$n - 1 \geq \ell, m \leq |\ell|, s = \pm \frac{1}{2}$

$\therefore$ 3, 3, 1, $-\frac{1}{2}$ is not permissible.

Sol 13: (D) $E \propto \frac{1}{\lambda}$

$$\Rightarrow \frac{E_1}{E_2} = \frac{\lambda_2}{\lambda_1} = 2$$

Sol 14: (B) Orbital angular momentum of an orbital is

$$L = \frac{h}{2\pi}\sqrt{\ell(\ell+1)}$$

For 2s, $\ell = 0$

$\therefore L = 0$

Sol 15: (A) For each value of 'ℓ'

We have $2\ell + 1$ orbitals

$\therefore$ No. of electrons = $\sum_{\ell=0}^{n-1} 2(2\ell+1)$

Sol 16: (D) $\lambda = \frac{h}{mv} = \frac{6.63 \times 10^{-27}}{6 \times 10^{-2} \times 1.4 \times 1.6 \times 10^2 \times \frac{5}{18}}$

$= 1.8 \times 10^{-27}$ m $= 1.8 \times 10^{-25}$ cm

Sol 17: (D) Given

$$R = 0.529 \times \frac{n^2}{Z} \text{ Å}$$

$$\Rightarrow R = \frac{0.529}{Z} \times 4$$

$$R' = \frac{0.529 \times 3^2}{Z} = \frac{R}{4} \times 9 = \frac{9R}{4} = 2.25R$$

Sol 18: (A) $\lambda_1 = R_H \times 1^2 \left[\frac{1}{1^2} - \frac{1}{2^2}\right]$

$$\lambda_2 = R_H \times 2^2 \left[\frac{1}{2^2} - \frac{1}{4^2}\right] = RH \left[\frac{1}{1^2} - \frac{1}{2^2}\right]$$

$$\Rightarrow \lambda_1 : \lambda_2 = 1 : 1$$

Sol 19: (C) $n_1 + n_2 = 4$

$n_2^2 - n_1^2 = 8$

$\Rightarrow (n_2 + n_1)(n_2 - n_1) = 8$

$\Rightarrow n_2 - n_1 = 2$

$\Rightarrow n_1 = 1$

$n_2 = 3$

$$\frac{1}{\lambda} = \frac{1}{R_H \times Z^2} \times \frac{1}{\left[\frac{1}{1} - \frac{1}{3^2}\right]} = \frac{9}{32R_H}$$

Sol 20: (A) Brackett series $n_1 = 4$

n_2 can be from 9 to 5

∴ No. of lines in Brackett series are 9 − 4 = 5

Sol 21: (B) Bohr used quantum theory for quantising the angular momentum of electron in atom.

Sol 22: (C) $\lambda = \frac{h}{mv}$

$$= \frac{6.6 \times 10^{-34}}{2 \times 10^{-1} \times 5 \times \frac{5}{18} \times 1.6} = 10^{-33} \left(\frac{6.6 \times 18}{3.2 \times 25}\right)$$

It's in the order of 10^{-30} s.

Sol 23: (D) Paschen series → $n_1 = 3$

Longest wavelength → 3 → 4, Shortest 3 → ∞

$$\Rightarrow \frac{1}{m'} = R_H \times 4^2 \left[\frac{1}{3^2} - \frac{1}{\infty^2}\right]$$

$$\Rightarrow \frac{1}{m} = R_H \times Z^2 \left[\frac{1}{3^2} - \frac{1}{4^2}\right]$$

$$\Rightarrow \frac{m'}{m} = \frac{2^2}{4^2} \frac{7/144}{1/9} = \frac{7}{64}$$

$$\Rightarrow m' = \frac{7m}{64}$$

Sol 24: (B) Y → $_9F^{19}$[since protons = 2s + 1, mass = 18 + 1]

⇒ X = $_{11}Na^{23}$ → 11 protons, 12 neutrons

23 g ⇒ 12 N neutrons

$$4.6 \text{ g} \rightarrow \frac{12 \times 4.6}{23} = 2.4 \text{ moles of neutron}$$

Sol 25: (C) Energy of photon

$$E = \frac{hC}{\lambda} = \frac{6.6 \times 10^{-34} \times 3 \times 10^8}{3.1 \times 10^{-8}} = 6.3 \times 10^{-18} \text{ J}$$

Work function= 12.8 × 1.6 × 10^{-19} = 2.04 × 10^{-18} J

$KE_{max} = 4.26 \times 10^{-18}$ J

$$V = \sqrt{2KE_{max}/m_e} = 2.18\sqrt{2} \times 10^6 \text{ m/s}$$

Sol 26: (B) By de-Broglie hypothesis

$$\lambda = \frac{h}{p} \Rightarrow |\Delta\lambda| = \frac{h}{p^2}.|\Delta p| \text{....(i)} \left[\because \frac{d\lambda}{dp} = \frac{-h}{p^2}\right]$$

and by Uncertainity principle

$$\Delta x.\Delta p = \frac{\hbar}{2} \Rightarrow \Delta p \Rightarrow \frac{h}{4\pi\Delta x} \text{ and give } \Delta x = \frac{7}{22} \text{ nm}$$

Put this in (i)

We get minimum Δλ as

$$|\Delta\lambda| = \frac{h}{p^2}.\frac{1}{4\pi\Delta x} = \frac{h^2}{4\pi p^2 \Delta x}$$

Sol 27: (B) For $3p_y$

ψ is not independent of θ, ϕ

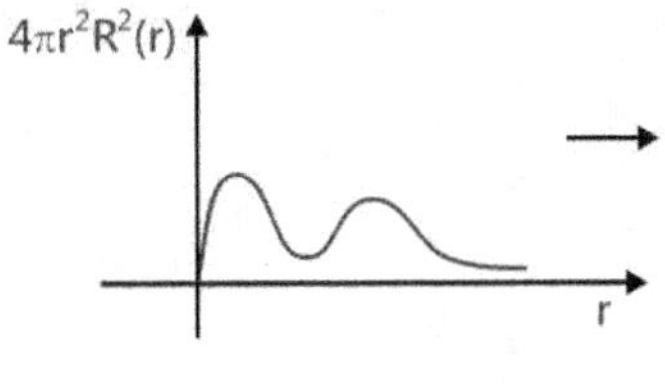

→ 2 maxima

2 → nodal plane

m can be 1 or 0 or −1.

Sol 28: (A) $\lambda_{radio} > \lambda_{micro}$

$r_n \propto n^2 \Rightarrow r_1 < r_2 < r_3 < r_4$

$E \propto -\frac{1}{n}$

$\therefore E_1 < E_2 < E_3 < E_4$ and $V \propto \frac{Z}{n}$

$\therefore Be^{+3} > Li^{+2} > He^{2+} > H$

Sol 29: (C) $L = \frac{nh}{2\pi}$

$L_n - L_{n-1} = \frac{h}{2\pi}$

ΔE doesn't depend on potential at ∞.

$KE \propto \frac{1}{r^2}$

∴ KE decreases on moving away from nucleus.

Previous Years' Questions

Sol 1: (A) James Chadwick discovered neutron ($_0n^1$).

Sol 2: (D) The radius of an atom is of the order of 10^{-8} cm

Sol 3: (A) NO^+ C_2^{2-} CN^- N_2

$14e^-$ $14e^-$ $14e^-$ $14e^-$

Sol 4: (B) N^{3-}, F^- and Na^+ (These three ions have e^- = 10, hence they are isoelectronic)

Sol 5: (D) $r_H = 0.529 \frac{n^2}{z}$ Å

For hydrogen ; n = 1 and z = 1 therefore

r_H = 0.529 Å

For Be^{3+} : Z = 4 and n = 2 Therefore

$$r_{Be^{3+}} = \frac{0.529 \times 2^2}{4} = 0.529 \text{Å}$$

Sol 6: (C) $\frac{1}{\lambda} = R\left[\frac{1}{n_1^2} - \frac{1}{n_2^2}\right]$

$$\frac{1}{\lambda} = 1.097 \times 10^7 m^{-1}\left[\frac{1}{1^2} - \frac{1}{\infty^2}\right]$$

$\therefore l = 91 \times 10^{-9}$

We known 10^{-9} = 1 nm So, λ = 91 nm

Sol 7: (C) $\Delta E = 1.312 \times 10^6 \left[\frac{1}{1^2} - \frac{1}{2^2}\right]$

$= 1.312 \times 10^6 \times \frac{3}{2} = 0.984 \times 10^6$ d

$= 9.84 \times 10^5$ J/mol.

Sol 8: (C) $E = E_1 + E_2$

$$\frac{hc}{\lambda} = \frac{hc}{\lambda_1} + \frac{hc}{\lambda_2} \Rightarrow \frac{1}{\lambda} = \frac{1}{\lambda_1} + \frac{1}{\lambda_2}$$

$$\frac{1}{355} = \frac{1}{680} + \frac{1}{\lambda_2} \Rightarrow \lambda_2 = 742.76 nm$$

Sol 9: (B) As

$$\lambda = \frac{h}{m\upsilon} = \frac{6.63 \times 10^{-34}}{1.67 \times 10^{-27} \times 1 \times 10^3} = 3.97 \times 10^{-10} m$$

$= 0.397 \times 10^{-9}$ m = 0.40 nm.

Sol 10: (C) $\Delta x + \Delta p = \frac{h}{4x}$

$$\Delta x \times [m\Delta\upsilon] = \frac{h}{4x}; \Delta u = \frac{600 \times 0.005}{100} = 0.03$$

$$\text{So, } \Delta x\, [9.1 \times 10^{-31} \times 0.03] = \frac{6.6 \times 10^{-34}}{4 \times 3.14}$$

$$\Delta x = \frac{6.6 \times 10^{-34}}{4 \times 3.14 \times 9.1 \times 0.03 \times 10^{-31}} = 1.92 \times 10^{-3} m$$

Sol 11: (B) is the correct option because it has the maximum value of n + *l*

Sol 12: (D) (4) and (5) belong to 3d-orbital which are same energy.

Sol 13: (B) The electronic configuration is Cr (chromium element in the ground state) = $1s^2\, 2s^2\, 2p^6\, 3s^2\, 3p^6\, 3d^5\, 4s^1$

Sol 14: (A) For 4f orbital electron n = 4

l = 3 (Because 0, 1, 2, 3)

s, p, d, f

m = + 3, +2, +1, 0, −1, −2, −3; s = +1/2

Sol 15: (B) 4p(2) 4s(3) 3d(4) 3p

According to (n + l) rule, increasing order of energy

(4) < (2) < (3) < (1)

Sol 16: (C) $\Delta x . m\, \Delta v = \frac{h}{4\pi}$

$$\Delta x = \frac{h}{4\pi\, m\, \Delta v}$$

$$\Delta v = 600 \times \frac{0.005}{100} = 0.03$$

$$\Rightarrow \Delta x = \frac{6.625 \times 10^{-34}}{4 \times 3.14 \times 9.1 \times 10^{-31} \times 0.03} = 1.92 \times 10^{-3}\text{ m}$$

Sol 17: (B) $\lambda = \frac{h}{mv} = \frac{6.63 \times 10^{-34}}{1.67 \times 10^{-27} \times 10^{3}} \equiv 0.40\text{ nm}$

Sol 18: (A) Energy required for 1 Cl_2 molecule

$= \frac{242 \times 10^{3}}{N_A}$ Joules.

This energy is contained in photon of wavelength 'λ'.

$$\frac{hc}{\lambda} = E \Rightarrow \frac{6.626 \times 10^{-34} \times 3 \times 10^{8}}{\lambda}$$

$$= \frac{242 \times 10^{3}}{6.022 \times 10^{23}}$$

$\lambda = 4947\ \overset{0}{A} \approx 494$ nm

Sol 19: (B) $IE_{He^+} = 13.6\, Z^2_{He^+}\left[\frac{1}{1^2} - \frac{1}{\infty^2}\right] = 13.6\, Z^2_{He^+}$, where $(Z_{He^+} = 2)$

Hence, $13.6 \times Z^2_{He^+} = 19.6 \times 10^{-18}$ J atom^{-1}.

$$(E_1)_{Li^{+2}} = -13.6\, Z^2_{Li^{+2}} \times \frac{1}{1^2}$$

$$= -13.6\, Z^2_{He^-} \times \left[\frac{Z^2_{Li^{+2}}}{Z^2_{He^+}}\right]$$

$$= -19.6 \times 10^{-18} \times \frac{9}{4} = -4.41 \times 10^{-17}\text{ J/atom}$$

Sol 20: (C) $(E_n)_H = -13.6 \frac{1^2}{n^2}$ eV

n = 2 ⇒ E_2 = −3.4 eV

JEE Advanced/Boards

Exercise 1

Sol 1: 1 eV = 1.6 × 10^{-19} J

Let λ be the wavelength of a photon of 1 eV energy.

$$E = \frac{hc}{\lambda}$$

$$\Rightarrow \lambda = \frac{hc}{E} = \frac{6.6 \times 10^{-34} \times 3 \times 10^{8}}{1.6 \times 10^{-19}}$$

= 12.3 × 10^{-7} m

= 1.23 × 10^{-6} m

Sol 2: 1st line in Lyman series of H

$$\frac{1}{\lambda} = R_H \times 1^2 \left[\frac{1}{1^2} - \frac{1}{2^2}\right]$$

In He^{2+}

$$\frac{1}{\lambda} = R_H \times Z^2 \left[\frac{1}{n_1^2} - \frac{1}{n_2^2}\right]$$

$$\Rightarrow \left[\frac{1}{n_1^2} - \frac{1}{n_2^2}\right] \times 2^2 = \frac{1}{1^2} - \frac{1}{2^2}$$

$$\Rightarrow \frac{1}{n_1^2} - \frac{1}{n_2^2} = \frac{1}{2^2} - \frac{1}{4^2}$$

⇒ n_1 = 2, n = 4

IE_2 of He = 13.6 × Z^2 eV/atom

= 13.6 × 4 × 1.6 × 10^{-19}

= 8.67 × 10^{-18} J/atom

1st of He^{2+} → r = 0.529 × $\frac{n^2}{Z}$ Å

n = 1, Z = 2

∴ r = 0.264 Å

Sol 3: $\lambda = \frac{h}{mv}$ and $eV = \frac{1}{2}mv^2$

$$\Rightarrow \lambda = \frac{h}{\sqrt{2meV}} \Rightarrow r = \frac{h^2}{2me\lambda^2}$$

$$= \frac{(6.6\times10^{-34})^2}{2\times9.1\times10^{-31}\times1.6\times10^{-18}\times(9\times10^{-12})^2}$$

$= 10^5 \times 0.0186$

$= 1.86 \times 10^3$ Volts.

Sol 4: According to Heisenberg's Uncertainity principle, the position and momentum of a moving particle cannot be found exactly.

But according to Bohr, electrons move in a stationary circular orbit with a fixed velocity which contradicts Heisenberg principle.

According to Bohr,

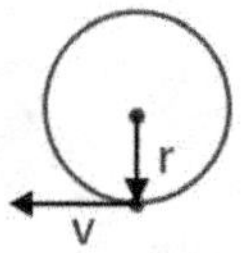

if $\bar{r}$ is fixed, we can find $\bar{v}$ exactly.

Sol 5: Let it move with a velocity v.

From conservation of energy

$\Delta KE + \Delta PE = 0$

$$\Rightarrow \frac{1}{2}mv^2 = \frac{1}{4\pi\epsilon_0}\cdot\frac{q_1q_2}{r}$$

$$\Rightarrow \frac{1}{2} \times 4 \times 1.67 \times 10^{-27} \times v^2$$

$$= \frac{9\times10^{11}\times29\times2\times(1.6\times10^{-18})^2}{10^{-13}}$$

$$\Rightarrow v = \sqrt{4012} \times 10^5 \text{ m/s} = 6.34 \times 10^6 \text{ m/s}$$

Sol 6: Velocity of electron in nth orbit of H atom is = $\frac{2.18\times10^6}{n}$ m/s

if n = 3

$v = 0.72 \times 10^6$ m/s

no. of revolutions per sec = $\frac{v}{2\pi r}$

$$= \frac{0.72\times10^6 \text{ m/s}}{2\times3.14\times0.529\times3^2\times10^{-10}} \text{ M}$$

$= 2.40 \times 10^{14}$ /sec

Sol 7: $E \propto \frac{Z^2}{n^2}$

Minimum energy of photon needed to remove electron in 3rd orbit of He+ is

$$-21.76 \times 10^{-19} \times \frac{2^2}{3^2} \text{ J}$$

$$\therefore \frac{hc}{\lambda} = +21.76 \times 10^{-19} \times \frac{4}{9} \text{ J}$$

$$\Rightarrow \lambda = \frac{6.6\times10^{-34}\times3\times10^8}{21.76\times4\times10^{-19}} \times 9 = 2.04 \times 10^{-7} \text{ m}$$

Sol 8: Bohr's theory has some failures

→ It can't explain the splitting of spectral lines in magnetic fields.

→ When the positions of electrons are recorded practically, the obtained results show that electrons don't move in a fixed and quantised orbit but they move in 3-dimensional orbitals which spread but nucleus to infinity.

But probability is high at Bohr's orbits compared to other regions

Sol 9: An atomic orbital is a mathematical function that describes the wave-like behaviours' of either one electron or a pair of electrons in an atom.

It gives us the probability of finding an electron at a certain point around the nucleus.

Each orbital in an atom is characterised by a unique set of values called quantum numbers they are

n → principle quantum number

l → Azimuthal quantum number

m → Magnetic quantum number

Sol 10: The emission spectrum of atomic hydrogen is divided into a number of spectral series with wavelength given by Rydberg's formula.

These observed spectral lines are due to electron moving between energy levels in the atom

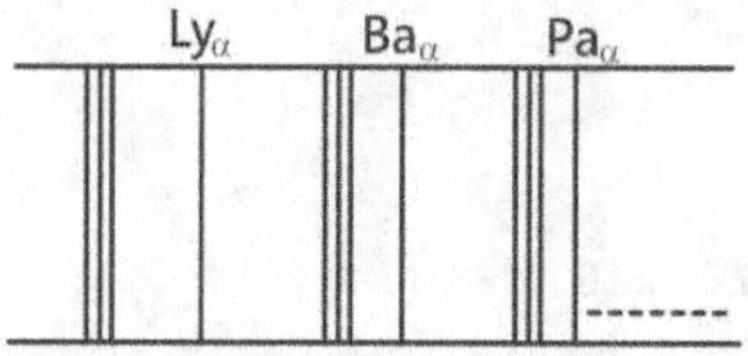

Sol 11: $\frac{1}{\lambda} = R_H Z^2 \left[\frac{1}{n_1^2} - \frac{1}{n_2^2}\right]$

$$\frac{1}{6500\,\text{Å}} = R_H(1)\left(\frac{1}{2^2} - \frac{1}{3^2}\right)$$

In Balmer, H_α - $2 \rightarrow 3$

H_β - $2 \rightarrow 4$

$$\frac{1}{\lambda'} = R_H(1)\left(\frac{1}{2^2} - \frac{1}{4^2}\right)$$

$$\Rightarrow \frac{\lambda'}{6500} = \frac{\frac{1}{4} - \frac{1}{9}}{\frac{1}{4} - \frac{1}{16}} = \frac{\frac{5}{36}}{\frac{3}{16}} = \frac{20}{27}$$

$\lambda' = 4814$ Å

Sol 12: $\frac{1}{\lambda} = R_H \times Z^2 \left[\frac{1}{n_1^2} - \frac{1}{n_2^2}\right]$

$$\text{Frequency} = \frac{C}{\lambda} = CR_H Z^2 \left[\frac{1}{n_1^2} - \frac{1}{n_2^2}\right]$$

$$= 3 \times 10^8 \times 1.09 \times 10^7 \left(1 - \frac{1}{9}\right) = 2.9 \times 10^{15} \text{ Hz}$$

Sol 13: $\Delta E = E_2 - E_1$

$$= \frac{-13.6}{2^2} - \left(\frac{-13.6}{1^2}\right) = -3.4 + 13.6 = 10.2 \text{ eV}$$

$$\text{Energy of X-ray} \rightarrow \frac{hC}{\lambda} = \frac{6.6 \times 10^{-34} \times 3 \times 10^8}{3 \times 10^{-8}}$$

$= 6.6 \times 10^{-18}$ J

$\Delta E = Z^2 \times 10.2 \text{ eV} = Z^2 \times 10.2 \times 1.6 \times 10^{-19}$ J

and $\Delta E = \frac{hc}{\lambda} \Rightarrow Z^2 \times 10.2 \times 1.6 \times 10^{-19} = 6.6 \times 10^{-18}$

$\Rightarrow Z^2 = 4 \Rightarrow Z = 2$ i.e. He

Sol 14: The shape of an atomic orbital is the locus where there is a significant probability of finding an electron of that orbital. For example, if we take an s-orbital $\rightarrow$ there is a good probability of finding electrons around a spherical surface of certain radius and p-orbital has a shape of a dumbbell.

Sol 15: For any orbital

Spherical radial nodes = $n - \ell - 1$

Angular nodes/nodal planes = ℓ

Total nodes = $n - 1$

Sol 16: Atomic number of chromium. Actual chromium configuration is $1s^2\, 2s^2\, 2p^6\, 3s^2\, 3p^6\, 3d^5\, 4s^1$ but expected is [Ar] $4s^2\, 3d^4$. The deviation can be explained by 2 reasons

1 $\rightarrow$ No. of exchange pairs. More the number of exchange pairs, more is the stabilisation energy.

2 $\rightarrow$ Spherical symmetry in both 4s and 3d orbitals.

So, it will be more stable which is not the case in $4s^2\, 3d^4$.

Sol 17: As we know

Energy of photon = Work function $\rightarrow KE_{max}$

$$\Rightarrow \frac{hc}{\lambda_1} = h\nu_0 + KE_1 \qquad \text{...(i)}$$

$$\frac{hc}{\lambda_2} = h\nu_0 + KE_2 \qquad \text{...(ii)}$$

Multiply (i) by (ii) and subtract from (ii)

$$\frac{2hc}{\lambda_1} - \frac{hc}{\lambda_2} = 2h\nu_0 - h\nu_0$$

$$\Rightarrow 3 \times 10^8 \left(\frac{2}{2.2 \times 10^{-7}} - \frac{1}{1.9 \times 10^{-7}}\right) = \nu_0$$

$$\Rightarrow \nu_0 = 3 \times 10^{15}\left(\frac{1}{1.1} - \frac{1}{1.9}\right) = 1.14 \times 10^5 \text{ Hz}$$

$$\lambda_0 = \frac{c}{\nu_0} = \frac{3 \times 10^8}{1.14 \times 10^{15}} = 2.61 \times 10^{-7} \text{ m}$$

Sol 18: By conservation of energy $E_0 = E_1 + E_2$

$$\frac{hc}{\lambda_0} = \frac{hc}{\lambda_1} + \frac{hc}{\lambda_2} \text{ where } \lambda_0 = 300 \text{ nm}$$

$\lambda_1 = 760$ nm

$$\Rightarrow \frac{1}{\lambda_0} = \frac{1}{\lambda_1} + \frac{1}{\lambda_2} \Rightarrow \frac{1}{\lambda_2} = \frac{1}{\lambda_0} - \frac{1}{\lambda_1}$$

$$\Rightarrow \lambda_2 = \frac{\lambda_0 \lambda_1}{\lambda_1 - \lambda_0} = \frac{300(760)}{760 - 300} = 495.6 \text{ nm}$$

Sol 19: By De-Broglie hypothesis

$$\lambda = \frac{h}{mv} = \frac{6.6\times10^{-34}}{0.15\times50} = 8.8 \times 10^{-35}\text{ m}$$

It is extremely low even compared to size of atoms. So, it's not observable.

Sol 20: According to conservation of energy $\frac{hc}{\lambda} = \omega_0 + KE_{max}$

when $\nu_1 = 1.6 \times 10^{16}$ Hz

$h\nu_1 = \omega_0 + KE_1$ .....(i)

when $\nu_2 = 10^{16}$ Hz and $KE_1 = 2KE_2$

$h\nu_2 = \omega_0 + KE_2$ .....(ii)

Multiply (ii) by 2 subtract (i) from it.

$2h\nu_2 - h\nu_1 = 2\omega_0 + 2KE_2 - (\omega_0 + KE_1) = \omega_0$

and $\omega_0 = h\nu_0$

$\Rightarrow 2h\nu_2 - h\nu_1 = h\nu_0$

$\Rightarrow \nu_0 = 2\nu_2 - 2\nu_1$

$= 2 \times 10^{16} - 1.6 \times 10^{16}$

$= 0.4 \times 10^{16}$

$= 4 \times 10^{15}$ Hz

Sol 21: The magnetic moment of an ion is

$M = \sqrt{n(n+2)}$ B.M. where n is the number of unpaired electron in the ion.

$\sqrt{n(n+2)} = \sqrt{35}$

$\Rightarrow n(n + 2) = 35 = 5 \times 7$

$\Rightarrow n = 5$

So, X^{3+} has 5 unpaired electron in 3d series.

□ |1|1|1|1|1|

$\therefore$ Atomic number = 18 + 5 + 3= 26

Sol 22: Bond energy of $I_2 \rightarrow$ 240 kJ/mole for 1 molecule

$$= \frac{240\times10^{3}}{6\times10^{23}} = 4 \times 10^{-19}\text{ J}$$

Energy of a photon $= \frac{hc}{\lambda}$

$$= \frac{6.66\times10^{-34}\times3\times10^{8}}{4.5\times10^{-7}} = 4.44 \times 10^{-19}\text{ J}$$

$\therefore$ KE of 2 Iodine atoms

$= 4.44 \times 10^{-19}\text{ J} - 4 \times 10^{-19}\text{ J}$

$= 0.44 \times 10^{-19}$ J

$\therefore$ KE of a single atom $= \frac{4.4\times10^{-20}}{2} = 2.2 \times 10^{-20}$ J

Sol 23: Energy of the incident photon, $E = \frac{hc}{\lambda}$

$$= \frac{6.6\times10^{-34}\times3\times10^{8}}{2.53\times10^{-7}} = 7.82 \times 10^{-19}\text{ J}$$

$$= \frac{7.82\times10^{-19}}{1.6\times10^{-19}}\text{ eV} = 4.89\text{ eV}$$

$\therefore$ Work function $= \frac{hc}{\lambda} - KE_{max}$

$= 4.89 - 0.24 = 4.65$ eV

Sol 24: 1 g atm $\rightarrow$ 1 N atoms

Lowest energy of visible region $\rightarrow$ Lowest energy line of Balmer series i.e. 3 $\rightarrow$ 2

$$E = E_3 - E_2 = \frac{-13.6}{n_1^2} - \left(\frac{-13.6}{n_2^2}\right)$$

$$\Rightarrow 13.6\left(\frac{1}{2^2} - \frac{1}{3^2}\right) = 13.6\left(\frac{5}{36}\right) = 1.88\text{ eV}$$

Total energy $= 6 \times 10^{23} \times 1.88$ eV

$= 6 \times 10^{23} \times 1.88 \times 1.6 \times 10^{-19}$ eV

$= 18.2 \times 10^{4}$ J

= 182 kJ per 1.0 gram atom

Sol 25: For X

$\lambda = 9.87$ Å. Moseley's law $= \sqrt{\nu} = k(Z - \alpha)$

$$\nu = \frac{c}{\lambda} = \frac{3\times10^{8}}{9.87\times10^{-10}}$$

$= 3.03 \times 10^{17}$

$\sqrt{\nu} = \sqrt{30.3\times10^{16}}$

$= 5.51 \times 10^{8}$

$\Rightarrow 5.51 \times 10^{8} = 4.9 \times 10^{7}(2 - 0.75)$

$\Rightarrow Z - 0.75 = 11.245$

$\Rightarrow Z \approx 12.$

For Y

$\lambda = 2.29$ Å

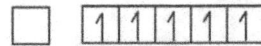

$$v = \frac{c}{\lambda} = \frac{3\times10^8}{2.29\times10^{-10}}$$

$= 1.31 \times 10^{18}$

$\sqrt{v} = 1.14 \times 10^9$

$\Rightarrow 1.14 \times 10^9 = 4.9 \times 10^7 (2 - 0.75)$

$\Rightarrow 2 - 0.75 = 23.35$

$\Rightarrow Z \approx 24$

Sol 26: $\frac{1}{\lambda} = R\left[\frac{1}{n_1^2} - \frac{1}{n_2^2}\right]$ and given that, its Balmer series so $n_1 = 2$

$$\Rightarrow \frac{1}{\lambda} = R\left[\frac{1}{2^2} - \frac{1}{n_2^2}\right]$$

$$\Rightarrow \frac{1}{\lambda} \times \frac{1}{R} = \frac{1}{2^2} - \frac{1}{n_2^2}$$

$$\Rightarrow \frac{912\text{Å}}{4344\text{Å}} = \frac{1}{2^2} - \frac{1}{n_2^2}$$

$$\Rightarrow 0.209 = \frac{1}{4} - \frac{1}{n_2^2}$$

$$\Rightarrow \frac{1}{n_2^2} = 0.0400$$

$\Rightarrow n_2^2 = 25$

$\Rightarrow n_2 = 5$

Sol 27: Change in energy of electron = Energy of photon

$E = E_3 - E_2$

$= -2.41 \times 10^{-12} - (-5.42 \times 10^{-12})$

$= 3.01 \times 10^{-12}$ erg

$= 3.01 \times 10^{-19}$ J

$$\lambda = \frac{hc}{E} = \frac{6.6\times10^{-34}\times3\times10^8}{3.01\times10^{-19}}$$

$\approx 6.6 \times 10^{-7}$ m

$= 6.6 \times 10^3$ Å

Sol 28:

$r = r_0.A^{1/3} = 1.4\times10^{-13}.A^{1/3}\text{cm}$

$= 1.4\times10^{-13}\times19^{1/3} = 5.07\times10^{-13}\text{cm}$

Vol of F. Nucleus $= \frac{4}{3}\pi r^3 = 7.17\times10^{-37}\text{cm}^3$

$$\text{Density} = \frac{\text{Mass}}{\text{Volume}} = \frac{19}{6.02\times10^{23}\times7.17\times10^{-37}}$$

$$= 0.44\times10^{14}\text{gm cm}^{-3}$$

Sol 29: From Heisenberg's Uncertainty principle

$$\Delta x.\Delta p \geq \frac{\hbar}{2}$$

$\Rightarrow x.m\,\Delta V \geq \frac{h}{4\pi}$ $\Delta V = 2 \times 0.01$ m/s

$$\therefore \Delta x = \frac{6.6\times10^{-34}}{4\times3.14\times10^{-3}\times10^{-2}\times2} = \frac{6.6\times10^{-29}}{12.56\times2}$$

$\approx 3 \times 10^{-30}$ m

Though, there is some uncertainty in the position, it is extremely negligible, i.e. we cannot observe such deviations.

Sol 30: By Heisenberg's uncertainty principle and $\Delta V = 2$ m/s

$$\Delta x.\Delta p \geq \frac{\hbar}{2}$$

$\Rightarrow \Delta x \times 9.109 \times 10^{-31} \times 2 = \frac{6.6\times10^{-34}}{4\times3.14}$

$\Rightarrow \Delta x = 0.028 \times 10^{-3}$

$\approx 2.8 \times 10^{-5}$ m

Here, the uncertainty in position is quite high compared to size of an atom.

$\therefore$ From the above 2 problems, we can see that uncertainty in position for macroparticles is negligible, but there, is high uncertainty in case of micro particles.

Exercise 2

Single Correct Choice Type

Sol 1: (C) Bohr's model says that electron move in a stationary state with fixed velocities. So, it contradicts uncertainty principle.

Sol 2: (B) Statement-II seems to be irrelevant.

Comprehension Type

Paragraph 1:

Sol 3: (A, B) $E = 13.6\left[\frac{1}{n_1^2} - \frac{1}{n_2^2}\right]$ eV

$$= 13.6\left[\frac{1}{1^2}-\frac{1}{3^2}\right] \times 1.6 \times 10^{-19}\ J$$

$$\approx 13.6 \times \frac{8}{9} \times 1.6 \times 10^{-19}\ J$$

$$= 1.93 \times 10^{-18}\ J$$

Sol 4: (C) No. of lines = $\frac{(\Delta n)(\Delta n+1)}{2}$ where $\Delta n = n_1 - n_2$.

In this case

$n_1 = 6, n_2 = 3$

$\therefore$ lines = $\frac{3(4)}{2} = 6.$

Sol 5: (B) 1st line in Lyman

$$\frac{1}{\lambda} = R \times 1^2 \times \left[\frac{1}{1^2}-\frac{1}{2^2}\right] \Rightarrow \lambda = \frac{4}{3R}$$

2nd line in Balmer

$$\frac{1}{\lambda} = R \times 1^2\left[\frac{1}{2^2}-\frac{1}{4^2}\right] \Rightarrow \lambda = \frac{16}{3R}$$

$$\lambda = \frac{12}{3R} = \frac{4}{R}$$

Sol 6: (C) Wave number = $\frac{1}{\lambda}$

$n_1 + n_2 = 4$

$n_2 - n_1 = 2$

$\Rightarrow n_1 = 1$

$n_2 = 3$

$$\frac{1}{\lambda} = R \times Z^2\left[\frac{1}{1^2}-\frac{1}{3^2}\right] \text{ and } Z = 3$$

$$= R(9)\left(\frac{8}{9}\right) = 8R$$

Paragraph 2:

Sol 7: (B) $\Delta KE = \Delta PE$

$$\Rightarrow \frac{1}{2}mv^2 = \frac{1}{4\pi\epsilon_0}\frac{q_1q_2}{r}$$

$$\Rightarrow \frac{1}{2} \times 4 \times 1.6 \times 10^{-27} \times v^2$$

$$= \frac{9\times10^{11}\times29\times2\times(1.6\times10^{-19})^2}{10^{-13}}$$

$$\Rightarrow v = \sqrt{4012} \times 10^5\ m/s$$

$$= 6.33 \times 10^6\ m/s$$

$$= 6.33 \times 10^8\ cm/sec$$

Sol 8: (A) Na, K, Cs can't be used as they are 1A group elements, they have only $1e^-$ in outer shell. So, they have less density and IE. So, Pt is suitable.

Sol 9: (C) This question is out of scope.

Though the value of N after several calculations comes out as

$$N(\theta) = \frac{nt}{4r^2}\left(\frac{2Z}{2k}\right)^2\left(\frac{e^2}{4\pi\epsilon_0}\right)^2\frac{1}{\sin^4\theta/2}$$

$\therefore N \propto \frac{1}{\sin^4\theta/2}$ (need not be included in the syllabus)

Sol 10: (C) In Rutherford's α-ray scattering experiment, most of the α-particles doesn't undergo any deflection. Few particles deviate through some angle and negligible no. of α-particles deflect for more than 90°.

Match the Columns

Sol 11: A → s; B → r; C → q; D → p

(A) 2nd orbit in He^+ $\rightarrow r = 0.529 \times \frac{2^2}{2}$ Å

$$\rightarrow E = -13.6 \times \frac{2^2}{2^2}\ eV$$

$$\lambda = \frac{h}{mv} = \frac{h}{\sqrt{2mE}}$$

$$= \frac{6.6\times10^{-34}}{\sqrt{2\times9.1\times10^{-31}\times13.6\times1.6\times10^{-19}}} = \sqrt{\frac{150}{13.6}}\ Å$$

(B) 3rd orbit in H-atom, $V \propto \frac{Z}{n}$

$$V = 2.18 \times \frac{Z}{n} = \frac{2.18}{3}\ m/s$$

(C) 1st orbit Li^{2+} ion $\rightarrow E = -13.6 \times \frac{Z^2}{n^2} = (-13.6)\ 9\ eV$

(D) 2nd orbit, Be^{3+} ion $\rightarrow r = 0.529 \times \frac{n^2}{Z}$

$$= 0.529 \times \frac{2^2}{Z} = 0.529\ eV$$

Sol 12: A → p; B → p, q, s; C → p, r; D → q, s

(A) 4s, 5p, 6d, ... show similar ψ

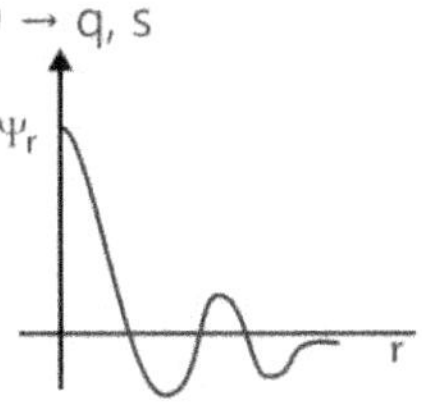

4s [since 3 maxima/minima locally]

$5p_x$, $6d_{xy}$ are not symmetric for all values of θ and ϕ.

4s, 5p, 6d, ...

(B) $\psi^2_r 4\pi r^2$

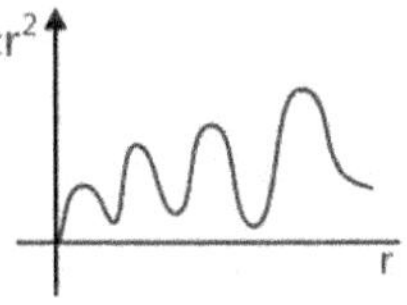

(C) $\psi(\theta, \phi)$ = k (i.e. independent of θ, ϕ)

only for circularly symmetric orbitals.

4s, 3s

(D) At one angular nodes for any orbital, no. of angular nodes = ℓ.

$5p_x$, $6d_{xy}$

Assertion Reasoning Type

Sol 13: (D) In Li^{2+}, if n_1 = 2, and n_2 > 2, then the line may not be in visible region, as energy of that line is 9 times the corresponding line H atom (which is in visible region). So, Statement-I is false.

Previous Years' Questions

Sol 1: (D) Neutron has no charge, hence, e/m is zero for neutron. Next, α-particle (He^{2+}) has very high mass compared to proton and electron, therefore very small e/m ratio. Proton and electron have same charge (magnitude) but former is heavier, hence has smaller value of e/m.

e/m : n < α < p < e

Sol 2: (C) X-rays is electrically neutral, not deflected in electric or magnetic fields.

yz-plane, a nodal plane

Sol 3: (A) Nodal plane is an imaginary plane on which probability of finding an electron is minimum. Every p-orbital has one nodal plane:

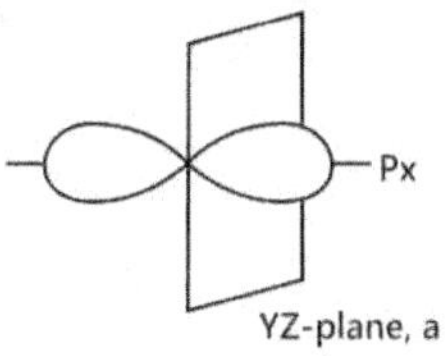

YZ-plane, a nodal plane

Sol 4: (C) $1s^7$ violate Pauli exclusion principle, according to which an orbital cannot have more than two electrons.

Sol 5: (B) Expression for Bohr's orbit is $r_n = \frac{a_0 n^2}{Z} = a_0$

When n = 2, Z = 4.

Sol 6: (A) The number of radial nodes is given by expression ($n - l - 1$)

For 3s, number of nodes = 3 − 0 − 1 = 2

For 2p, number of nodes = 2 − 1 − 1 = 0

Sol 7: (A, C) Alpha particles passes mostly undeflected when sent through thin metal foil mainly because

(A) It is much heavier than electrons.

(C) Most part of atom is empty space.

Sol 8: (A, D) Then these electrons in the 2p orbital must have same spin, no matter up spin or down spin.

Sol 9 (C) Statement-I is correct, Be ($1s^2$, $2s^2$) has stable electronic configuration, removing an electron requires more energy than the same for B ($2p^1$). Reason is incorrect (Aufbau principle)

Sol 10.1: (B) S_1 is spherically symmetrical state, i.e., it corresponds to an s-orbital. Also, it has one radial node.

Number of radial nodes = $n - l - 1$

$\Rightarrow n - 0 - 1 = 1$

(i) n = 2, i.e., S_1 = 2s orbital

Sol 10.2: (C) (ii) Ground state energy of electron in H-atom (E_H)

$$E_H = k\frac{Z^2}{n^2} = k(Z = 1, n = 1)$$

For S_1 state of Li^{2+}.

$$E = \frac{k(3)^2}{2^2} = \frac{9}{4}k = 2.25k$$

Sol 10.3: (B) (iii) In S_2 state, E(Li^{2+}) = k (given)

$$\Rightarrow k = \frac{9k}{n^2}$$

$\Rightarrow n = 3$

Since, S_2 has one radial node.

$3 - l - 1 = 1$

$\Rightarrow l = 1$

Sol 11: A → r; B → q; C → p; D → s

(A) $V_n = \frac{1}{4\pi \epsilon_0}\left(\frac{Ze^2}{r}\right)$

$K_n = \frac{1}{8\pi \epsilon_0}\left(\frac{Ze^2}{r}\right) \Rightarrow \frac{V_n}{K_n} = -2(R)$

(B) $E_n = -\frac{Ze^2}{8\pi \epsilon_0 r} \propto r^{-1}$

$\Rightarrow x = -1(Q)$

(C) Angular momentum = $\sqrt{l(l+1)}\frac{h}{2\pi}$ = 0 in 1s orbital.

(D) $r_n = \frac{a_0 n^2}{Z} \Rightarrow \frac{1}{r_n} \propto Z(S)$

Sol 12: A → q; B → p, q, r, s; C → p, q, r; D → p, q, r

(A) Orbital angular momentum

(L) = $\sqrt{l(l+1)}\frac{n}{2\pi}$ i.e,L depends on azimuthal quantum number only.

(B) To describe a one electron wave function, the quantum numbers n, l and m are needed. Further to abide by Pauli exclusion principle, spin quantum number (s) is also needed.

(C) For shape, size and orientation, only n, l and m are needed.

(D) Probability density (ψ^2) can be determined if n, l and m are known.

Sol 13: When n = 3, l = 0, 1, 2, i.e., there are 3s, 2p and 3d orbital. If all these orbitals are completely occupied as

[↑↓] [↑↓|↑↓|↑↓] [↑↓|↑↓|↑↓|↑↓|↑↓]

Total 18 electrons, 9 electrons with s = + $\frac{1}{2}$ and 9 with s = $-\frac{1}{2}$

Alternatively: In any nth orbit, there can be a maximum of $2n^2$ electrons. Hence, when n = 3, number of maximum

Electrons – 18. Out of these 18 electrons, 9 can have spin –1 /2 and remaining nine with spin = + 1/2

Sol 14: Energy of photon

$\frac{hc}{\lambda}J = \frac{hc}{e\lambda}eV = \frac{6.625\times10^{-34}\times3\times10^{8}}{300\times10^{-9}\times1.602\times10^{-19}} = 4.14eV$

For photoelectric effect to occur, energy of incident photons must be greater than work function of metal. Hence, only Li, Na, K and Mg have work functions less than 4.14V.

Sol 15: At radial node, ψ_2 must vanish i.e.

$\psi_{2r}^2 = 0 = \left[\frac{1}{4\sqrt{2\pi}}\right]^2\left(2-\frac{r_0}{a_0}\right)^2 e^{-\frac{r_0}{a_0}}$

$\Rightarrow 2-\frac{r_0}{a_0} = 0$

$\Rightarrow r_0 = 2a_0$

(b) $\frac{h}{mv} = \frac{6.625\times10^{-34}}{100\times10^{-3}\times100} = 6.625\times10^{-35}m$

= 6.625×10^{-25} Å (negligibly small)

Sol 16: $mvr = \frac{nh}{2\pi}$

$v = \frac{nh}{2\pi mr} = \frac{6.625\times10^{-34}}{2\times3.14\times9.1\times10^{-31}\times0.529\times10^{-10}}$

= 2.18×10^6 ms^{-1}.

(b) $\lambda = \frac{h}{mv} = \frac{6.625\times10^{-34}}{9.1\times10^{-31}\times2.18\times10^{6}}$

= 0.33×10^{-9} m = 3.3 Å

(c) Orbital angular momentum (L) = $\sqrt{l(l+1)}\frac{h}{2\pi}$

$= \sqrt{2}\left(\frac{h}{2\pi}\right)$ [∵ for p-orbital, l = 1]

Sol 17: Since, $\lambda = \frac{h}{mV} = \frac{h}{\sqrt{2M\,K.E.}}$ (since KE. ∝ T)

$\Rightarrow \lambda \propto \frac{1}{\sqrt{MT}}$

For two gases,

$\frac{\lambda_{He}}{\lambda_{Ne}} = \sqrt{\frac{M_{Ne}}{M_{He}}\frac{T_{Ne}}{T_{He}}} = \sqrt{\frac{20}{4}\times\frac{1000}{200}}$

$= \sqrt{25} = 5$

Sol 18: Single electron species don't follow the $(n+l)$ rule but multi electron species do.

Ground state of $H^- = 1s^2$

First excited state of $H^- = 1s^1, 2s^1$

Second excited state of $H^- = 1s^1, 2s^0, 2p^1$

Sol 19: (C) As per Bohr's postulate,

$$mvr = \frac{nh}{2\pi}$$

$$\text{So, } v = \frac{nh}{2\pi mr}$$

$$KE = \frac{1}{2}mv^2$$

$$\text{So, } KE = \frac{1}{2}m\left(\frac{nh}{2\pi mr}\right)^2$$

$$\text{Since, } r = \frac{a_0 \times n^2}{z}$$

So, for 2nd Bohr orbit

$$r = \frac{a_0 \times 2^2}{1} = 4a_0$$

$$KE = \frac{1}{2}m\left(\frac{2^2h^2}{4\pi^2m^2 \times (4a_0)^2}\right)$$

$$KE = \frac{h^2}{32\pi^2 ma_0^2}$$

3. PERIODIC TABLE AND PERIODICITY

1. INTRODUCTION

You must have visited a library. There are thousands of books in a large library. In spite of this, if you ask for a particular book, the library staff can locate it easily. How is it possible? In a library, the books are classified into various categories and subcategories. They are arranged on the shelves accordingly. Therefore, locating the books becomes easy. Same is the story with chemical elements. A large number of elements and compounds are known today. But a systematic classification of these elements has made their study possible and easy. The well organized and tabulated classification of elements, as we know it today, is called the **Periodic Table**. It not only helps to locate, identify and characterize the element and its properties but also points out the directions in which new investigations are made.

2. GENESIS OF PERIODIC CLASSIFICATION

In 18th century, the number of elements was limited. In 19th century, scientists began to seek ways to classify elements because of their rapidly increasing number. They started recognizing patterns in properties and began to develop classification schemes. Some such early attempts of classification are described below.

2.1 Prout's Hypothesis

The atomic weights of all elements are simple multiples of atomic weight of hydrogen. Prout gave this hypothesis on the basis of Dalton's atomic theory and the atomic weights of some elements known at that time. But this hypothesis could not last longer, because there are some atomic weights which are fractional not in whole number.

2.2 Dobereiner's Triads

It was first attempt towards classification of elements. He arranged similar elements in groups of three elements called triad and the atomic mass of the middle elements of the triad is approximately the arithmetic mean of the other two.

Table 13.1: Example of dobereiner's triad

Triad				Mean of at. Mass of (I) and (III) element
(i)	Li^7	Na^{23}	K^{39}	$\frac{7+39}{2}=23.00$
(ii)	$Cl^{35.5}$	Br^{80}	I^{127}	$\frac{35.5+127}{2}=81.25$
(iii)	P^{31}	As^{75}	Sb^{120}	$\frac{31+120}{2}=75.50$
(iv)	S^{32}	Se^{79}	Te^{127}	$\frac{127+32}{2}=79.50$
(v)	Ca^{40}	Sr^{88}	Ba^{137}	$\frac{40+137}{2}=88.50$

Merits: After Dobereiner, Chemists focused on chemicals in groups having similar physical and chemical properties.

Demerits: All the known elements did not follow this rule .Law of triads was rejected as some triads nearly had same atomic masses, e.g., (Fe, Co, Ni),(Ru, Rh, Pd),(Os, Ir, Pt)

2.3 Newland's Rule of Octaves

When the lighter elements are arranged in order of their increasing atomic weights, then every eighth element is similar to the first element in its properties, similarly as the eighth note of a musical scale is similar to the 1st one. e.g. Na, 8th element resembles in their properties with Li. Similarly K, the 8th element with Na, and so on.

	do	**re**	**mi**	**pha**	**sol**	**la**	**si**
Symbol of element	Li	Be	B	C	N	O	F
	7	9	11	12	14	16	19
Symbol of element	Na	Mg	Al	Si	P	S	Cl
	23	24	27	28	31	32	35.5

It is clear from the above table that sodium is the eighth element from lithium, whose properties resemble that of lithium.

This type of classification was limited up to only 20 elements.

Demerits

(i) Law of octave worked quite well for lighter elements but failed with heavier elements.

(ii) Properties of elements were not taken into account and the elements were arranged in the order of their increasing atomic masses.

(iii) No places were left for unknown elements and so, many elements occupied wrong positions. Thus, resulted in the rejection of the attempt.

2.4 Lothar Meyer's Volume Curve

The graphs of atomic volumes against weights are known as Lothar Meyer's volume curves.

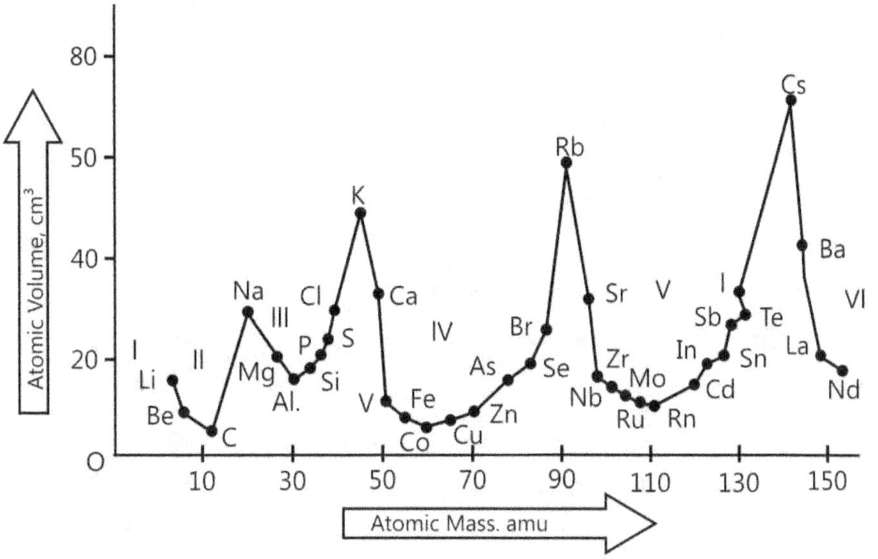

Figure 13.1: Lothar Meyer's volume curve

Features of curve

(a) Alkali metals having the largest atomic volumes occupy the maxima of the curve.

(b) Alkaline earth metals lie at about the mid points on the descending portions of the curve.

(c) The halogens occupy positions on ascending portions of the curve before inert gases.

(d) The transition metals occupy the minima of the curve.

Demerits: It lacked practical utility as it is not easy to remember the position of different elements on the curve.

2.5 Mendeleev's Periodic Law

According to Mendeleev's periodic law, the physical and chemical properties of elements are periodic functions of their atomic weights.

Merits of Mendeleev's periodic table

(a) Study of elements and their compounds becomes easy and systematic, as by knowing the property of one element in a group, then the properties of the other elements present in the same group can easily be predicted.

(b) Helps in the discovery of new elements. As Mendeleev left some blank spaces for some unknown elements and further, predicted the properties of these elements e.g. ekaluminium, ekasilicon.

(c) Correction of doubtful atomic mass.

(d) Correction in the valency of some elements.

(e) Correction in the position of some elements.

(f) Classification of elements then known, was done for the first time and the elements having similar properties were kept in the same group.

(g) It encouraged research and led to discovery of newer elements.

(h) Mendeleev had even predicted the properties of many elements not discovered at that time. This helped in the discovery of these elements.

For example: Mendeleev predicted the properties of the following elements.

(i) Eka-boron - This was later called scandium (Sc)

(ii) Eka- aluminium - This was later called gallium (Ga)

(iii) Eka-silicon - This was later called germanium (Ge)

Limitations of Mendeleev's periodic table

(a) The position of hydrogen was found to be anomalous due to its resemblance with the 1st group alkali metals and also with the 7th group halogens in their properties.

(b) Position of isotopes: Isotopes must have different positions but they were placed in the same group.

(c) Position of isobars: They were placed in different groups.

(d) Dissimilar elements were placed together in the same group like K and Cu in 1st group.

(e) Similar elements were placed in different groups.

(f) Some higher atomic weight elements were placed before the lower atomic weight elements

(g) e.g. Ar^{40} precedes K^{39}, $Co^{58.9}$ precedes $Ni^{58.7}$, $Te^{127.6}$ precedes I^{127}.

(h) Position of metals and non-metals: Both were placed together in the same group.

(i) Diagonal relationship could not be explained.

(j) Position of lanthanides and actinides was not properly specified.

(k) No proper position to VIII group elements.

(l) There was no indication whether lanthanides and actinides were associated with group IIIA or group IIIB.

(m) Position of Isobars- These elements had different groups when mass remained the same.

(n) Lot of stress was given to the valence of elements.

2.6 Modern Periodic Law and Modern Periodic Table

Mosley: Proved that the square root of frequency (f) of the rays, which are obtained from a metal on showering high velocity electrons is proportional to the nuclear charge of the atom.

This can be represented by the following expression.

$\sqrt{f} = a(Z-b)$ where Z is nuclear charge on the atom and a and b are constants.

The nuclear charge on an atom is equal to the atomic number.

Modern Periodic Table

According to modern periodic law, "The properties of elements are the periodic function of their atomic numbers".

Period-The details about the seven periods are as follows:-

Period	**Atomic number**		**Number of elements**
	From	**to**	
First	H (1)	He (2)	2
Second	Li (3)	Ne (10)	8
Third	Na (11)	Ar (18)	8
Fourth	K (19)	Kr (36)	18
Fifth	Rb (37)	Xe (54)	18
Sixth	Cs (55)	Rn (86)	32 (including lanthanides)
Seventh	Fr (87)	Ha (105)	19 (including actinides)

Group: The modern periodic table has 18 vertical columns and according to CAS system there are 16 groups having the following number of elements.

Group		**Number of Elements**
(a)	I A group	7 (H, Li, Na, K Rb, Cs, Fr) Alkali metals
(b)	II A group	6 (Be, Mg, Ca, Sr, Ba, Ra) Alkaline earth metals
(c)	III A group	5 (B, Al, Ga, In, Tl) Boron family
(d)	IV A group	5 (C, Si, Ge, Sn, Pb) Carbon family
(e)	V A group	5 (N, P, As, Sb, Bi) Nitrogen family
(f)	VI A group	5 (O, S, Se, Te, Po) Oxygen family (Chalcogen)
(g)	VII A group	5 (F, Cl, Br, I, At) Halogen family
(h)	Zero group	6 (He, Ne, Ar, Kr, Xe, Rn) Inert elements
		32 (Sc, Y, La, Ac & 14 lanthanide elements & 14 actinide elements.)
(i)	III B group	These are elements of IIIB group, which could not be accommodated in one column and therefore written separately outside the periodic table.

(j)	IV B group	4 (Ti, Zr, Hf, Rf)
(k)	V B group	4 (V, Nb, Ta, Db)
(l)	VI B group	3 (Cr, Mo, W)
(m)	VII B group	3 (Mn, Tc, Re)
(n)	VIII (3) group	9 (Fe, Co, Ni, Ru, Rh, Pd, Os, Ir, Pt)
(o)	I B group	3 (Cu, Ag, Au)
(p)	II B group	3 (Zn, Cd, Hg)

Advantages of the Long Form of the Periodic Table

(a) The table is based on a more fundamental property i.e. atomic number.

(b) It correlates the position of elements with their electronic configuration more clearly.

(c) The completion of each period is more logical. In a period, as the atomic number increases, the energy shells are gradually filled up until an inert gas configuration is reached.

(d) It eliminates the even and odd series of IV, V and VI periods of Mendeleev's periodic table.

(e) The position of VIII group is also justified in this table. All the transition elements have been brought to the middle as the properties of transition elements are intermediate between s-and p-block elements.

(f) Due to the separation of two sub-groups, dissimilar elements do not fall together. One vertical column accommodates elements with same electronic configuration thereby showing same properties.

(g) The table completely separates metals and non-metals. Non-metals are present in upper right corner of the periodic table.

(h) There is a gradual change in properties of the elements with increase in their atomic numbers i.e., periodicity of properties can be easily visualized. The same properties occur after the intervals of 2, 8, 8, 18, 18 and 32 elements which indicates the capacity of various periods of the table.

(i) The greatest advantage of this periodic table is that this can be divided into four blocks namely s-, p-, d- and *f*-block elements.

(j) This arrangement of elements is easy to remember and reproduce.

Defects of the Long Form of the Periodic Table

(a) The position of hydrogen is still disputable as it was there in MENDELEEV periodic table in group IA as well as IVA & VIIA.

(b) Helium is an inert gas but its configuration is different from that of the other inert gas elements

(c) Lanthanide and actinide series could not be adjusted in the main periodic table and therefore they had to be provided with a place separately below the table.

To Locate Group and Period if Atomic Number is given

Locate period: Write electronic configuration of each element for which the atomic number is given. The number of outermost shell suggests the period to which it belongs in the periodic table.

Locate it's Block and Group

Group can be located after knowing the block of an element as follows:

For s-block elements,

gp. no. = No. of s-electrons in valence shell

For p-block elements,

gp. no. = No. of s-electrons + p-electrons in valence shell + 10

For d-block elements,

gp. no. = No. of (n – 1)d + ns electrons

Illustration 1: How many elements from the following atomic number are p-block elements?

83, 79, 42, 64, 37, 54 34

Use the following data for predicting answer – (Atomic Number of noble gases are given in the bracket) **(JEE MAIN)**

He[2] Ne[10] Ar[18] Kr[36]

Xe[54] Rn[86]

Sol: 83 – *p*-block $Xe^{54}\ 6s^2\ 5d^{10}\ 4f^{14}\ 6p^3$

79 – d-block $Xe^{54}\ 6s^1\ 4f^{14}\ 5d^{10}$

42 – d-block $Kr^{36}\ 5s^1\ 4d^5$

64 – f-block $Xe^{54}\ 6s^2\ 5d^1\ 4f^7$

37 – s-block $Kr^{36}\ 5s^1$

54 – p-block Xe

34 – p-block $Ar^{18}\ 4s^2\ 3d^{10}\ 4p^4$

Illustration 2: (a) Write the electronic configuration of the elements given below:

A (At. No. = 9), B (At. No. = 12), C (At. No. = 29), D (At. No. = 54) and E (At. No. = 58)

(b) Also predict the period, group number and block to which they belong. **(JEE MAIN)**

Sol: (a) Electronic configuration of the element A, B, C, D and E are as follows:

Element	**At. No.**	**Electronic configuration**
A	9	$1s^2\ 2s^2\ 2p^5$
B	12	$1s^2\ 2s^2\ 2p^6\ 3s^2$
C	29	$1s^2\ 2s^2\ 2p^6\ 3s^2\ 3p^6\ 3d^{10}\ 4s^1$
D	54	$1s^2\ 2s^2\ 2p^6\ 3s^2\ 3p^6\ 3d^{10}\ 4s^2\ 4p^6\ 4d^{10}\ 5s^2\ 5p^6$
E	58	$1s^2\ 2s^2\ 2p^6\ 3s^2\ 3p^6\ 3d^{10}\ 4s^2\ 4p^6\ 4d^{10}\ 5s^2\ 5p^6\ 6s^2\ 5d^1\ 4f^1$

(b) **Element A:** Receives the last electron in 2p-orbital, therefore, it belongs to **p-block**.

Group number = 10+ No. of electrons in the valence shell = 10 + 7 = **17**.

Period =Principal quantum number of the valence shell = **2nd**.

Element B: Receives the last electron in 3s-orbital, thus, it belongs to **s-block**.

Group number = No. of electrons in the valence shell=**2**

Period = Principal quantum number of the valence shell = **3rd**

Element C: Receives the last electron in the 3d-orbital, thus, it belongs to **d-block**.

Group number = No. of electrons (penultimate shell +valence shell) = 10 + 1 = **11**

Period = Principal quantum number of the valence shell = **4th**

Element D: Receives its last electron in the 5p-orbital, thus, it belongs to **p-block.**

Group number = 10 + No. of electrons in the valence shell = 10 + 8 = **18**

Period of the element =Principal quantum number of the valence shell = **5th**

Element E: Receives its last electron in the 4f-orbital, thus, it belongs to **f-block**

It may be noted here, that, the filling of 4f-orbital occurs only when one electron has already entered the 5d-orbital. Therefore, element E belongs **to f-block** and not to **d-block**. Since it belongs to lanthanide series, there is no such group number of its own but is usually considered to lie in **group 3**.

Period = Principal quantum number of the valence shell = **6th**.

Illustration 3: Elements A, B, C, D and E have the following electronic configurations.

A: $1s^2\ 2s^2\ 2p^1$ **B: $1s^2\ 2s^2\ 2p^6\ 3s^2\ 3p^1$** **C: $1\ s^2\ 2s^2\ 2p^6\ 3s^2\ 3p^3$**

D: $1s^2\ 2s^2\ 2p^6\ 3s^2\ 3p^5$ **E: $1s^2\ 2s^2\ 2p^6\ 3s^2\ 3p^6\ 4s^2$**

Which among these will belong to the same group in the periodic table? **(JEE ADVANCED)**

Sol: We know that elements having similar valence electronic configuration belong to the same group of the periodic table. Therefore, elements **A** and **B** having three electrons in the valence shell, i.e. $2s^2\ 2p^1$ and $3s^2\ 3p^1$ respectively belong to the same group, i.e., group 13 of the periodic table.

3. NAMING OF ELEMENTS HAVING ATOMIC NUMBER GREATER THAN 100

(a) The name is derived directly from the atomic number of the elements using the following numerical roots:

Table 13.2: Naming of elements having atomic number greater than 100

Digit	Name	Abbreviation
0	nil	n
1	un	u
2	bi	b
3	tri	t
4	quad	q
5	pent	p
6	hex	h
7	sept	s
8	oct	o
9	oct	e

(b) The roots are put together in the order of the digits which make up the atomic number and are terminated by 'ium' to spell out the name. The final 'n' of 'enn' is removed when it occurs before 'nil' and the final 'i' of 'bi' and of 'tri' when it occurs before 'ium'

(c) The symbol of the element is composed of the initial letters of the numerical roots which make up the name.

Illustration 4: Eka-aluminium and eka-silicon were the names given by Mendeleev for the unknown elements gallium and germanium respectively. A recently discovered element was first named as eka-mercury. What is its atomic number? Write its group number, electronic configuration, IUPAC name. **(JEE MAIN)**

Sol: The element which comes after mercury in the periodic table is called eka-mercury. Its various parameters are:

Z = 80 + 32 = 112 IUPAC name: Uub

Official name Cn(copernicium) E.C. = [Rn] $5f^{14}\ 8d^{10}\ 7s^2$

4. CLASSIFICATION OF PERIODIC TABLE BASED ON BLOCKS

s-block Elements: Elements of groups 1 and 2 including He in which the last electron enters the s-orbitals of the valence shell are called s-block elements. There are only **14 s-block elements** in the periodic table.

Characteristics:

(a) The electronic configuration of outermost shell of s-block elements is ns^1 (alkali metals; group1) or ns^2 (alkaline earth metals; group 2)

(b) The valence of group I elements is +1 and those of group II elements is +2.

(c) These are soft metals having low melting points and boiling points.

(d) Most of these form ionic compounds on account of their lower ionization energy.

(e) Most of these metals (except Be & Mg) and their salts imparts characteristic colour to the flame e.g., sodium imparts a golden yellow colour; potassium imparts violet colour to the flame.

(f) These are highly reactive elements and are strong reducing agents.

(g) All are good conductors of heat and electricity.

p-block Elements: Elements of groups 13-18 in which the last electron enters the p-orbitals of the valence shell are called p-block elements.

Characteristics:

(a) The electronic configuration of the outermost shell of p-block elements (group 13, 14, 15, 16, 17 and 18) is $ns^2\ np^{1\text{-}6}$.

(b) These elements include metals and non-metals with a few metalloids. The metallic character, however, decreases along the period but increases down the group.

(c) These possess relatively higher ionization energy which tends to increase along the period but decreases down the group.

(d) Most of them form covalent compounds.

(e) Most of these elements show negative (except some metals) as well as positive oxidation states (except F).

(f) The oxidizing power of these elements increases along the period but decreases down the group.

d-Block Elements: There are three complete series and one incomplete series of d-block elements. These are: 1st or 3d-transition series which contains ten elements with atomic number 21-30 ($_{21}Sc$-$_{30}Zn$).

2nd or 4d-transition series which contains ten elements with atomic numbers 39-48($_{39}Y$-$_{48}Cd$).

3rd or 5d transition series which contains ten elements with atomic numbers 57 and 72-80

($_{57}La$, $_{72}Hf$-$_{80}Hg$).4th or 6d transition series which is incomplete at present and contains only nine elements. These are

$_{89}Ac$, $_{104}Rf$, $_{105}Ha$, Unh (Seaborgium, Z = 106), $_{107}Bh$ (Bohrium), $_{108}Hs$ (Hassium), $_{109}Mt$ (Meitnerium), Ds (Darmstadtium, Z= 110) and Cn (Copernicium, Z = 112) or Ekamercury. The element, Z = 111 has not been discovered yet. Thus, in all, there are 39 d-block elements.

Characteristics:

(a) The electronic configuration of outermost shell of d-block elements is ns^{0-2} followed with (n–1) $s^2p^6d^{1-10}$.

(b) All (except Hg) are hard, ductile metals with high melting and boiling points.

(c) All of these are good conductors of heat and electricity.

(d) Their ionization energies are higher than s-block elements but lesser than p-block elements.

(e) Most of the transition metals form coloured ions (Zn^{2+}, Hg^{2+}, Cd^{2+} are colourless.)

(f) These elements show variable oxidation states.

(g) Most of these elements possess catalytic activity.

(h) Metals and their ions are generally paramagnetic due to the presence of unpaired electrons.

(i) Most of the transition metal ions possess the tendency to form complex ions.

(j) Most transition metals form alloys.

f -block Elements: f-Block elements are also called inner-transition elements. In these elements, the f-subshell of the inner-penultimate is progressively filled up. There are two series of f-block elements each containing 14 elements. The fourteen elements from $_{58}Ce$ - $_{71}Lu$ in which, 4f-subshell is progressively filled up are called lanthanides or rare elements. Similarly, the fourteen elements from $_{90}Th$ – $_{103}Lr$ in which, 5f-subshell is progressively filled up are called actinides.

Characteristics:

(a) The electronic configuration of outermost shell of f-block elements is ns^2, followed with $(n-2)f^{1-14}$, $(n-1)\ d^{0-2}$.

(b) All are metals.

(c) Lanthanoids are also known as **rare earth elements** whereas most of the members of actinoid series are known as **transuranic elements** (made artificially).

(d) These show variable valency.

(e) These form coloured ions.

(f) Actinoids are radioactive.

(g) These also form complexes.

5. POSITION OF METALS AND NON METALS IN PERIODIC TABLE

Metals, Non Metals and Metalloids in Periodic Table

(a) Trends in metallic character in Periodic table

(i) The metallic character increases down the group and decreases along the period.

(ii) The non-metallic character decreases down the group and increases along the period.

Note: All the non-metals and metalloids belong to p-block (except H and He).

6. EFFECTIVE NUCLEAR CHARGE AND SHIELDING

In a polyelectronic atom, the internal electrons repel the electrons of the outermost orbit. This results in the decrease in the nuclear attraction on the electrons of the outermost orbit.

Therefore, only a part of the nuclear charge is effective on the electrons of the outermost orbit. Thus, the inner electrons protect or shield the nucleus and thereby decrease the effect of nuclear charge towards the electrons of the outermost orbit.

Thus, part of the nuclear charge works against outer electrons, and is known as effective nuclear charge $Z^{*} = Z - S$

Z^* = effective nuclear charge, S = shielding constant and Z = nuclear charge

A scientist named Slater, determined the value of shielding constant and put forward some rules which are listed below:

1. The shielding effect or screening effect of each electron of 1s orbital is 0.30.

2. The shielding effect of each electrons of ns and np i.e. electron of the outermost orbit, is 0.35.

3. The shielding effect of each electron of s, p or d orbitals of the penultimate orbit (n – 1) is 0.85.

4. The shielding effect of each electron of s, p, d or f orbital of the inner penultimate orbit (n – 2) and below this is 1.0

Table 13.3: Z* for II period elements

Element	Atomic number	Calculation of			$\frac{349}{35.3}$	$Z^* = Z - S$
		[0.35 ×No. of nth electrons)-1]	[0.85×No. of (n–1)th electrons]	[1.0×No. of inner electrons]		
Li	3	—	0.85 × 2	—	1.70	1.30
Be	4	0.35 × 1	0.85 × 2	—	2.05	1.95
B	5	0.35 × 2	0.85 × 2	—	2.40	2.60
C	6	0.35 × 3	0.85 × 2	—	2.75	3.25
N	7	0.35 × 4	0.85 × 2	—	3.10	3.90
O	8	0.35 × 5	0.85 × 2	—	3.45	4.55
F	9	0.35 × 6	0.85 × 2	—	3.80	5.20
Ne	10	0.35 × 7	0.85 × 2	—	4.15	5.85

Illustration 5: What is the screening constant for the last electron in Sc? **(JEE MAIN)**

Sol: $1s^2\ 2s^2\ 2p^6\ 3s^2\ 3p^6\ 3d^1\ 4s^2$

Last electron is in 4s orbital.

The shielding effect for one electron of 4s = 0.35

Electrons of 3rd shell = 9; their contribution = 9 × 0.85

Contribution of 2nd and 1st shell = 10 × 1

Total = 0.35 + 9 × 0.85 + 10 × 1 = 18

$\frac{349}{35.3} = 18$

$z^* = z - S = 21 - 18 = 3$

7. TRENDS IN PHYSICAL PROPERTIES

7.1 Atomic Radius

It refers to the distance between the centre of the nucleus of the atom to the outermost shell containing electrons. Since absolute value of the atomic size cannot be determined, it is usually expressed in terms of the following operational definitions.

(a) Covalent Radius

(i) Normally, this term is used for non-metals.

(ii) It is defined as half of the distance between two successive nuclei of two covalently

(iii) bonded atoms in a molecule.

Covalent radius = $\frac{1}{2}$ × Internuclear distance between two covalently bonded like atom(d)

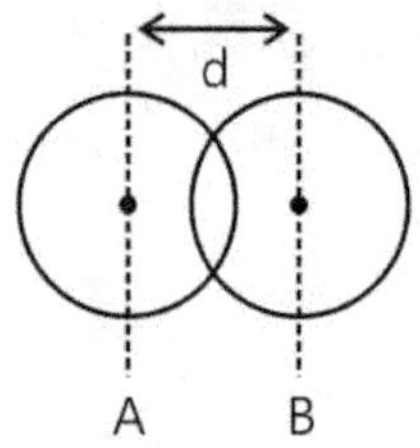

Figure 13.2-a: Diagrammatic representation of covalent radius

In other words, covalent radius is the radius of a spherical atom that leads to observed bond lengths when the spheres are just touching each other.

Covalent radius may be of following types

Single bond covalent radius

Double bond covalent radius

(b) Vander Waals Radius

(i) Van der Waals radius is defined as half of the internuclear separation of two non-bonded atoms of the same element on their closest possible approach. The term is used for non- metals (covalent compound) and noble gases.

(ii) It is half of the distance between two successive nuclei of two covalently bonded molecules of like atoms or two successive molecules of inert gases.

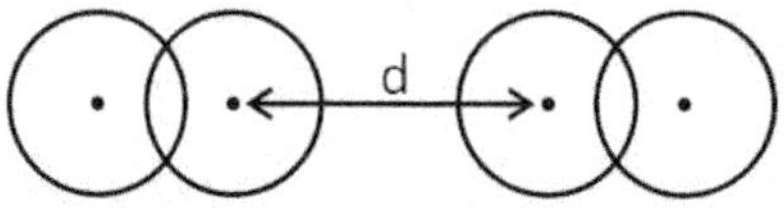

Figure 13.2-b: Diagrammatic representation of vander waal radius

Vander Waals radius = $\frac{1}{2}$ × Internuclear distance between two successive nuclei of two covalent molecules (d)

(c) Crystal Radius or Metallic Radius

(i) The term is usually used for metals.

(ii) It is defined as half of the distance between two successive nuclei of two adjacent metal atoms in the metallic closed packed crystal lattice.

(d) Ionic Radius

(i) This term is used in case of ions.

(ii) It is the distance of outermost shell of an anion or cation from its nucleus. In other words, it is defined as the effective distance from the nucleus of the ion which is under influence in an ionic bond.

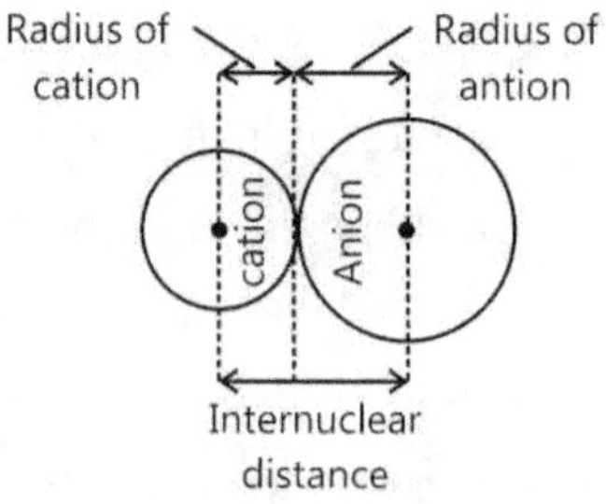

Figure 13.2-c: Diagrammatic representation of ionic radius

NOMORECLASS CONCEPTS

It is evident that van der Waals radius is greater than Covalent radius of an atom because Vander Waals forces of attraction are weaker than covalent bond forces.

Trends of Atomic Radius

(a) Along the period: On moving across a period, atomic radii decreases because effective nuclear charge increases.

(b) Down the group: On moving down a group, atomic radii increases, because number of orbits increases.

Factors Affecting Atomic Radii

(a) **Effective nuclear charge:** As the effective nuclear charge increases, the attractive force between nucleus and valence electron increases. Thus, across a period, atomic size/atomic radii decreases.

So, Atomic radii $\propto \frac{1}{Z_{eff}}$

(b) **Size of valence shell:** Atomic radii is the measure of radius of valence shell. As the value of n (principal quantum no.) increases, for an orbit, its size increases, thus down a group, atomic radii increases.

(c) **Multiplicity of bond:** Covalent radii decreases, as the multiplicity of bond increases.

For example, in case of carbon

	C – C	C = C	Δ_{A-B}
D_{c-c} (Å)	1.54	1.34	1.20
r_c (Å)	0.77	0.67	0.60

(d) **Percentage ionic character in bond:** Covalent radii of an atom in a bond depends upon % of ionic character. Increase in ionic character % leads in shortening of bond, decreasing the atomic radii.

(e) **Cationic Radii:** Size of cation is always lesser than its parent atom and greater the charge on cation, smaller is its ionic radii. E.g. Fe > Fe^{+} > Fe^{2+} > Fe^{3+} (decreasing ionic radii)

Formation of cation involves loss of electron. Thus, effective nuclear charge increases, pulling the remaining electrons more tightly towards the nucleus.

(f) **Anionic Radii:** Size of an anion is always larger than its parent atom. Formation of an anion involves gain of electrons by an atom and so, effective nuclear charge decreases. Thus, the valence shell electrons are less tightly held by the nucleus.

NOMORECLASS CONCEPTS

Lanthanide contraction also plays an important role in deciding the trends of atomic radii accounting for the similar atomic radii of palladium and Pt.

Illustration 6: Which one of the following pairs would have a large size? Explain.

(i) K or K^+ (ii) Br or Br^- (iii) O^{2-} or F^- (iv) Li^+ or Na^+ (v) P or As (vi) Na^+ or Mg^{2+}

(JEE MAIN)

Sol:

(i) Due to higher effective nuclear charge, K^+ has smaller atomic size than K.

(ii) Due to lower effective nuclear charge, the size of Br^- is greater than that of Br.

(iii) O^{2-} and F^- are isoelectronic species. Since effective nuclear charge of O^{2-} is lower than that of F^-, therefore, O^{2-} has a greater atomic size than F^-.

(iv) Li^+ and Na^+ both belong to group 1. Because of a greater number of shells (2, in case of Na^+ and 1, in case of Li^+), Na^+ has a bigger atomic size than Li^+.

(v) As has four shells while P has three. Therefore, atomic size of As is greater than that of P.

(vi) Na^+ and Mg^{2+} are isoelectronic cations. Therefore, due to lower effective nuclear charge, the ionic radius of Na^+ is greater than that of Mg^{2+}.

Illustration 7: Arrange the following in order of increasing radii?

(i) I, I^+, I^- (ii) C, N, Si, P (iii) O^{2-}, N^{3-}, S^{2-} F^- **(JEE MAIN)**

Sol: The size of a neutral, positive and negative species is compared.

(i) Size of a cation is always smaller while that of an anion is always bigger than the neutral atom, i.e. $I^+ < I < I^-$

(ii) C and N lie in 2nd period while Si and P lie below them in the 3rd period. Since elements in the 3th period have higher atomic size than those in the 2nd period. Therefore, atomic radii of Si and P are higher than those of C and N respectively. Since, atomic radii decreases across a period due to higher nuclear charge, therefore, C has higher atomic radius than N and Si has higher atomic radius than P. Thus, the overall order of increasing atomic radii is: N < C < P < Si.

(iii) Among isoelectronic ions, the size of anions increases as the nuclear charge decreases: $F^- < O^{2-} < N^{3-} < S^{2-}$.

Illustration 8: Select from each group the species which has the smallest radius stating appropriate reasons.

(i) O, O^-, O^{2-} (ii) P^{3+}, P^{4+}, P^{5+} **(JEE ADVANCED)**

Sol: Ionic radius decrease on loss of electrons since the nuclear charge increases.

(i) O has the smallest radius. The anion is larger than its parent atom. Also, the anion of the same atom with higher negative charge is bigger in size as compared to the anion with smaller negative charge as proton to electron ratio decreases. Thus, attraction between valence shell electrons and nucleus decreases. Hence, the electron cloud expands.

(ii) The ionic radius decreases as more electrons are ionized off .Thus, valency increases. So, the correct order is $P^{5+} < P^{4+} < P^{3+}$.

Illustration 9: Arrange the following ions in the increasing order of their size: Be^{2+}, Cl^-, S^{2-}, Na^+, Mg^{2+}, Br^-?

(JEE ADVANCED)

Sol: Ionic radius $\propto \dfrac{1}{\text{Nuclear charge}}$

$$Be^{2+} < Mg^{2+} < Na^+ < Cl^- < S^{2-} < Br^-$$

Be^{2+} is smaller than Mg^{2+} ,as, Be^{2+} has one shell whereas, Mg^{2+} has two shells. Mg^{2+} and Na^+ are isoelectronic species.

Cl^- is smaller than Br^- as Cl^- .has three shells whereas Br^- has four shells.

NOMORECLASS CONCEPTS

- In isoelectronic ions, size decreases with increase in atomic number.

  For e.g. $N^{3-} > O^{2-} > F^{-} > Na^{+} > Mg^{2+} > Al^{3+}$

  Although, Ne is also isoelectronic to them but its atomic radius is 1.60Å. It should not be compared to other radii because noble gases do not form ionic compounds and their radii are simply Van der Waals radii.

- In the end of the period, the atomic radii of inert gases are exceptionally higher because they do not form molecules and their radii are simply Van der Waals radii.

7.2 Ionization Potential

It is the amount of energy required to remove the most loosely bound electron from an isolated gaseous atom i.e.

$$M(g) + IE \rightarrow M^{+}(g) + e^{-}$$

The amount of energies required to remove the first, second, third etc. electrons from an Isolated gaseous atom are called successive ionization energies and are designated as IE_1, IE_2,

IE_3 etc. It may be noted that IE_2 is always greater than IE_1 .Thus, the order is- $IE_3 > IE_2 > IE_1$

NOMORECLASS CONCEPTS

The removal of a second electron is relatively more difficult because after the removal of the first electron, remaining electrons in the cation are more effectively pulled by the nucleus due to increased effective nuclear charge, thus, $IE_3 > IE_2 > IE_1$

Factors affecting Ionisation Potential

(a) Number of shells: With the increase in the number of shells, the atomic radius increases i.e. the distance of outermost shell electron from the nucleus increases and hence the ionization potential decreases.

(b) Effective Nuclear Charge: Atomic size decreases with the increase in effective nuclear charge because, higher the effective nuclear charge, stronger will be the attraction of the nucleus towards the electron of the outermost orbit and higher will be the ionization potential.

(c) Shielding Effect: The electrons of the inner orbits repel the electrons of the outermost orbit due to which the attraction of the nucleus towards the electrons of the outermost orbit decreases and thus the atomic size increases and the value of ionization potential decreases.

(d) Stability of half-filled and fully filled orbitals: The atoms whose orbitals are half-filled (p^3, d^5, f^7) or fully-filled $(s^2, p^6, d^{10}, f^{14})$ have greater stability than the other. Therefore, they require greater energy to remove an electron. However, stability of fully filled orbitals is greater than that of the half-filled orbitals.

(e) Penetration power: In any atom, the s-orbital is nearer to the nucleus in comparison to p, and f orbitals. Therefore, greater energy is required to remove an electron from s-orbital than from p, d and f orbitals. The order is as follows- s > p > d > f

Periodic Trends in Ionisation Potential

In a Period: The value of ionization potential normally increase across a period, because effective nuclear charge increases and the atomic size decreases.

Exceptions: In the second period, ionization potential of Be is greater than that of B, and in the third period, ionization potential of Mg is greater than that of Al due to the high stability of fully filled orbitals. In the second period, ionization potential of N is greater than O and in the third period, ionization potential of P is greater than that of S, due to the stability of half-filled orbitals.

In a Group: The value of ionization potential normally decreases down the group because both, atomic size and shielding effect increase.

Exception: The value of ionization potential remains almost constant from Al to Ga in the III A group. (B > Al, Ga > In).

In IV B group i.e. Ti, Zr and Hf, the I.P. of Hf is higher than that of Zr due to Lanthanide contraction. Thus the I.P. of IV B group varies as Ti > Zr < Hf.

Ionisation Potential of Transition Elements: In transition elements, the value of ionization potential has changes very little across a period. This is because, the outermost orbit remains the same but electrons get filled up in the (n-1)d orbitals resulting in very little increase in the values of ionization potential.

In transition element series, the first ionization potential normally increases with increase in atomic number on going from left to right, but this periodicity is not uniform. The value of ionization potential of transition elements depends on two important factors-

(a) The value of ionization potential increases with increase in effective nuclear charge.

(b) The value of ionization potential decreases with increase in shielding effect when number of electrons increase in (n – 1) orbitals.

  (i) In the first transition element series, the first ionization potential normally increases on going from left to right from Sc to Cr because shielding effect is much weaker in comparison to effective nuclear charge. The value of first ionisation potential of Fe, Co and Ni remains constant, because shielding effect and effective nuclear charge balance each other. The value of ionization potential shows a slight increase from Cu to Zn because they have fully filled s and d orbitals. The value of first ionisation potential of Mn is maximum because it has maximum stability due to fully filled s and half-filled orbitals.

Inner Transition Elements: The size of inner transition elements is greater than that of d block elements. Therefore, the value of ionization potential of f- block elements is smaller than that of d- block elements. Due to the almost constant atomic size of f- block elements in a period, the value of their ionisation potential remains more constant than that of d- block elements.

NOMORECLASS CONCEPTS

Ionization energy and the stable oxidation states of elements:

(i) When difference in two successive IE values (Δ^2) for an atom is approximately 10-15 eV or less, then, the higher ox. state will be more stable.

E.g.

For Al, $I.E_1 = 6.0$ eV. $I.E_2 = 18.8$ eV,

$I.E_3 = 28.4$ eV and $I.E_4 = 120$ eV,

Then, for Al(II); $\Delta I.E._{(1.2)} = 12.8$ eV

Al (III); Δ $I.E._{(2.3)} = 9.6$ eV

Al (IV); Δ $I.E._{(3.4)} = 91.6$ eV

NOMORECLASS CONCEPTS

So, Al (III) is more stable than Al (I) or Al (II).

(ii) If the value of $\propto$ is greater than 16.0 eV, then, the lower ox. state will be more stable.

E.g., For Na, $I.E._1$ = 5.1 eV; $I.E_2$ = 47.3 eV

$\therefore \Delta IE_{(1.2)} = 42.4$ eV

So, Na(I) is formed and not Na(II).

Illustration 10: Calculate the energy required to convert all the atoms of magnesium to magnesium ions present in 24 mg of magnesium vapours? First and second ionization enthalpies of Mg are 737.76 and 1450.73 kJ mol^{-1} respectively. **(JEE MAIN)**

Sol: According to the definition of successive ionization enthalpies,

$$Mg(g)+\Delta_iH_1 \rightarrow Mg^+(g)+e^-(g) \quad ; \quad \Delta_iH_1 = 737.76 \text{ kJ mol}^{-1}$$

$$Mg^+(g)+\Delta_iH_2 \rightarrow Mg^{2+}(g)+e^-(g) \quad ; \quad \Delta_iH_2 = 1450.73 \text{ kJ mol}^{-1}$$

Δ Total amount of energy needed to convert Mg (g) atom into Mg^{2+}(g) ion = $\Delta_iH_1 + \Delta_iH_2$

= (737.76 + 1450.73) kJ m = 2188.49 kJ mol^{-1}

24 mg of Mg = $\frac{24}{1000}$ g = $\frac{24}{1000 \times 24}$ mole = 10^{-3} mole

Therefore, amount of energy needed to ionize 10^{-3} mole of Mg vapours = 2188.49 × 10^{-3} = 2.188 kJ.

Illustration 11: The Δ_iH_1 and Δ_iH_2 of Mg(g) are 740 and 1450 kJ mol^{-1}. Calculate the percentage of Mg^+(g) and Mg^{2+} (g) if 1 g of Mg(g) absorbs 50 kJ of energy. **(JEE MAIN)**

Sol: Δ_iH_1 and Δ_iH_2 is the ionization enthalpy i.e. heat energy used to remove the loosely bound electron.

No. of moles of Mg vapours present in 1 g = 1/24 = 0.0417

Energy absorbed in the ionization of 0.0417 mole of Mg(g) to Mg^+(g) = 0.0417 × 740

= 30.83 kJ; Energy left unused = 50 – 30.83 = 19.17 kJ

Now, 19.17 kJ will be used to ionize Mg^+(g) to Mg^{2+}(g)

Na Mg Al (Metals) No. of moles of Mg^+(g) converted into Mg^{2+}(g) = 19.17/1450 = 0.0132

No. of moles of magnesium ions left as Mg^+(g) = 0.0417–0.0132 = 0.0285

Si (Metalloid) % of Mg^+(g) = (0.0285/0.0417) × 100 = 68.35% and % of Mg^{2+}(g) = 100 – 68.38 = 31.65%

Illustration 12: The first (Δ_iH_1) and the second (Δ_iH_2) ionization enthalpies (kJ mol^{-1}) of a few elements designated by Roman numerals are shown below: **(JEE ADVANCED)**

Element	Δ_iH_1	Δ_iH_2
I	2372	5251
II	520	7300

III	900	1760
IV	1680	3380

Which of the above elements is likely to be (a) a reactive metal (b) a reactive non-metal (c) a noble gas (d) a metal that forms a stable binary halide of the formula AX_2(X = halogen).

Sol:

(i) Since Element II has a very low Δ_iH_1 but a very high Δ_iH_2, therefore, it has only one electron in the valence shell and hence is likely to be a reactive metal (i.e., an alkali metal).

(ii) Since the Δ_iH_1 of Element IV is very high and its Δ_iH_2 is not so high (actually almost double), IV is likely to be a reactive non-metal (i.e. a halogen).

(iii) Among the elements listed, Δ_iH_1 of element I is the highest and its, Δ_iH_2 is also not so high, therefore, it must be a noble gas.

(iv) The Δ_iH_1 of element III is higher than that of element II, but unlike element II, its Δ_iH_2 is only about twice its Δ_iH_1, therefore, it is likely that element III has two electrons in the valence shell (i.e., alkaline earth metal). As such it will form a stable binary halide of the formula AX_2 where A is the metal and X is the halogen.

Illustration 13: From each set, choose the atom which has the largest ionization enthalpy and explain your answer.

(i) F, O, N (ii) Mg, P Ar (iii) B, Al, Ga **(JEE MAIN)**

Sol: Largest ionization enthalpy is the highest amount of energy needed to remove the valence electron due to a stronger nuclear charge and a smaller atomic size.

(i) F, O, N-All belong to 2nd period. Among these, F has the highest Δ_iH_1 because of its smallest size and highest nuclear charge.

(ii) Mg P, Ar-All lie in the 3rd period. Among these, Ar has the highest Δ_iH_1 because it has stable inert gas configuration.

(iii) B, Al, Ga-All lie in group 13. B has the highest Δ_iH_1 due to its smallest size.

Illustration 14: Compare qualitatively the first and second ionization potentials of copper and zinc. Explain the observation. **(JEE ADVANCED)**

Sol:

	IE_1 kJ mol^{-1}	IE_2 kJ mol^{-2}
Cu	744	1961
Zn	906	1736

IE_1 of copper is less than that of zinc, because removal of electron takes place from $4s^1$ (attaining a more stable configuration $3d^{10}$) whereas, in case of zinc, it is from completely filled $4s^2$ (attaining the configuration $4s^1$)

IE_2 of copper is higher than zinc, because the removal of 2nd electron from a stable configuration (d^{10}) requires higher energy.

Illustration 15: The first four successive ionization energies for an element are (6.113, 11.871, 50.908, 67.01) respectively. What is the number of valence shell electrons? **(JEE ADVANCED)**

Sol: The difference in second and third ionization is very large. Therefore, the no. of valence shell electrons should be 2. The element would attain the noble gas configuration after losing these 2 electrons.

7.3 Electron Affinity

It is the amount of energy released when a neutral isolated gaseous atom accepts an electron to form a gaseous anion.

$X(g) + E \rightarrow X^-(g) + EA$

Similarly, second and third electron can be added to form gaseous dinegative and trinegative ions. The energy changes accompanying the addition of first, second, third etc. electrons to neutral isolated gaseous atoms are called successive electron affinities and are designated as EA_1, EA_2, EA_3, etc.

Since an atom has a natural tendency to accept an electron, therefore, the first electron affinity (EA_1) is always taken as positive. However, the addition of second electron to the negatively charged ion is opposed by coulombic repulsion. Hence, energy has to be supplied for the addition of second electron.

Thus, second electron affinity (EA_2) of an element is taken as negative.

For example,

$O(g) + e^- \rightarrow O^-(g);\ EA_1 = +141\ kJ\ mol^{-1}$ $\qquad \Delta H_{EA_1} = -141\ kJ/mol$

$O^-(g) + e^- \rightarrow O^{2-}(g);\ EA_2 = -780\ kJ\ mol^{-1}$ $\qquad \Delta H_{EA_2} = +780\ kJ/mol$

Factors Affecting Electron Affinity

Atomic size or atomic radius: When the atomic size/radius increases, the electrons entering the outermost orbit is more weakly attracted by the nucleus and the value of electron affinity is lower.

Effective Nuclear charge: When effective nuclear charge is more, then, the atomic size is less. Then, the atom can easily gain an electron and possess a higher value of electron affinity.

Stability of Fully-Filled and Half-Filled orbitals: The stability of the configuration having fully-filled orbitals (p^6, $d^{10}f^{14}$) and half—filled orbitals (p^3, d^5, f^7) is relatively higher than that of other configurations. Hence, such type of atoms have a lesser tendency to gain an electron, therefore, their electron affinity values will be very low or zero.

Trends in Electron Affinity: In a period, atomic size decreases with the increase in effective nuclear charge and hence, increases the electron affinity.

Exception:

(a) Ongoing from C^6 to N^7 in the second period, the values of electron affinity decrease instead of increasing. This is because there are half-filled ($2p^3$) orbitals in the outermost orbit of N, which are more stable. On the other hand, the outermost orbit in C has a $2p^2$ configuration.

(b) In the third period, the value of electron affinity of Si is greater than that of P. This is because the electronic configuration of the outermost orbit in P atom is $3p^3$, which being half-filled, is relatively more stable. The values of electron affinity of inert gases are zero, because their outermost orbit has fully-filled p orbitals.

(c) In a period, the value of electron affinity goes on decreasing on going from group IA to group IIA. The value of electron affinity of the elements of group IIA is zero because ns orbitals are fully-filled and such orbitals have no tendency to accept electrons.

In a Group: The values of electron affinity normally decrease down a group because the atomic size increases, decreasing the actual attractive force of the nucleus.

Exceptions:

(a) The value of the electron affinity of F is lower than that of Cl, because the size of F is very small and compact and the charge density is high on the surface. Therefore, the incoming electron/s experience more repulsion in comparison to Cl accounting for the highest value of Cl in the periodic table.

(b) The values of electron affinity of alkali metals and alkaline earth metals can be regarded as zero, because they do not have the tendency to form anions by accepting electron/s.

NOMORECLASS CONCEPTS

(a) Cl has the highest electron affinity (3.7)

(b) Higher the EA of an element, easier is the addition of an electron.

(c) Following are some important observations derived from the general trend of electron affinity:

(i) More the tendency to gain an electron, more is the non-metallic nature. Therefore, non-metallic nature increases along the period but decreases down the group.

(ii) More the tendency to form an anion, more is the tendency to show ionic bonding. Therefore, tendency of non- metals to show ionic bonding increases along the period but decreases down the group.

(iii) More the tendency to get reduced, more is the oxidizing nature. Therefore, oxidizing power increases along the period but decreases down the group

Order of Oxidizing power- F > Cl > Br > I.

Illustration 16: The electron gain enthalpy of chlorine is –349 kJ mol^{-1}. How much energy in kJ is released when 3.55 g of chlorine is converted completely into Cl^- ion in the gaseous state? **(JEE MAIN)**

Sol: According to the definition of electron gain enthalpy, $Cl(g) + e^-(g) \rightarrow Cl^-(g) + 349\ kJ\ mol^{-1}$

$C \equiv C$ Energy released when 1 mole (=35.5 g) of chlorine atoms change completely into $Cl^-(g)$ = 349 kJ

Energy released when 3.55 g of chlorine atoms change completely into $Cl^-(g) = \frac{349}{35.3} \times 3.55 = 34.9$ kJ

Illustration 17: Which of the following pairs of elements would have more negative electron gain enthalpy? Explain (i) N or O (ii) S or O (iii) C or Si **(JEE ADVANCED)**

Sol: (i) The electron gain enthalpy of O is highly negative while that of N is slightly positive.

Reason: The electronic configuration of N is quite stable ($1s^2\ 2s^2\ 2p_x^1\ 2p_y^1\ 2p_z^1$) since it has exactly half-filled 2p-orbitals and hence has no tendency to accept an extra electron. In other words, energy has to be supplied to add an extra electron. Thus, electron-gain enthalpy of N is slightly positive. In contrast, the electronic configuration of O ($1s^2\ 2s^2\ 2p_x^2\ 2p_y^1\ 2p_z^1$) is not so stable but it has a higher nuclear charge and smaller atomic size than N and hence, it has a higher tendency to accept an extra electron. In other words, electron gain enthalpy of O is highly negative.

(ii) S has more negative electron gain enthalpy than O.

Reason: The size of O is much smaller than that of S. As a result, the electron-electron repulsions in the smaller 2p-subshell of O are comparatively more than those present in the bigger 3p-subshell of S. Therefore, S has a higher tendency to accept an additional electron than O.

(iii) C has a more negative electron gain enthalpy than Si.

Reason: This is because C-atom has a smaller size than Si-atom. (Note that the electron-electron repulsions in these atoms are not very large because they contain only 4 electrons in the outermost shell.)

Illustration 18: (i) Arrange the following elements in order of decreasing electron gain enthalpy: B, C, N, O.

(JEE ADVANCED)

Sol: N has +ve electron gain enthalpy while, all others have –ve electron gain enthalpies. While moving from B $\rightarrow$ C $\rightarrow$ O, size decreases leading to –ve electron gain enthalpy in the same order. Thus, the overall decreasing order of electron gain enthalpies is N, B, C, O.

7.4 Electronegativity

The tendency of an atom to attract the shared pair of electrons of the covalent bond towards itself is called electronegativity of that atom.

Factors Affecting Electronegativity

Atomic size: Electronegativity of a bonded atom decreases with increase in size since the attractive force on the valence electrons decreases and hence electronegativity decreases.

Hybridisation state of atom: Electronegativity increases with increase in the s-character of the hybrid orbital. This is because, the s-orbital is nearer to the nucleus and thus, suffers greater attraction leading to increased electronegativity. The number of covalent bonds present between two bonded atoms is known as its bond order. With increase in the bond order, the bond distance decreases, effective nuclear charge increases and thus electronegativity increases. Increasing order of electronegativity is as follows: $C-C < C=C < C\equiv C$. When effective nuclear charge is high, the nucleus will attract the shared electrons with greater strength to give high electronegativity.

Oxidation number: The electronegativity value increases with the increase in oxidation number since the radius decreases with the increase in oxidation number.

The increasing order of electronegativity is as follows: $Fe < Fe^{+2} < Fe^{+3}$

Electronegativity does not depend on stability of fully-filled or half-filled orbitals because it is simply the capacity of the nucleus to attract a bonded pair of electrons.

Trends in Electronegativity: Atomic size decreases across a period. Thus, electronegativity increases. Atomic size increases down a group decreasing the electronegativity.

F has maximum electronegativity value in the periodic table, while Cs has minimum.

According to the Pauling scale, the electronegativity value of F is 4.0, O is 3.5 N is 3.0 and Cl is 3.1

Exceptions

(a) The elements of group IIB i.e. Zn, Cd and Hg show increase in electronegativity value down the group.

(b) The elements of group IIIA, i.e Al to Ga show increase in electronegativity value down the group.

(c) The elements of group IVA, Si onwards, show no change in electronegativity value down the group.

Measurment of Electronegativity

Pauling Scale: If two atoms, A and B, having different electronegativity values, get bonded to form a molecule, AB, then the bond between A and B in A–B will have both covalent and ionic properties.

Δ_{A-B} = Observed bond energy – Energy of 100% covalent or $\Delta_{A-B} = D - E_{A-B}$

Where D = Observed bond energy

E_{A-B} = Bond energy of pure covalent bond of A – B

The value of E_{A-A} and E_{B-B} is $E_{A-B} = \frac{1}{2}[E_{A-A} + E_{B-B}]$

$= 0.208\sqrt{\Delta_{A-B}} = X_A - X_B$ where, $X_A > X_B$ or $0.043 \times \Delta_{A-B} = (X_A - X_B)^2$

Mulliken Electronegativity Scale: Mulliken suggested that the value of electronegativity of an element as an average of the values of ionization potential and electron affinity of the element.

$$X_M = \frac{I.P. + E.A}{2} \text{ (in eV)}$$

Where X_M = Electronegativity value as given by Mulliken

$$X_P = \frac{X_M}{2.8} = \frac{I.P. + E.A}{5.6}$$

Where X_p = Electronegativity value as given by Pauling or $X_p = 0.336\,(X_M - 0.615)$

Allred–Roschow's Scale

$$X_{AR} = \frac{Z_{eff}.e^2}{r^2} \qquad X_p = 0.359\,\frac{Z_{eff}.e^2 + 0.744}{r^2}$$

$Z_{eff} = Z - s^2, p^6, d^{10}, f^{14}$ where Z = Nuclear charge

ΔIE = Shielding constant

Or $X_p = 0.359\,X_{AR} + 0.744$

Sanderson's Scale: In Sanderson scale, the stability ratio of an atom itself has been regarded as its electronegativity.

$$X_s \text{ or S.R.} = \frac{\text{Average electron density of an atom}}{\text{Electron density of the isoelectronic iner gas}}$$

This is related to Pauling scale as follows-

$$\sqrt{X_P} = \frac{0.2}{\text{S.R. or } X_s} \times 0.77$$

Applications of Electronegativity

(a) Nomenclature: Name of the more electronegative element is written at the end and 'ide' is suffixed to it. The name of the less electronegative element is written before the name of more electronegative element of the formula.

E.g. Correct formula	Name
(a) IBr	Iodine bromide
(b) OF_2	Oxygen difluoride
(c) Cl_2O	Dichlorine oxide

(b) Nature of bond: If the electronegativity difference between the two elements is 1.7 or more, then, an ionic bond is formed between them. Whereas, if it is less than 1.7, then, covalent bond is formed.

(HF is an exception in which a bond is covalent, although, difference of electronegativity is 1.9)

(c) Metallic and non-metallic nature: Low EN shows metallic nature and high EN shows non-metallic nature

Hydrolysis of AX – where A = Other element

and X = Halogen

If electronegativity of X > Electronegativity of A, then, on hydrolysis, product will be HX.

Example In BCl_3, EN of Cl > EN of B

(d) Partial Ionic character in covalent bonds: Partial ionic character is generated in covalent compounds due to the difference in electronegativities.

Hanny and smith: Calculates percentage of ionic character from the electronegativity difference. Percentage of ionic character = $16(X_A - X_B) + 3.5\,(X_A - X_B)^2$

$= 16\Delta + 3.5\Delta^2$

$= (0.16\Delta + 0.035\Delta^2) \times 100$

Here, X_A is electronegativity of element A.

X_B is electronegativity of element B.

$\Delta = X_A - X_B$

(e) Bond length: $d_{A-B} = r_A + r_B - 0.09(X_A - X)_B$

or $d_{A-B} = \frac{1}{2}(D_{A-A} + D_{B+B}) - 0.09(X_A - X)_B$

Here, $X_A > X_B$

(f) Bond energy and stability: Bond strength and stability of A–B increases on increase in difference of electronegativities of atoms A and B. Therefore, H–F > H–Cl > H–Br >H–I

(g) Acidic strength of hydrides: Bond energy (Strength) $\propto$ stability of molecule.

Order of stability of hydrohalides is HF > HCl > HBr > HI

So, Order of their acidic strength will be – HF < HCl < HBr < HI

Down the 5A group

NH_3	*Thermal	stability decreases
PH_3	*Basic	character decreases
AsH_3	*Acidic	character increases

In PH_3 and AsH_3, the difference in the electronegativites of X_A and X_B is very less, so their bond energy decreases and hence acidic character (losing H^+ ion) increases.

(h) Nature of oxides: If the difference of the two electronegativities (X_O–X_A) is 2.3 or more than 2.3, then the oxide will be basic in nature. Similarly, if value of X_O–X_A is lower than 2.3, then the compound will be first amphoteric and then acidic in nature.

Oxide	Na_2O	MgO	Al_2O_3	SiO_2	P_2O_5	SO_3	Cl_2O_7
(X_O–X_A)	2.6	2.3	2.1	1.8	1.5	1.1	0.5
Nature	Strongly basic	Basic	Amphoteric	weakly acidic	Acidic	Strong acidic	Strongest acidic

(i) Nature of hydroxides: According to Gallis, if electronegativity of A in a hydroxide (AOH) is more than 1.7, then, it will be acidic in nature whereas, it will be basic in nature, if electronegativity is less than 1.7.

For example	NaOH	and	ClOH
Electronegativity (X_A)	0.9		3.00
Nature	Basic		Acidic

If the value is more than X_O–X_H, then that hydroxide will be basic, otherwise, it will be acidic.

Illustration 19: Among the following, how many elements have lower electronegativity than oxygen atom-F, Cl, Br, I, H, S, P, K, Ca **(JEE MAIN)**

Sol: F is the most electronegative element. Order of electronegativity- F > O > N = Cl > ...... Except F, all will have electronegativity less than oxygen.

Illustration 20: What is the difference in the electronegativity of two atoms, when the percentage ionic character is 19.5%? **(JEE ADVANCED)**

Sol: Hanny – Smith equation

Percentage of ionic character $= \left(16\Delta + 3.5\ \Delta^2\right)$

Δ = difference in electronegativity

$19.5 = 16\Delta - 3.5\ \Delta^2$

$3.5\ \Delta^2 + 16\ \Delta - 19.5 = 0$

Solving quadratic, we get, $\Delta = 1$

8. TRENDS IN CHEMICAL PROPERTIES

Valency/valence, also known as **valence number**, is the number of valence bonds a given atom has formed, or can form, with one or more atoms.

Table 13.4: Variation of valence in a group:

Group	1	2	13	14	15	16	17	18
Number of valence electron/s	1	2	3	4	5	6	7	8
Valence	1	2	3	4	3,5	2,6	1,7	0,8

Variation of valence in a period: On moving along the period, the number of valence electrons increases from 1 to 8. Consequently, the valence of the elements with respect to hydrogen increases from 1 to 4 up to group IV and then decreases to 1 as shown in the table. However, valence with respect to oxygen increases from one to seven along the period.

Variation of valence in group: On moving down the group, the number of valence electrons remains the same. Therefore, all the elements in a group have the same valence. For example, elements of group I have valency 1 and elements of group II have valency 2.

Variation of valence in transition elements: Transition metals show variable valence of 1, 2 or 3 as they can use electrons from their outermost as well as penultimate shell, during chemical reactions as energy difference between them is small.

Some anomalous properties of second period elements

Consider the elements of II period.

II period: Li Be B C N O F Ne

The elements of gp. 1, 2 (Li & Be) and of gp. 13–17 (B to F) differ in many respects from other members of their group. (These points of difference will be later studied in detail.)Some anomalous properties of 2nd period elements are given below with their explanation.

(a) Covalence: The maximum covalency of 2nd period elements is four while, other members may also show higher covalency. e.g., BF_4^- exists but $[BF_6]^{3-}$ is known. Similarly OF_2 is known but OF_4, OF_6 are not while, SF_4 are SF_6 are known, N is never pentavalent etc.

Explanation: These elements have only two shells in their atom and the valence shell contain 4 orbitals only (one 2s and three 2p) so, a maximum of four bonds can be formed. In the 3rd period and onwards, the valence shell contain empty d–orbitals also. So, covalency may be more than four.

(b) $p\pi - p\pi$ **Multiple bonding:** Bonds like C=C, C≡C, N≡N, C=O etc. exist due to $p\pi-p\pi$ multiple bonding. These elements are smaller in size and mostly electronegative in their respective groups thus forming multiple bonds.

(c) **Diagonal relationship:** Diagonal relationship between elements of II & III periods.

Li Be B C

Na Mg Al Si

Diagonal relationship between these elements can be explained on the basis of approximately similar charge/size ratio of diagonally related elements.

9. PERIODIC PROPERTIES OF ELEMENTS

9.1 Periodicity Along the Period

(a) **Ionization enthalpy:** Increases along the period (with exception).

(b) **Electron gain enthalpy:** Increases along the period (with exception).

(c) **Electronegativity:** Increases along the period.

(d) **Atomic radius:** Decreases along the period.

(e) **Ionic radius:** The radii of isoelectronic ions decrease with increase in atomic number.

(f) **Atomic volume:** Volume occupied by 1 g-atom of an element in solid state is Atomic Volume.

Atomic volume decreases up to metals and then increases:

	Na	Mg	Al	Si	P	S	Cl	Ar
Atomic volume (cm^3)	24	14	10	12	17	16	19	23.7

(g) **Melting point, Boiling point**: Increases along the period for metals.

	Na	Mg	Al
M.P (°C)	98	649	660
B.P (°C)	883	1100	1800

(h) **Density:** Increases along the period.

(i) **Reducing behavior:** Decreases along the period.

Na Mg Al (Strong reductants) Si P S Cl (S, Cl: Strong oxidants)

(j) **Metallic character:** Decreases along the period.

Na Mg Al (Metals) Si (Metalloid) P S Cl (Non–metals)

Increasing non-metallic nature →
Decreasing metallic nature

(k) **Electropositive character:** Decreases along the period.

(l) Nature of oxides: The basic character of oxides decreases and acidic character increases along the period.

Elements: Na Mg Al Si P S Cl

Oxides: $\underbrace{Na_2O \quad MgO}_{\text{Basic}}$ $\underset{\text{Amphoteric}}{Al_2O_3}$ $\underbrace{SiO_2 \; P_2O_5 \; SO_3 \; Cl_2O_7}$

(m) Nature of hydrides: The basic character of hydrides decreases along the period.

(n) Valency: (i) Valency with respect to oxygen increases from one to seven along the period (table 1.9). (ii) Valency with respect to hydrogen increases first from one to four & then decreases to one (table 1.9)

Table 13.5: Valence of Elements

Group	I	II	III	IV	V	VI	VIII
Valency with respect to oxygen	1	2	3	4	5	6	7
Formula of the oxide	R_2O	RO	R_2O_3	RO_2	R_2O_5	RO_3	R_2O_7
Valency with respect to hydrogen	1	2	3	4	3	2	1
Formula of the hydride	RH	RH_2	RH_3	RH_4	RH_3	RH_2	RH

9.2 Periodicity along the Group

(a) Ionisation enthalpy : Decreases

(b) Electron gain enthalpy : Decreases

(c) Electronegativity : Decreases

(d) Atomic radii : Increases

(e) Ionic radii : Increases

(f) Atomic volume : Increases

(g) M.P./B.P : Decreases

(h) Density : Increases

(i) Oxidant-Reductant nature :

* Reducing nature of metals : Increases

* Oxidizing nature of non-metals : Decreases

(i) Metallic character : Increases

(j) Electro positive character : Increases

(k) Basic character of oxide : Increases

(l) Basic character of hydride : Decreases

NOMORECLASS CONCEPTS

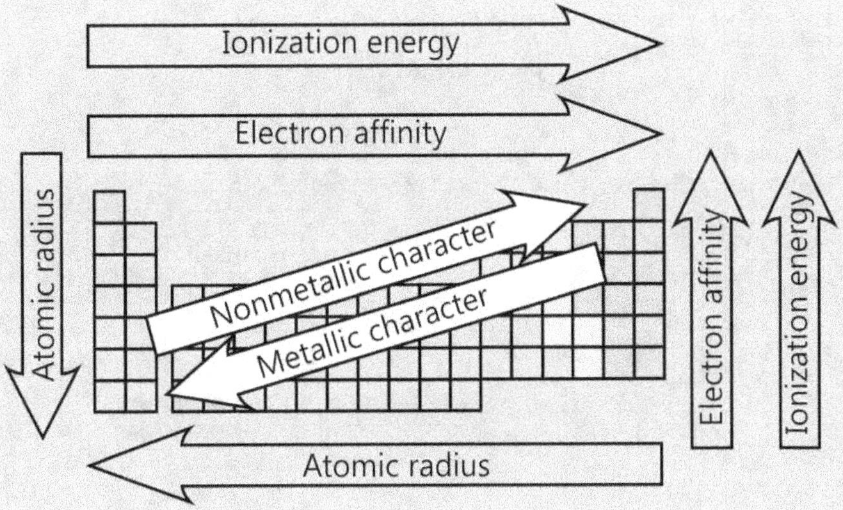

Figure 13.3: Variation of properties across the table

Some Important Things to Remember

1. Liquid radioactive element: Francium (Fr)
2. Rarest element in earth crust: Astatine (At)
3. Most poisonous metal: Plutonium (Pu)
4. Element with the maximum number of natural isotopes: Tin(Sn)
5. First man-made element: Technetium (Tc)
6. The size of the largest atom (Cs) is approximately 4.4 times to that of the smallest (H) atom.
7. Out of 17 non-metals known, 11 are gases, one is liquid (Br) and 5 are solids (C, P, S, Si and I)
8. Element with highest M.P.: Carbon (C in diamond)
9. Metal with highest M.P.: Tungsten (W)
10. Metal with maximum density: Iridium(Ir) (22.61 g/cc)
11. Metal with minimum density: Lithium(Li)
12. Lightest element: Hydrogen(H)
13. Most acidic oxide: Dichlorine heptoxide (Cl_2O_7)

POINTS TO REMEMBER

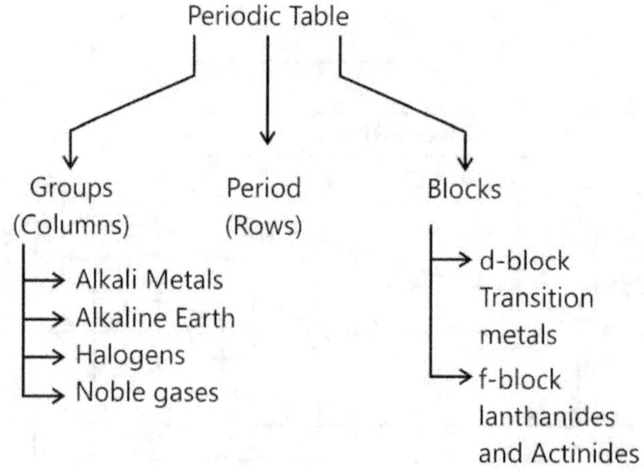

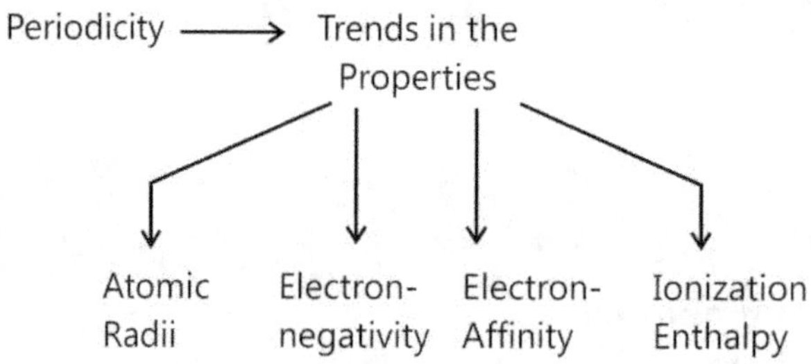

JEE Main/Boards

Exercise 1

Q.1 The electron affinity of each one of Be and Mg is zero. Give reason.

Q.2 Explain the following

(i) The process $O(g) + e^- \rightarrow O^-(g) + 141$ kJ mol^{-1} gives a positive electron affinity though

$\Delta E = -141$ kJ mol^{-1} (exothermic).

(ii) The process $O^-(g) + e^- + 700$ kJ $mol^{-1} \rightarrow O^{2-}(g)$ gives a negative electron affinity though ,

$\Delta E = +700$ kJ mol^{-1} (endothermic).

Q.3 There is an irregular trend in the ionization energies of group 13 elements. Explain.

Q.4 Mg^{2+}, O^{2-}, Na^+, F^-, N^{3-} (Arrange in decreasing order of ionic size)

Q.5 Why Ca^{2+} has a smaller ionic radius than K^+.

Q.6 Arrange in decreasing order of atomic size: Na, Cs, Mg, Si, Cl.

Q.7 Why the first ionization energy of carbon atom is greater than that of boron atom whereas, the reverse is true for the second ionization energy.

Q.8 (IE_1) of Be is greater than $(IE)_1$ of B but $(IE)_2$ of B is greater than that of Be. Explain.

Q.9 EA of Cu is 123 kJ mol^{-1} but that of Zn is –87 kJ mol^{-1}. Explain.

Q.10 In the preparation of hydrocarbon by Kolbe's electrolysis, generally RCOONa or RCOOK are taken but not RCOOLi. Explain.

Q.11 The IE do not follow a regular trend in II and III periods with increasing atomic number. Why ?

Q.12 Explain why a few elements such as Be (+0.6), N (+0.3) & He(+0.6) have positive electron gain enthalpies while majority of elements have negative values.

Q.13 Which bond in each pair is more polar

(A) P–Cl or P–Br (B) S–Cl or S–O

(C) N–O or N–F

Q.14 LiCl is hydrated but NaCl is always anhydrous. Explain.

Q.15 Explain physical property of third period?

Q.16 Arrange the following according to the instructions given against each.

(i) S, O, Se, C (Increasing order of atomic radius)

(ii) Ca, Al, O, N (Increasing paramagnetism)

(iii) F, Cl^-, Br^-, I^- (Increasing polarizability)

(iv) Na, Mg and Al (Increasing electropositive character)

(v) Na, P, Si, Al, S (Decreasing order of IE)

(vi) I^-, Br^-, Cl^- (Decreasing order of ionic size)

(vii) C,F, Li, O, Cs (Increasing order of IE_1)

(viii) O, F, Be, C, N (Decreasing order of electron affinity)

(ix) Mg^{2+}, O^{2-}, Na^+, F^-, N^{3-} (Decreasing order of ionic size)

(x) Na^+, I^-, Mg^{2+}, Rb^+, Cl^- (Decreasing order of ionic radii)

Q.17 Explain the following:

(i) Fluorine is the most electronegative and Cesium is the most electropositive element.

(ii) The first ionization energy of carbon atom is greater than that of boron atom whereas, the reverse is true for the second ionization energy.

(iii) Electron affinities of halogens are high.

(iv) The formation of F^-(g) from F(g) is exothermic whereas that of O^{2-}(g) from O^-(g) is endothermic.

(v) Transition and inner transition metals have variable oxidation states.

(vi) Zero group elements are chemically inert.

(vii) The second ionization potential of an element is higher than its first ionization potential.

(viii) The radius of an anion is greater than the parent atom while that of a cation is lesser than the atom.

Q.18 State giving reasons which one will have a higher value

(i) IE_1 or F or Cl

(ii) EA of O or O^-

(iii) ionic radius of K^+ or Cl^-

Q.19 From among the elements, choose the following: Cl, Br, F, Al, C, Li, Cs & Xe.

(i) The element with highest electron affinity.

(ii) The element with lowest ionization potential.

Q.20 In the ionic compound KF, the K^+ and F^- ions are found to have practically radii, about 1.34 Å each. What do you predict about the relative covalent radii of K and F?

Q.21 Explain applications of electronegativity

Q.22 Define atomic size and ionic size?

Q.23 Which oxide is more basic, MgO or BaO? Why?

Q.24 The basic nature of hydroxides of group 13 (III–A) decreases progressively down the group. Comment.

Q.25 Based on location in P.T., which of the following would you expect to be acidic & which basic.

(i) CsOH (ii) IOH

(iii) $Sr(OH)_2$ (iv) $Se(OH)_2$

(v) FrOH (vi) BrOH

Q.26 Compare the following giving reasons Acidic nature of oxides: CaO, CO, CO_2, N_2O_5, SO_3.

Q.27 Write the drawback of Mendeleev table?

Q.28 Write the postulates of modern periodic table?

Q.29 Explain electron affinity of periodic tables?

Q.30 Explain factors affecting ionization potential?

Q.31 From among the elements, choose the following: Cl, Br, F, Al, C, Li, Cs & Xe.

(i) The element whose oxide is amphoteric.

(ii) The element which has smallest radii.

(iii) The element whose atom has 8 electrons in the outermost shell.

Exercise 2

Single Correct Choice Type

Q.1 When the following five anions are arranged in order of decreasing ionic radius, the correct sequence is:

(A) Se^{2-}, I^-, Br^-, O^{2-}, F^- (B) I^-, Se^{2-},O^{2-}, Br^-, F^-

(C) Se^{2-}, I^-, Br^-, F^-,O^{2-} (D) I^-, Se^{2-},Br^-, O^{2-}, F^-

Q.2 Which of the following is wrong regarding the stability of the ions Ge,Sn and Pb

(A) $Ge^{2+} < Sn^{2+} < Pb^{2+}$ (B) $Ge^{4+} > Sn^{4+} > Pb^{4+}$

(C) $Sn^{4+} > Sn^{2+}$ (D) $Pb^{2+} < Pb^{4+}$

Q.3 The law of triads is not applicable on

(A) Cl, Br, I (B) Na, K, Rb

(C) S, Se, Te (D) Ca, Sr, Ba

Q.4 The atomic volume was choosen as the basic of periodic classification of elements by

(A) Niels Bohr (B) Mendeleev

(C) Lothar meyer (D) Newlands

Q.5 The majority of gaseous elements in the periodic table are placed

(A) At bottom left hand side

(B) At top right hand side

(C) Below the main table

(D) Along side d block elements

Q.6 The electronic configuration

$1s^2, 2s^2, 2p^6, 3s^2, 3p^6, 3d^{10}, 4s^2, 4p^6, 4d^{10}, 5s^2$ is for:

(A) f-block element (B) d-block element

(C) p-block element (D) s-block element

Q.7 Fluorine has the highest electronegativity among the ns^2 np^5 group on the Pauling scale, but the electron affinity of fluorine is less than that of chlorine because:

(A) The atomic number of fluorine is less than that of chlorine

(B) Fluorine being the first member of the family behaves in an unusual manner

(C) Chlorine can accommodate an electron better than fluorine by utilizing its vacant 3d orbital

(D) Small size, high electron density and an increased electron repulsion makes addition of an electron to fluorine less favourable than that in the case of chlorine.

Q.8 The last electron in each normal elements of period is filled in

(A) The same energy sublevel

(B) The same enrgy level

(C) The same orbital

(D) Relation between I_x and I_z is uncertain

Q.9 The greater stability of the lower oxidation state in heavier p block metals in the consequence of

(A) Electronic transition within p-orbitals

(B) Electronic transition from s to p-orbitals

(C) Inert pair effect

(D) Expansion of octet

Q.10 Oxidation number of p-block elements is [Excluding inert gases]

(A) Equal to group number

(B) Group number +2

(C) Between the range [Group no...(Group no. 8)]

(D) Number of unpaired electrons in the valence shell

Q.11 The correct order of second ionization potential of carbon, nitrogen, oxygen and fluorine is:

(A) C > N > O > F (B) O > N > F > C

(C) O > F > N > C (D) F > O > N > C

Q.12 Which statement is wrong:

(A) 2nd ionization energy shows jump in alkali metals

(B) 2nd electron affinity for halogens is zero

(C) Maximum electron affinity exists for F

(D) Maximum ionization energy exists for He

Q.13 Which of the following is the configuration of second excited state of the element isoelectronic with O_2 or P^- or Cl^+

(A) $[Ne]3s^2 3p_x^1 3p_y^1 3p_z^1$

(B) $[Ne]3s^2$

(C) $[Ne]3s^1 3p_z^1 3p_y^1 3p_z^1 3d^1_{xy} 3d^1_{yz}$

(D) $[Ne] 3s^1 3p_x^1 3p_z^1 3d^1_{xy}$

Q.14 Metallic radii of transition elements

(A) First increase, then decrease periodically

(B) First decrease, then remain almost constant

(C) First increase, then remaining almost constant

(D) First increase, then increase periodically

Q.15 For the formation of a covalent bond, the difference in the value of electro negativities should be:

(A) Equal to or less than 1.7 (B) More than 1.7

(C) 1.7 or more (D) None of these

Q.16 Which one of the following is the correct order of interactions?

(A) Covalent < hydrogen bonding < Vander Waal's < dipole-dipole

(B) Van der Waal's < hydrogen bonding< dipole-dipole < covalent

(C) Van der Waal's < dipole-dipole < hydrogen bonding < covalent

(D) Dipole-dipole < Van der Waal's < hydrogen

Q.17 Properties of the elements of which of the following pairs do not resemble?

(A) Li and Mg (B) Be and Al

(C) Mg and Al (D) B and Si

Q.18 Which of the following statement is not true?

(A) The atoms have no tendency to accept electrons in empty higher energy levels

(B) The atoms have no tendency to accept electrons in empty high energy sublevels

(C) The alkali metals have no tendency to accept electrons

(D) The atoms with exactly half-filled electronic configurations have no tendency to accept electrons.

Q.19 IP_1 and IP_2 of Mg are 178 and 348 kcal mol^{-1}. The energy required for the reaction, $Mg \rightarrow Mg^{2+} + 2e^-$

(A) +170 kcal (B) +526 kcal

(C) $E - 2E_2$ (D) $(E_1 - E_2)/2$

Q.20 Electronic configuration of an element of atomic weight 40 is 2, 8, 8, 2 which of the following statement regarding this element is not correct

(A) It belongs to the second group of the P.T

(B) It has 20 neutrons

(C) The formula of its oxide is MO_2

(D) It has 20 Protons

Q.21 There are four elements P, Q, R and S: their configuration are also given. Show that which element will have highest value of IE_2?

(A) $P = [He]2s^2$ (B) $Q = [He]\ 2s^2 2p^2$

(C) $R = [He]\ 2s^2 2p^1$ (D) $S = [He]\ 2s^1 2p^3$

Q.22 The ionization potential of nitrogen is greater than that of oxygen because

(A) Nitrogen is an inert element

(B) The outermost shell of nitrogen has half-filled orbitals

(C) The radius of nitrogen is more than that of oxygen

(D) The radius of oxygen is more than that of nitrogen

Q.23 Be and Mg have zero value of electron affinities, because

(A) Be and Mg have $[He]\ 2s^2$ and $[Ne]\ 3s^2$ configuration respectively

(B) 2s and 3s orbitals are filled to their capacity

(C) Be and Mg are unable to accept electron

(D) All the above are correct

Q.24 If I_1 and I_3 etc. represent the successive ionization potentials of an atom then the correct order is.

(A) $I_1 > I_2 > I_3$ (B) $I_1 < I_2 > I_3$ (C) $I_1 < I_2 < I_3$ (D) $I_2 > I_1 > I_3$

Q.25 For I.P, which order is wrong

(1) F > O (2) O > N (3) S > P (4) Be > B

Code is –

(A) 1, 2, 3 (B) 2, 3 (C) 1, 4 (D) 1, 2, 4

Q.26 Alkali metals do not form dipositive ions, because

(A) The difference in the first and second I.P. is more than 16 eV

(B) The difference in the first and second I.P. is less than 11 eV

(C) Alkali metals have one electron in their ultimate energy level

(D) Oxidation state of alkali metals is +1

Q.27 Electronic configuration of X^{+2} and Y^{+3} are: X^{+2} = [Ar] $3d^8$, Y^{+3} = [Ar] $3d^3$. What are the atomic number of X^0 and Y^0 respectively.

(A) 28, 24 (B) 28, 25 (C) 28, 26 (D) 28, 27

Q.28 The electronegativity values of C, N, O and F

(A) Increase from carbon to fluorine

(B) Decrease from carbon to fluorine

(C) Increase up to oxygen and minimum at a fluorine

(D) Is minimum at nitrogen and then increase continuously

Q.29 The valency in the II period from left to right

(A) Increases

(B) Decreases

(C) First increases, then decreases

(D) First decreases, then increases

Q.30 Which of the following represents incorrect relation of electronegativity

(A) C > O > N (B) C < O >N

(C) O > C < N (D) O > N > C

Q.31 In a period, elements are arranged in strict sequence of

(A) Decreasing charges in the nucleus

(B) Increasing charges in the nucleus

(C) Constant charges in the nucleus

(D) Equal charges in the nucleus

Q.32 In the periodic table, the metallic character of elements

(A) Decreases across a period and down a group

(B) Decreases across a period and increases down a group

(C) Increases across a period and down a group

(D) Increases across a period and decreases down a group

Q.33 The screening effect of inert electrons of the nucleus causes

(A) A decrease in the ionization potential

(B) An increase in the ionization potential

(C) No effect on the ionization potential

(D) An increase in the nuclei attraction of the electrons

Q.34 The statement that is not correct for the periodic classification of elements is

(A) The properties of elements are the periodic functions of their atomic numbers

(B) Non-metallic elements are lesser in number than metallic elements

(C) The first ionization energies along a period vary in a regular manner with increase in the atomic number

(D) For transition elements, the d-sub-shells are filled with electrons monotonically with an increase in atomic number

Q.35 Match List I with List II and select the correct answer using the codes given below the lists

List I	List II
(i) $1s^2\ 2s^2\ 2p^6\ 3s^2\ 3p^6\ 4s^2$	(p) In
(ii) $1s^2\ 2s^2\ 2p^6\ 3s^2\ 3p^6\ 3d^{10}\ 4s^1$	(q) In
(iii) $1s^2\ 2s^2\ 2p^6\ 3s^2\ 3p^6\ 3d^{10}\ 4s^2\ 4p^6 4d^{10}$	(r) Ca
(iv) $1s^2\ 2p^2\ 2p^6\ 3s^2\ 3d^{10}\ 4s^2\ 4s^2\ 4p^{10} 5s^2 5p^1$	(s) Cu

Codes:	i	ii	iii	iv
(A)	p	q	r	s
(B)	p	r	q	s
(C)	r	s	q	p
(D)	p	s	r	q

Q.36 Match List-I with List-II and select the correct answer from the codes given below the lists

List I	List II
(A) ns^2, np^5	(p) Chromium
(B) (n–)d^{10}, ns^1	(q) Copper
(C) (n–1)d^{10}, ns^2, np^6	(r) Krypton
(D) (n–1)d^{10}, ns^2 np^6	(s) Bromine (n = 4)

Q.37 Match List I with List II and select the correct answer from the codes given below the lists

List I	List II
(A) Highest ionization potential	(p) Technitium
(B) Highest electronegativity	(q) Lithium
(C) Artificial element	(r) Helium
(D) High reducing ability	(s) Fluorine

Q.38 The correct order of relative basic character of NaOH, $Mg(OH)_2$ and $Al(OH)_3$ is–

(A) $Al(OH)_3 > Mg(OH)_2 > NaOH$

(B) $Mg(OH)_2 > NaOH > Al(OH)_3$

(C) $NaOH > Mg(OH)_2 > Al(OH)_3$

(D) $Al(OH)_3 > NaOH < Mg(OH)_2$

Q.39 Match List–I with List–II and select the correct answer from the codes given below the lists

List I	List II
(i) Increasing atomic size	(p) Cl < O < F
(ii) Decreasing atomic radius	(q) Li < Be < B
(iii) Increasing electronegativity	(r) Si < Al < Mg
(iv) Decreasing effective nuclear charge	(s) Na> N > F

Codes:	i	ii	iii	iv
(A)	r	s	p	q
(B)	s	q	r	p
(C)	p	q	r	s
(D)	q	p	s	r

Previous Years Questions

Q.1 Which of the following statement is correct with respect to the property of element with an increase in atomic number in the carbon family (group 14) ***(2004)***

(A) Atomic size decrease

(B) Ionisation energy increase

(C) Metallic character decrease

(D) Stability of +2 oxidation state increase

Q.2 The first ionization potentials in electron volts of nitrogen and oxygen atoms are respectively given by ***(1987)***

(A) 14.6, 13.6 (B) 13.6, 14.6

(C) 13.6, 13.6 (D) 14.6, 14.6

Q.3 The elements which occupy the peaks of ionization energy curve, are ***(2000)***

(A) Na, K, Rb, Cs (B) Na, Mg, Cl, I

(C) Cl, Br, I, F (D) He, Ne, Ar, Kr

Q.4 Arrange F, Cl, O, N in the decreasing order of electronegativity

(A) O > F > N > Cl (B) F > N > Cl > O

(C) Cl > F > N > O (D) F > O > N > Cl

Q.5 Ionic radii of ***(1999)***

(A) $Ti^{4+} < Mn^{7+}$ (B) $^{35}Cl^- < {}^{37}Cl^-$

(C) $K^+ > Cl^-$ (D) $P^{3+} > P^{5+}$

Q.6 In which block, 106th element belongs

(A) s-block (B) p-block

(C) d-block (D) f-block

Q.7 The first $(\Delta_1 H_1)$ and second $(\Delta_1 H_2)$ ionization enthalpies (in kJ mol^{-1}) and the electron gain enthalpy $(\Delta_{eg}H)$ (in kJ mol^{-1}) of the elements I, II, III, IV and V are given below

Element	$\Delta_1 H_1$	$\Delta_1 H_2$	$\Delta_{eg}H$
I	520	7300	–60
II	419	3051	–48
III	1681	3374	–328
IV	1008	1846	–295
V	2372	5251	+48

The most reactive metal and the least reactive non-metal of these are respectively ***(2010)***

(A) I and V (B) V and II

(C) II and V (D) V and II

(E) V and III

Read the assertion and reason carefully to mark the correct option out of the options given below:

(A) If both assertion and reason are true and the reason is the correct explanation of the assertion.

(B) If both assertion and reason are true but reason is not the correct explanation of the assertion.

(C) If assertion is true but reason is false.

(D) If the assertion and reason both are false.

(E) If assertion is false but reason is true.

Q.8 Assertion: Positive ions will be wider than parent atoms ***(1999)***

Reason: Nuclear charge pulls them closer.

Q.9 Assertion: More is the electron affinity greater is the reducing character. ***(2000)***

Reason: Reducing character depends on number of electrons gained.

Q.10 Assertion: Ground state configuration of Cr is $3d^5, 4s^1$. ***(2004)***

Reason: A set of half filled orbitals containing one electron each with their spin parallel provides extra stability to the system.

Q.11 Assertion: I.E. Of $_7N$ is more that of $_8O$ as well as $_6C$. ***(2005)***

Reason: This is due to difference in reactivity towards oxygen.

Q.12 Which one of the following constitutes a group of the isoelectronic species? ***(2008)***

(A) C_2^{2-}, O_2^-, CO, NO (B) $NO^+, C_2^{2-}, CN^-, N_2$

(C) $CN^-, N_2, O_2^{2-}, C_2^{2-}$ (D) N_2, O_2^-, NO^+, CO

Q.13 Which one of the following pairs of species have the same bond order? ***(2008)***

(A) CN^- and NO^+ (B) CN^- and CN^+

(C) O_2^- and CN^- (D) NO^+ and CN^+

Q.14 In which of the following arrangements, the sequence is not strictly according to the property written against it? ***(2009)***

(A) $CO_2 < SiO_2 < SnO_2 < PbO_2$: increasing oxidising power

(B) $HF < HCl < HBr < HI$: increasing acid strength

(C) $NH_3 < PH_3 < AsH_3 < SbH_3$: increasing basic strength

(D) $B < C < O < N$: increasing first ionization enthalpy.

Q.15 The set representing the correct order of ionic radius is: ***(2009)***

(A) $Li^+ > Be^{2+} > Na^+ > Mg^{2+}$

(B) $Na^+ > Li^+ > Mg^{2+} > Be^{2+}$

(C) $Li^+ > Na^+ > Mg^{2+} > Be^{2+}$

(D) $Mg^{2+} > Be^{2+} > Li^+ > Na^+$

Q.16 The bond dissociation energy of B – F in BF_3 is 646 kJ mol^{-1} whereas that of C-F in CF_4 is 515 kJ mol^{-1}. The correct reason for higher B-F bond dissociation energy as compared to that of C- F is: ***(2009)***

(A) Smaller size of B-atom as compared to that of C- atom

(B) Stronger σ bond between B and F in BF_3 as compared to that between C and F in CF_4

(C) Significant $p\pi - p\pi$ interaction between B and F in BF_3 whereas there is no possibility of such interaction between C and F in CF_4.

(D) Lower degree of $p\pi - p\pi$ interaction between B and F in BF_3 than that between C and F in CF_4.

Q.17 The correct sequence which shows decreasing order of the ionic radii of the elements is ***(2010)***

(A) $Al^{3+} > Mg^{2+} > Na^+ > F^- > O^{2-}$

(B) $Na^+ > Mg^{2+} > Al^{3+} > O^{2-} > F^-$

(C) $Na^+ > F^- > Mg^{2+} > O^{2-} > Al^{3+}$

(D) $O^{2-} > F^- > Na^+ > Mg^{2+} > Al^{3+}$

Q.18 Among the following the maximum covalent character is shown by the compound: ***(2011)***

(A) $SnCl_2$ (B) $AlCl_3$

(C) $MgCl_2$ (D) $FeCl_2$

Q.19 Which one of the following order represents the correct sequence of the increasing basic nature of the given oxides? ***(2011)***

(A) $MgO < K_2O < Al_2O_3 < Na_2O$

(B) $Na_2O < K_2O < MgO < Al_2O_3$

(C) $K_2O < Na_2O < Al_2O_3 < MgO$

(D) $Al_2O_3 < MgO < Na_2O < K_2O$

Q.20 The molecule having smallest bond angle is: ***(2012)***

(A) NCl_3 (B) $AsCl_3$ (C) $SbCl_3$ (D) PCl_3

Q.21 Which of the following on thermal decomposition yields a basic as well as an acidic oxide? ***(2012)***

(A) $NaNO_3$ (B) $KClO_3$ (C) $CaCO_3$ (D) NH_4NO_3

Q.22 Which of the following represents the correct order of increasing first ionization enthalpy for Ca, Ba, S, Se and Ar? ***(2013)***

(A) Ca < S < Ba < Se < Ar

(B) S < Se < Ca < Ba < Ar

(C) Ba < Ca < Se < S < Ar

(D) Ca < Ba < S < Se < Ar

Q.23 The first ionisation potential of Na is 5.1 eV. The value of electron gain enthalpy of Na^+ will be: ***(2013)***

(A) -2.55 eV (B) -5.1 eV

(C) -10.2 eV (D) +2.55 eV

Q.24 The ionic radii (in Å) N^{3-}, O^{2-} and F^- are respectively: ***(2015)***

(A) 1.36, 1.40 and 1.71

(B) 1.36, 1.71 and 1.40

(C) 1.71, 1.40 and 1.36

(D) 1.71, 1.36 and 1.40

JEE Advanced/Boards

Exercise 1

Q.1 Use the following system of naming elements in which first alphabets of the digits are written collectively,

0	1	2	3	4	5	6	7	8	9
nil	uni	bi	tri	quad	pent	hex	sept	oct	enn

to write the three-letter symbols for the elements with atomic number 101 to 109.

Q.2 Arrange the following in increasing order of the property as indicated:

(i) Pb, Pb^{2+} and Pb^{4+}(size)

(ii) Mg, Al, Si and Na(ionisation potential)

(iii) MgO, SrO, Rb_2O, NiO, Cs_2O (basic character)

(iv) Be(OH), $Mg(OH_2$, $Ca(OH)_2$, $Ba(OH)_2$ (basicity)

(v) Cl^-, K^+, Ca^{2+}, Ar (ionization energy)

(vi) As, F, S, Cl (Electronegativity)

(vii) LiCl, LiBr, LiI (Ionic character)

(viii) Li^+, Na^+, K^+, Rb^+, Cs^+ (mobility of hydrated ions)

(ix) Li, Na, K, Rb and Cs (hydrated radii)

(x) Be^{2+}, Mg^{2+}, Ca^{2+}, Sr^{2+}, Ba^{2+} (hydration of ions)

Q.3 Why are inert gases mono-atomic?

Q.4 Why does the third period contains 8 elements and not 18?

Q.5 Arrange the following in order of increasing ionic radius:

(i) Cl^-, P^{3-}, S^{2-}, F^- (ii) Al^{3+}, Mg^{2+}, Na^+, O^{2-}, F^-

(iii) Na^+, Mg^{2+}, K^+.

Q.6 If internuclear distance between Cl atoms in Cl_2 is 10 Å & between H atoms is H_2 is 2 Å, then calculate internuclear distance between H and Cl (Electronegativity of H = 2.1 & Cl = 3.0).

Q.7 The As-Cl bond distance in $AsCl_3$ is 2.20 Å. Estimate the SBCR (single bond covalent radius) of As.

(Assume EN of both to be same and radius of Cl = 0.99 Å.)

Q.8 The Pt–Cl distance has been found to be 2.32 Å in several crystalline compounds. If this value applies to both of the compounds shown in figure, what is Cl–Cl distance in (i) and (ii)

(i)

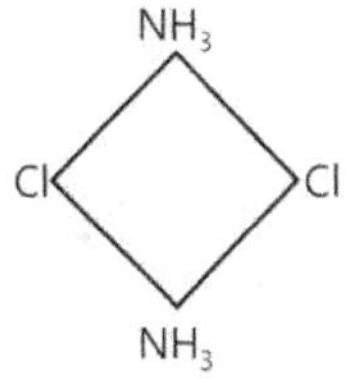

(ii)

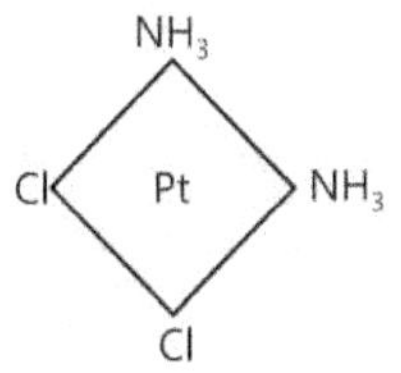

Q.9 K^+ and F^- have identical radius about 1.34 Å. What should be the atomic radius of K and F atoms.

Q.10 The radii of noble gases are greater than that of the radii of precedent halogens.

Q.11 The covalent radii (Å) do not increases regularly from B to Tl (B = 0. 80 Å, Al = 1.25 Å; Ga = 1.25 Å, In = 1.50 Å ; Tl = 1.55 Å) How do you account for this anomalous behavior.

Q.12 Calculate E.N. of chlorine atom on Pauling scale if I.E. of Cl^- is 4eV & of E.A. of Cl^+ is +13.0 eV.

Q.13 Why in isoelectronic species, the ionic / atomic radii decrease with increase in atomic numbers?

Q.14 The ionisation energy of the coinage metals fall in the order Cu > Ag < Au.

Q.15 The first ionization energy of carbon atom is greater than that of boron where as the reverse is true for second ionization energy. Explain.

Q.16 Define ionization potential?

Q.17 The I[st] ionization energy of potassium is less than that of Cu but the reverse is true for IInd ionisation energy.

Q.18 The sums of first and second ionization energies and those of third fourth ionisation energies in (kJ mol^{-1}) of nickel and platinum are

	$(IE)_1 + (EI)_2$	$(IE)_3 + (IE)_4$
Ni	2.49	8.80
Pt	2.66	6.70

Based on this information; write

(i) The most stable oxidation states of Ni and Pt

(ii) Name of one of the metals which can more easily form compounds in its +4 oxidation state.

Q.19 The first ionization enthalpy of magnesium is higher than of sodium. On the other hand, the second ionization enthalpy of sodium is very much higher than that of magnesium. Explain.

Q.20 Na^+ has higher value of ionization energy than Ne, though both have same electronic configuration. Explain?

Exercise 2

Single Correct Choice Type

Q.1 Two p-block elements x (outer configuration ns^2 , np^3) and z (outer configuration ns^2 np^4) occupy neighbouring positions in a period. Using this information, which of the following is correct with respect to their ionization potential I_x and I_z.

(A) $I_x > I_z$ (B) I_z and I_x (C) $I_z = I_x$ (D) $I_z > I_x$

Q.2 True position of lanthanides are:

(A) After III B group and in the 6th period

(B) After III B group in the 3rd period

(C) After VI B group in the 3rd period

(D) After III B group in the 4th Period

Q.3 The main cause of diagonal relationship between Be and Al is

(A) Similarity in ionic sizes

(B) Similar ionic potentials

(C) Similar electronegativity

(D) Similar atomic volume

Q.4 Electronic configuration of four elements are: a = $1s^2$; $2s^2$, $2p^1$; b = $1s^2$; $2s^2$, $2p^2$; c = $1s^2$; $2s^2$, $2p^5$; d = $1s^2$; $2s^2$, $2p^6$; $3s^1$. Which one of these would most readily form diatomic molecule?

(A) a (B) b (C) c (D) d

Q.5 EN of the element A is E_1 and IP is E_2. Hence EA will be:

(A) $2E_1 - E_2$ (B) $E_1 - E_2$ (C) $E_1 - 2E_2$ (D) $E_1 + E_2$

Q.6 | 1↿⇂ | | 1↿⇂ | 1 | 1 |

The above configuration would be of the species

(i) S (ii) Cl^+ (iii) p (iv) Ar^{-2}

Code is –

(A) i, ii (B) iiii, iv (C) i, iv (D) ii, iii

Q.7 Which of the following statements is correct for the addition of an electron to an isolated and gaseous uni-negatively charged oxygen O^- ion?

(A) The addition of electron cannot occur

(B) The addition of electron occurs with evolution of energy

(C) The addition of electron occurs with absorption of energy

(D) None of the above

Q.8 In the transformation $Na(s) \rightarrow Na^+(g)$, the energies involved are

(A) Ionization energy

(B) Sublimation energy

(C) Ionization energy and sublimation energy both

(D) None of the above

Q.9 The process of requiring absorption of energy is

(A) $F \rightarrow F^-$ (B) $Cl \rightarrow Cl^-$

(C) $O^- \rightarrow O^{-2}$ (D) $H \rightarrow H^-$

Q.10 The cynide CN^- & N_2 are isoelectronic. But in contrast to CN– , N_2 is chemically inert because of –

(A) Low bond energy

(B) Absence of bond polarity

(C) Unsymmetrical electron distribution

(D) Presence of more electron in bonding

Previous Year Questions

Q.1 The correct order of radii is ***(2000)***

(A) N < Be < B (B) $F^- < O^{2-} < N^{3-}$

(C) Na < Li < K (D) $Fe^{3+} < Fe^{2+} < Fe^{4+}$

Q.2 The set representing the correct order of first ionization potential is ***(2001)***

(A) K > Na > Li (B) Be > Mg > Ca

(C) B > C > N (D) Ge > Si > C

Q.3 Identify the least stable ion amongst the following ***(2000)***

(A) Li^+ (B) Be^- (C) B^- (D) C^-

Q.4 The statements that is/are true for the long form of the periodic table is/are ***(1988)***

(A) It reflects the sequence of filling the electrons in the order of sub-energy level s, p, d and f

(B) It helps to predict the stable valency states of the elements

(C) It reflects tends in physical and chemical properties of the elements

(D) It helps to predict the relative iconicity of the bond between any two elements

Q.5 Sodium sulphate is soluble in water whereas barium sulphate is sparingly soluble because ***(1989)***

(A) The hydration energy of sodium sulphate is more is more than its lattice energy

(B) The lattice energy of barium sulphate is more than its hydration energy

(C) The lattice energy has no role to play in solubility

(D) The hydration energy of sodium sulphate is less than its lattice energy

Q.6 The softness of group IA metals increases down the group with increasing atomic number. ***(1986)***

Q.7 In group IA of alkali metals, the ionization potential decreases down the group. Therefore, lithium is a poor reducing agent. ***(1987)***

Q.8 Arrange the following in order of their ***(1985)***

(i) Decreasing ionic size Mg^{2+}, O^{2-}, Na^+, F^-

(ii) Increasing first ionization energy Mg, Al, Si, Na

(iii) Increasing bond length F_2, N_2, Cl_2, O_2

Q.9 Arrange the following in the order of their increasing size: ***(1986)***

Cl^-, S^{2-}, Ca^{2+}, Ar

Q.10 Compare qualitatively the first and second ionization potentials of copper and zinc. Explain the observation. ***(1996)***

Q.11 Arrange the following ions in order of their increasing radii ***(1997)***

Li^+, Mg^{2+}, K^+, Al^{3+}.

Q.12 The species having bond order different from that in CO is *(2007)*

(A) NO^- (B) NO^+ (C) CN^- (D) N_2

Q.13 Statement-I: Pb^{4+} compounds are stronger oxidizing agents than Sn^{4+} compounds. *(2008)*

Statement-II: The higher oxidation states for the group 14 elements are more stable for the heavier members of the group due to 'inert pair effect'.

(A) Statement-I is True, Statement-II is True; Statement-II is correct explanation for Statement-I

(B) Statement-I is True, Statement-II is True; Statement-II is NOT a correct explanation for Statement-I

(C) Statement-I is True, Statement-II is False

(D) Statement-I is False, Statement-II is True

Q.14 Among the following, the number of elements showing only one non-zero oxidation state is: *(2010)*

O, Cl, F, N, P, Sn, Tl, Na, Ti

Q.15 The increasing order of atomic radii of the following Group 13 elements is *(2016)*

(A) Al < Ga < In < Ti (B) Ga < Al < Tl < In

(C) Al < In < Ga < Ti (D) Al < Ga < Tl < In

Important Questions

JEE Main/Boards

Exercise 1

Q.1 Q.9 Q.10 Q.14

Q.16 Q.17 (D) Q. 21

Exercise 2

Q.13 Q.16 Q.23 Q.26 Q.33

Q.35 Q.37

Previous Years' Questions

Q.7 Q.10

JEE Advanced/Boards

Exercise 1

Q.9 Q.11 Q.14

Exercise 2

Q.5 Q.10

Previous Years' Questions

Q.5 Q.6

Answer Key

JEE Main/Boards

Exercise 2

Single Correct Choice Type

Q.1 D	**Q.2** D	**Q.3** B	**Q.4** C	**Q.5** B	**Q.6** D	**Q.7** D
Q.8 B	**Q.9** C	**Q.10** C	**Q.11** C	**Q.12** C	**Q.13** C	**Q.14** B
Q.15 A	**Q.16** C	**Q.17** C	**Q.18** D	**Q.19** B	**Q.20** C	**Q.21** C
Q.22 B	**Q.23** D	**Q.24** C	**Q.25** B	**Q.26** A	**Q.27** A	**Q.28** A
Q.29 C	**Q.30** A	**Q.31** B	**Q.32** B	**Q.33** A	**Q.34** C	**Q.35** C
Q.36 C	**Q.37** A	**Q.38** C	**Q.39** A			

Previous Years' Questions

Q.1 D	**Q.2** A	**Q.3** D	**Q.4** D	**Q.5** D	**Q.6** C	**Q.7** C
Q.8 D	**Q.9** E	**Q.10** A	**Q.11** C	**Q.12** B	**Q.13** A	**Q.14** C
Q.15 B	**Q.16** C	**Q.17** D	**Q.18** B	**Q.19** D	**Q.20** C	**Q.21** C
Q.22 C	**Q.23** B	**Q.24** C				

JEE Advanced/Boards

Exercise 2

Single Correct Choice Type

Q.1 A	**Q.2** A	**Q.3** B	**Q.4** C	**Q.5** A	**Q.6** A	**Q.7** C
Q.8 C	**Q.9** C	**Q.10** B				

Previous Years' Questions

Q.1 B	**Q.2** B	**Q.3** B	**Q.4** B, C, D	**Q.5** A, B	**Q.6** True	**Q.7** False
Q.12 A	**Q.13** C	**Q.14** 2	**Q.15** B			

Solutions

JEE Main/Boards

Exercise 1

Sol 1: As we know, the electronic configuration of Be and Mg are [He] $2s^2$ and [Ne] $3s^2$ respectively. As we see, both of these elements have a fully filled s-orbital & making their configuration very stable compared to other configuration with incompletely filled orbitals. Therefore, they have very less tendency to gain an electron, which makes their electron affinity equal to zero.

Sol 2: The electronic configuration of oxygen is [He] $2s^22p^4$,which is not very stable. Since this is neither a half filled or a full filled configuration, therefore it has a natural tendency to accept an electron which makes it first electron affinity positive and the process exothermic but, after an electron is added, the atom becomes negatively charged and the addition of a second electron is opposed by coulombic repulsion. Hence energy has to be supplied for the addition of second electron. Thus, its second electron affinity is negative and the process is endothermic.

Sol 3: We know that the ionisation energy increases while going from left to right in a period. But, there is an exception while going from group 12 to group 13, which means that ionisation energy decreases from group 12 to group 13. This exception is explained by the high stability of group 12 elements due to fully filled orbitals which makes it difficult to remove the outermost electron compared to group 13 where there are no full filled orbitals. Hence, their ionization energy is smaller.

Sol 4: All these ions have the same electronic configuration, i.e. [Ne].So, difference in size will be due to electronic charge only. Now, we know that the ionic size decreases with increasing positive charge, because a loss of electron results in increased effective nuclear charge and hence the remaining electrons are held more firmly bound with the nucleus, hence decreasing ionic size. Also, ionic size increases with increasing ionic charge, because addition of an electron results in an increase in natural repulsion between electrons which decreases effective nuclear charge and leads to increase in ionic size. Therefore, the order of ionic size will be $N^{3-} > O^{2-} > F^- > Na^+ > Mg^{2+}$

Sol 5: Both Ca^{2+} and K^+ have the same electronic configuration, i.e. [Ar] and hence equal number of shells. The ionic radius decreases with increase in positive charge, because the loss of an electron results in an increase in effective nuclear charge and hence a decrease in radius as the remaining electrons are more firmly bound to the nucleus. If 2 electrons are removed, the effective nuclear charge will increase even more and the size of the ion will decrease further. Therefore, the ionic radius of Ca^{2+} is less than K^+

Sol 6: All the elements except Cs are period 3 elements and we know that atomic size decreases in going from left to right in a period due to increased effective nuclear charge, resulting from the increase in number of protons for same number of electronic shells. Also, the atomic radius of Cs will be bigger than the rest because of more number of shells is Cs. Hence order

Cs > Na > Mg > Si > Cl

Sol 7: Electronic configuration of carbon: [He] $2s^22p^2$

Electron configuration of Boron: [He] $2s^22p^1$

First ionization energy of carbon is more than Boron because of increased effective nuclear charge in carbon, making removal of electron difficult. After removal of electron:

Electronic configuration of C^+: [He] $2s^22p^1$

Electronic configuration of B^+: [He] $2s^2$

Now B^+ has a stable full filled orbital making its configuration very stable compared to incompletely filled configuration of C^+. Hence second ionization energy of Boron is more than Carbon

Sol 8: Electronic configuration of Be: [He] $2s^2$

Electronic configuration of B: [He] $2s^2$ $2p^1$ (IE), of Be > (IE) of B due to stable full filled orbital configuration of Be making removal of electron difficult.

After removal of electron

Electronic configuration of Be^+: [He] $2s^1$

Electronic configuration of B^+: [He] $2s^2$

Here, $(IE)_2$ of B > $(IE)_2$ of Be as now B^+ has a more stable configuration than Be^+ due to full filled orbitals which are not present is Be^+

Sol 9: Electronic configuration of Cu: [Ar] $3d^2\ 4s^2$

Electronic configuration of Zn: [Ar] $3d^{10}\ 4s^2$

Zn has a much more stable configuration than Cu because of fully filled d orbital compared to Cu. A stable configuration means that Zn has no tendency to accept an electron which makes its EA negative. Cu on the other hand can attain a stable configuration by adding an electron, hence its EA is positive.

Sol 10: For Kolbe's electrolysis, we want the acid to dissociate into ions easily during the reaction. But since the acidic strength of carbonic acids increases while going from top to bottom in a group, therefore, RCOOLi will be dissociated into ions to a very little extent compared to RCOONa or RCOOK, Hence RCOOLi is not used.

Sol 11: The general trend of IE is that it increases while going from left to right in a period. However IE is more for an element which has a stable configuration due to half-filled or full filled orbitals. Hence, IE of group 7 > IE of group 8 and IE of group 4 > IE of group 5 because group 4 and group 7 have full filled and half-filled orbital respectively.

Sol 12: Electronic configuration of Be: [He] $2s^2$

Electronic configuration of N: [He] $2s^2\ 2p^3$

Electronic configuration of He: [He]

We see, that each of those elements have a full filled or half-filled orbital which makes their configuration very stable and because of this, they have no tendency to accept an electron and hence they have negative E_A and hence positive value of electron gain enthalpy.

Sol 13: Polarity of the bond depends upon the electronegativity difference between the two elements. The more electronegativity difference will be the more polar the bond will be

(A) P – Cl > P – Br as $(EN)_{Cl} > (EN)_{Br}$

$= [(EN)_{Cl} - (EN)_P] > [(EN)_{Br} - (EN)_S]$

(B) S – O > S – Cl as $(EN)_O > (EN)_{Cl}$

$= [(EN)_O - (EN)_S] > [(EN)_{Cl} - (EN)_P]$

(C) N – F > N – O as $(EN)_F > (EN)_O$

$= [(EN)_F - (EN)_N] > [(EN)_F - (EN)_O]$

Sol 14: LiCl is hydrated because hydration energy of Li^+ is much higher than Na^+ due to small size of Li^+ ion. Hence, NaCl is amorphous because of low hydration enthalpy of Na^+.

Sol 15: Atomic Radius: The atomic radius decreases while going from left to right till because of increased effective nuclear charge which pulls the electrons closer to the nucleus and hence decreases atomic radius. Atomic radius of Ar is larger because Vander Waals radius is calculated for Ar as it does not form any strong bonds.

First Ionisation Energy: (IE), generally increases across the period from left to right due to increase in effective nuclear charge and poor screening of outer most electrons. However (IE) of Mg > (IE), of Al or Mg has a stable configuration because of full filled orbital and (IE), of P > (IE), of S as P has half-filled orbital making its configuration stable.

Electron Affinity (EA): Electron affinity increases across the period due to increased nuclear charge and decreasing atomic size. However EA of $S_A > E_A$ of P, due to half-filled orbital stable configuration of P which makes addition of electron difficult.

Electronegativity: It increases across the period due to the decrease in atomic size and hence increased effective nuclear charge.

Sol 16: (i) O < C < S < Se

Since, atomic radius increases while going down in a group O < S < Se. Also since atomic radius decreases across a period O < C .

(ii) Paramagnetism will increase with an increase in number of unpaired electrons.

∴ Ca (0 unpaired electron) Al (1 unpaired) < O (2 unpaired) < N (3 unpaired)

Ca < O < N

(iii) The order of polarisability will depend on the order of ionic size. The larger the ionic size, more will be polarizability.

∴ Order - $F^- < Cl^- < Br^- < I^-$

(iv) Al < Mg < Na

Since electropositive character decreases across a period.

(v) P > S > Si > Al > Na

I.E. increases across the period

I.E. of P > I.E. of S due to stable half-filled orbital configuration of P

(vi) $I^- > Br^- > Cl^-$

Ionic radius increases down the period.

(vii) Cs < Li < C < O < F

IE, increases across the period.

(viii) F > O > C > N > Be

Electron affinity increases across a period $(EA)_C > (EA)_N$ due to stable configuration of N (half-filled orbital)

(ix) $N^{3-} > O^{2-} > F^- > Na^+ > Mg^{2+}$

Ionic size decreases with increasing positive charge.

Positive charge means increase in effective nuclear charge and hence decrease in size negative charge lead to increase in repulsion between electrons thus increases atomic size

(x) $I^- > Rb^+ > Cl^- > Na^+ > Mg^{2+}$

Ionic radii decreases with increase in positive charge and with decrease in number of shells.

Sol 17: (i) We know that electronegativity increases across a period from left to right and decreases while going down in a group. Therefore F is the most electronegative and Cs is the least electronegative element.

(ii) Electronic configuration of C: [He] $2s^22p^2$

Electronic configuration of B: [He] $2s^22p^1$

(IE), of C > (IE) of B because of increased effective number charge in C compared to B

Electronic configuration of C^+: [He] $2s^22p^1$

Electronic configuration of B^+: [He] $2s^2$

$(IE)_2$ of B > $(IE)_2$ of C as B^+ has a stable electronic configuration due to fully filled s orbital which makes removal of electron difficult.

(iii) The outermost electronic configuration of halogens is of the form ns^2np^5. Thus, by adding one electron, they can attain noble gas configuration, which makes addition of electrons very favorable and hence results in high electron affinity.

(iv) The formation of $F^-(g)$ from F(g) enables fluorine to have a stable noble gas configuration. Hence, this process is exothermic.

For formation of $O^{2-}(g)$ from $O^-(g)$, an extra electron needs to be added to an already negative $O^-(g)$ which will be difficult due to coulombic repulsion between electrons making this process endothermic

(v) Transition and inner transition metals have variable oxidation states because ns and (n–1)d orbitals have very similar energy and there is not a huge jump in the amount of energy required to remove the third electron compared with the first and second (which are removed from ns orbitals)

(vi) Zero group elements (noble gases) are chemically inert as these have a very stable electronic configuration of ns^2np^6 and have no tendency to either accept or donate an electron.

(vii) $(IE)_2$ of an element is always greater than $(IE)_1$ as increased amount of energy is needed to remove an electron from an already positively charged ion due to increased nuclear charge.

(viii) Radius of cation is lesser than atom as removal of an electron leads to increase in effective nuclear charge and hence decrease in size.

Radius of anion is bigger than atom because addition of an electron results in increased coulombic repulsion between electrons causing decrease in effective nuclear charge and hence increase in size.

Sol : 18 (a) IE1 of F will be higher than that of Cl.

Reason: As we go down in a group, atomic radius increases and consequently the effective nuclear charge decreases and thus ionization decreases.

(b) EA of O^- is greater than O.

Reason: The second electron affinity of oxygen is particularly high because the electron is being forced into a small, very electron-dense space.

(c) Ionic radius of Cl^- will be greater than K^+.

Reason: Both are isoelectronic species thus ionic radius decreases with increase in nuclear charge.

Sol 19: (i) F because EA increases across the period and decreases down the group

(ii) Cs because IE decreases down the group and increases across the period.

Sol 20 Covalent radius of K > covalent radius of F

Since, size of K^+ < size of K and size of F^- > size of F

Sol 21: (i) Nomenclature:

Name of more electronegative element is written at the end and 'ide' is sufficed to it

Name of less electronegative element is written first in the formula. Eg. I, Br Iodine Bromine

(ii) If difference is EN > 1.7 ionic bond

If difference is EN < 1.7 covalent bond

(iii) Metallic and non-metallic nature

Low EN: metallic

high EN: nonmetallic

(iv) Percentage ionic character in covalent bonds =

$16(X_A - X_B) + 3.5\ (X_A - X_B)^2$

(v) Bond length

When electronegativity difference increases, bond length decreases

$d_{A-B} = r_A + r_B - 0.09\ |x_A - x_B|$

(vi) Bond strength and stability

It increases with increase in electronegativity difference

Eg. H–F > H–Cl > H–Br > H–I

(vii) Acidic strength of hydrides

It decreases with increase in electronegativity difference

H – I > H – Br > H – Cl > H – F

(ix) Acidic nature of oxides

If $X_0 - X_A \geq 2.3$ basic oxides

$X_0 - X_A < 2.3$ acidic oxides

(x) Nature of hydroxide

In AOH, if $(EN)_A > 1.7$, acidic

$(EN)_A < 1.7$, basic

Sol 22: Atomic size: It is the distance between the centre of the nucleus and the outermost shell containing electrons.

Ionic size: It is the distance between the centre of the nucleus and the outermost shell containing electrons in an ion.

Sol 23: BaO is more basic because $(EN)_{Ba} < (EN)_{Mg}$

Sol 24 This statement is wrong since the electronegativity decreases while going down in a group. Therefore, basic strength should increase.

Sol 25: (i) CsOH: basic ($(EN)_{Cs} < 1.7$)

(ii) IOH: acidic $(EN)_I > 1.7$)

(iii) $Sr(OH)_2$: basic ($(EN)_{Sr} < 1.7$)

(iv) $Se(OH)_2$: acidic ($(EN)_{Se} > 1.7$)

(v) FrOH: basic ($(EN)_{Fr} < 1.7$)

(vi) BrOH: acidic ($(EN)_{Br} > 1.7$)

Sol 26: Acidic nature: $N_2O_5 > SO_3 > CO_2$ <> CO > CaO

Acidic strength decreases with increase in electronegativity difference. Also, CO is neutral and CaO is basic.

Sol 27: (i) Anomalous position of hydrogen.

(ii) Isotopes were placed in same group despite having different properties.

(iii) Isobars are placed in different groups.

(iv) Metals and non-metals both placed together in same group.

(v) Some similar elements placed differently.

(vi) Position of lanthanides and actinides is not properly specified.

Sol 28: (i) The nuclear charge on an atom is equal to the atomic number.

(ii) The properties of elements are the periodic functions of their atomic number.

Sol 29: Electron affinity (EA) is the amount of energy released when a neutral gaseous atom accepts an electron to form gaseous anion

$X(g) + E \rightarrow X^-(g) + EA$

EA increase across the period in periodic table due to increased effective nuclear charge with the exception of elements having stable configurations (due to half or full filled orbitals) for which EA is high than adjacent elements. EA decreases down the group due to decrease in effective nuclear charge.

Sol 30: (i) **Number of shells:** IE decreases with increase in number of shells.

(ii) **Effective Nuclear Charge:** IE increases with increase in effective nuclear charge

(iii) **Shielding effect:** The more will be the shielding effect. Less will be the IE.

(iv) The elements which have half-filled or fully filled orbitals have higher IE due to their stable electronic configuration.

(v) **Penetrative Power:** In any atom, the s orbital is nearer to the nucleus than p, d or f orbitals. Therefore, greater energy is required to remove an electron from s orbital. Decreasing order of IE is:

s > p > d > f

Sol 31: (i) Al, due to its intermediate electronegativity.

(ii) F,due to highest effective nuclear charge and least number of shells.

(iii) Xe, It is a group zero element (a noble gas).

Exercise 2

Single Correct Choice Type

Sol 1: (D) I^- will have the largest ionic radius increase further, we know that ionic radius will down the group and decrease across a period. And for isoelectronic species, the ionic radius will increase with increase in negative charge

Sol 2: (D) In group 13, the stability of +2 oxidation state will increase and stability of +4 oxidation state will decrease down the group due to inert pair effect, which is negligible in Ge and most significant in Pb.

Sol 3: (B) We check the atomic radius of all the elements in the four options.

For option (B), we find that the atomic mass of the middle elements is not equal to the average of atomic mass of other 2 elements.

Sol 4: (C) Lothar Meyer used atomic volume for plotting Lothar Mayer's volume curves.

Sol 5: (B) Majority of low atomic number non metallic elements are gases which are placed at top right hand side.

Sol 6: (D) Since, here s orbital is filled last, it is a s-block element

Sol 7: (D) All the reasons given are correct.

Sol 8: (B) The last electron in each period is filled in np orbital which is the same shell, i.e. same energy level.

Sol 9: (C) This is a consequence of inert pair effect, according to which in heavier p-block elements, the s orbitals become very close to nucleus and thus become relatively inert and are not removed.

Sol 10: (C) p-block elements can display a range of oxidation states due to availability of electrons in both of s and p orbitals and also possibility of attaining noble gas configuration by adding electrons.

Sol 11: (C) We note that order of (IE) is F > N > O > C

Here (IE) of N > (IE) of O because N has a stable half-filled orbital configuration and also IE increases across a period.

But after removal of an electron O will have a stable, half-filled orbital configuration and therefore, its IE will be more than F^+. Hence order O > F > N > C

Sol 12: (C) Electron affinity is higher for Cl due to its low electron density.

$(IE)_2$ of alkali metals shows a jump because they have stable noble gas configuration in +1 oxidation state

$(IE)_2$ of halogens is zero because halogens have noble gas configuration in –1 oxidation state

He has maximum ionisation energy due to its small size and fully filled orbital configuration.

Sol 13: (C) The element isoelectronic with O_2 is sulphur (s).

Its configuration in group state [Ne] $3s^2 3p^4 3d^0$

Outer shell configuration in ground state

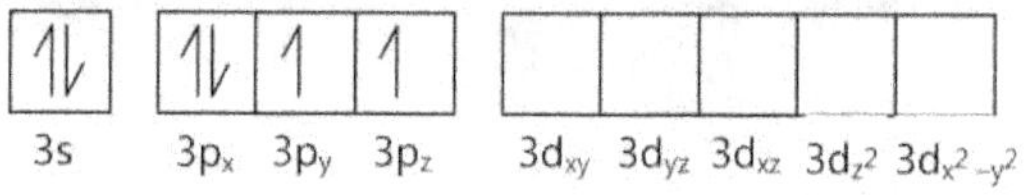

Outer shell configuration in second excited state (all electrons get unpaired

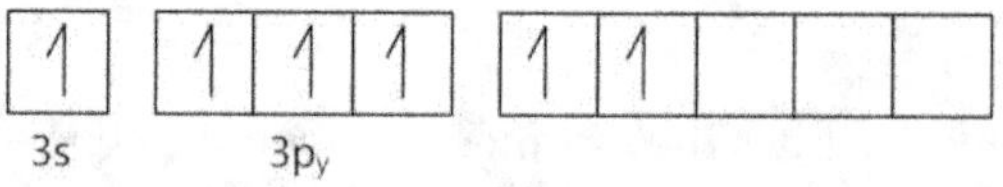

Sol 14: (B) Metallic radii of transition elements first decreases due to increase in effective nuclear charge, but after since the electrons are filled in (n – 1)d orbital, after a few elements, the nuclear charge on ns electrons remains almost constant due to extra stability.

Sol 15: (A) If difference in electronegativity > 1.7, ionic bond

Sol 16: (C) In covalent bond, there is actual bond formation therefore they are the strongest of allowed by hydrogen bonding. Vander waal's interactions are least strong because they are developed due to momentary dipole interactions.

Sol 17: (C) The properties of other three pairs resemble due to diagonal relationship between them (identical charge 1 size ratio)

Sol 18: (D) The added electron will have no tendency to go to the higher energy level or sublevel as this will further increase the energy of the system. Alkali metals have no tendency to accept electrons due to low effective nuclear charge.

Atoms with exactly half filled configuration will have low (but not zero) tendency to accept electrons.

Sol 19: (B) Energy required = $IP_1 + IP_2$

$= (178 + 3481 kcalmol^{-1}) = 526kcalmol^{-1}$

Sol 20: (C) Configuration $= 1s^2 2s^2 2b^2 3s^2 3p^6 4s^2$

Since last filled shell is 4 and it contains 2 electrons, 4th period and 2nd group

Number of neutrons= Atomic weight – Atomic Number = 40 -20 = 20

For this element this the most stable oxidation state will be +2 as in +2 oxidation state. It will attain a therefore formula of metal oxide = Mo

Sol 21: (C) (Given answer = a =wrong)

For R, after removal of an electron, a fully filled stable configuration is achieved which increases its $(IE)_2$ above all other 3 elements.

Sol 22: (B) Nitrogen has a configuration of $[He]2s^2\ 2p^3$, which is stable because of the half-filled p- orbital and thus, it has more IE than Oxygen.

Sol 23: (D) All the given reasons are correct.

Sol 24: (C) With the removal of an electron, the further removal of electron becomes increasingly difficult due to increase in effective nuclear charge and decrease in size. Hence $I_1 < I_2 < I_3$

Sol 25: (B) Only (B) and (C) order is wrong

F > O, since IP increases across the period

N > O, as N has a stable half filled configuration

P > S, as P has a stable half filled configuration

Be > B, as Be has a stable fully filled configuration.

Sol 26: (A) If difference between $(IP)_1$, and $(IP)_2$ > 16 eV, then element will not form a dipositive ion.

Sol 27: (A) Inconsistency in question, taking x^{-2}, no correct option

No. of electrons in X^{-2} = 18 + 8 = 26

Atomic number of X^0 = 26 – 2 = 24

No. of electrons in Y^{-3} = 18 + 3 = 21

Atomic number of Y = 21 + 3 = 24

Sol 28: (A) Electronegativity increases across a period.

Sol 29: (C) The valency first increases as the number of unpaired electrons increases, but then starts decreasing as electrons begin to get paired.

Sol 30: (A) Electronegativity of carbon is least.

Sol 31: (B) Elements are arranged on the basis of increasing atomic number (i.e. increased nuclear charge)

Sol 32: (B) Metallic character decreases across the period and increases down the group

Sol 33: (A) Due to screening effect of inner electrons, the effective nuclear charge on the valence electrons decreases which lowers the IP.

Sol 34: (C) IE does not vary in a regular manner as IE is higher for elements which have stable half-filled or fully filled configurations compared to their adjacent elements, which violates the general trend that IE increases across a period. All other statements are correct.

Sol 35: (C) Just check the electronic configuration for each elements

In : $1s^2 2s^2 2p^6 3s^2 3p^6 3d^{10} 4s^2 4p^6 4d^{10} 5s^2 5p^1$ (iv)

Pd : $1s^2 2s^2 2p^6 3s^2 3p^6 3d^{10} 4s^2 4p^6 4d^{10}$ (iii)

Ca : $1s^2 2s^2 2p^6 3s^2 3p^6 4s^2$ (i)

Cu : $1s^2 2s^2 2p^6 3s^2 3p^6 3d^{10} 4s^1$ (ii)

Sol 36: (C) Just check the configuration of each element, taking n = 4

Chromium : [Ar] $3d^5 4s^1$

Copper : [Ar] $3d^{10} 4s^1$

Krypton : [Ar] $3d^{10} 4s^2 4p^6$

Bromine : [Ar] $3d^{10} 4s^2 4p^5$

Sol 37: (A) Based on facts

Helium has highest ionization potential.

Fluorine has highest electronegativity.

Technitium is an artificial element.

Lithium has a very high reducing ability.

Sol 38: (C) Basic character decreases across a period.

Sol 39: (A) atomic radius, and Atomic size, decreases across a period electronegativity increases across a period. Effective nuclear charge increases across a period

Previous Years' Questions

Sol 1: (D) As we go down the group, inertness of ns^2 pair increase. Hence, tendency to exhibit +2 oxidation state increases and that of +4 oxidation state decreases.

Sol 2: (A) First I.E. of N > First I.E. of O.

Sol 3: (D) All the noble gases occupy the peaks of I.E. curve.

Sol 4: (D) Electronegativity increases on going from left to right in a period. Thus electronegativity of F > O > N > Cl.

Sol 5: (D) Nuclear charge per electron is greater in P^{5+}. Therefore its size is smaller.

Sol 6: (C) Element belongs to d-block is unnilhexium $(Unh)_{106}$.

Sol 7: (C) I represents Li

II represents K

III represents Br

IV represents I

V represents He

So, amongst these, II represents most reactive metal and V represents least reactive non-metal.

Sol 8: (D) Positive ions will be smaller than parent atoms.

Sol 9: (E) Assertion is false but reason is true.

More is the electron affinity, greater is the oxidizing character.

Sol 10: (A) Both assertion and reason are true and reason is the correct explanation of assertion.

I.E. of N is more than that of $_{x}O$ as well as $_{y}C$.

Sol 11: (C) Assertion is true but reason is false.

N is half-filled ($1s^2\ 2s^22p^3$) and therefore more stable and hence energy required to lose electron is greater.

Sol 12: (B) NO^+, C_2^{2-}, CN^- and N_2. All have fourteen electrons.

Sol 13: (A) Both are isoelectronic and have same bond order.

Sol 14: (C) Correct basic strength is

$NH_3 > PH_3 > AsH_3 > BiH_3$

Sol 15: (B) Follow the periodic trends

Sol 16: (C) Option itself is the reason

Sol 17: (D) For isoelectronic species higher the $\frac{Z}{e}$ ratio, smaller the ionic radius

$\frac{Z}{e}$ for $O^{2-} = \frac{8}{10} = 0.8$

$F^- = \frac{9}{10} = 0.9$

$Na^+ = \frac{11}{10} = 1.1$

$Mg^{2+} = \frac{12}{10} = 1.2$

$Al^{3+} = \frac{13}{10} = 1.3$

Sol 18: (B) Greater charge and small size of cation cause more polarization and more covalent is that compound.

Sol 19: (D) Across a period metallic strength decreases & down the group it increases.

Sol 20: (C) As the size of central atom increases lone pair bond pair repulsions increases so, bond angle decreases.

Sol 21: (C) $CaCO_3 \rightarrow \underset{Basic}{CaO} + \underset{Acidic}{CO_2}$

Sol 22: (C) Order of increasing

ΔH_{IE_1} : Ba < Ca < Se < S < Ar

Ba < Ca ; Se < S : On moving top to bottom in a group, size increases. So ionisation energy decreases.

Ar: Maximum value of ionisation energy, since it is an inert gas.

Sol 23: (B) $Na \rightarrow Na^+ + e^-$ I^{st} I.E.

$Na^+ + e^- \rightarrow Na$ Electron gain enthalpy of Na^+

Because reaction is reverse so then.

$\Delta H_{eg} = -5.1$ eV.

Sol 24: (C) These are isoelectronic species.

As negative charge increases, ionic radius increases.

JEE Advanced/Boards

Exercise 1

Sol 1: 101: Unnilunium

102: Unnilbium

103: Unniltrium

104: Unnilquadium

105: Unnilpentium

106: Unnilhexium

107: Unnilseptium

108: Unniloctium

109: Unnilennium

Sol 2: (i) $Pb^{4+} < Pb^{2+} < Pb$

Size decreases with increasing positive charge.

(ii) Na < Al < Mg < Si

IP increases across a period.

More for Mg because of its stable half filled orbital configuration.

(iii) NiO < MgO < Sr < Rb_2O < Cs_2O

Basic character increases down a group and decreases across a period.

Less for transition metal oxides

(iv) $Be(OH)_2 < Mg(OH)_2 < Ca(OH)_2 < Ba(OH)_2$

Basic character increases down the group.

(v) $Cl^- < Ar < K^+ < Ca^{2+}$

Same configuration in all.

IE is decided by effective nuclear charge

(vi) As < P < S < Cl

EN decreases down the group and increases across the period

(vii) LiI < LiBr < LiCl

Ionic character decreases with decrease in electronegativity difference.

(viii) $Li^+ < Na^+ < K^+ < Rb^+ < Cs^+$

Mobility of ions increases down the group

(ix) Cs < Rb < K < Na < Li

Size of hydrated radii decreases down the group

(x) $Ba^{2+} < Sr^{2+} < Ca^{2+} < Mg^{2+} < Be^{2+}$

Extent of hydration decreases down the group

Sol 3: Inert gases have an electronic configuration of ns^2np^6 which is an extremely stable configuration due to all fully filled orbitals. Hence, they have no tendency to combine with other atoms and change this configuration, so they are monoatomic.

Sol 4: The third period contains 8 elements as the energy of 3d orbital is more than 4s orbital. 3d orbital is filled during the fourth period making the number of elements in third period equal to 8.

Sol 5: (A) $F^- < Cl^- < S^{2-} < P^{3-}$

Ionic radius increase with increase in number of shells and negative charge

(B) $Al^{3+} < Mg^{2+} < Na^+ < F^- < O^{2-}$

Ionic radius decrease with increasing positive charge

(C) $Mg^{2+} < Na^+ < K^+$

Ionic radius decrease with increase in positive charge and increase with increase in number of shells.

Sol 6: We have,

Internuclear distance

$d_{H-Cl} = r_H + r_{Cl} - 0.09\,|\,r_H - r_{Cl}\,|$

Now, $r_{Cl} = \dfrac{10}{2}$ Å = 5 Å

$r_H = \dfrac{2}{2}$Å = 1 Å

$x_{Cl} - x_H = 3 - 2.1 = 0.9$

$\therefore\ d_{H-Cl} = 5 + 1 - 0.09 \times 0.9 = 5.919$

Hence, internuclear distance = 5.919 Å

Sol 7: We have

$d_{As-Cl} = r_{As} + r_{Cl} - 0.09 \mid EN_{Cl} - EN_{As} \mid$

Here, $d_{As-Cl} = 2.20$ Å

$r_{Cl} = 0.99$Å

$EN_{Cl} - EN_{e} = 0$

$\therefore r_{As} = d_{As-Cl} - r_{Cl} = 1.21$ Å

Hence, SBCR of As is 1.21 Å

Sol 8: (A) We have Pt–Cl distance = 2.32Å from the figure, we have

Cl–Cl distance = 2 × (Pt –Cl distance) = 4.64 Å

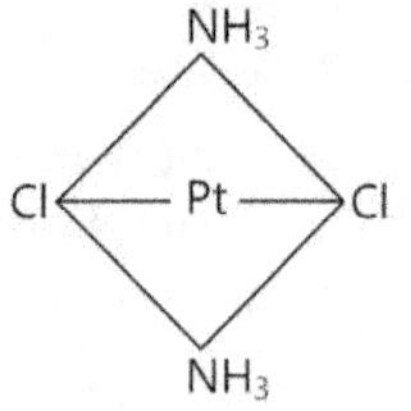

Hence Cl–Cl distance is 4.64 Å

(B) Here, since this is a square planar geometry. Therefore, Cl–Pt–Cl angle will be equal to 90°

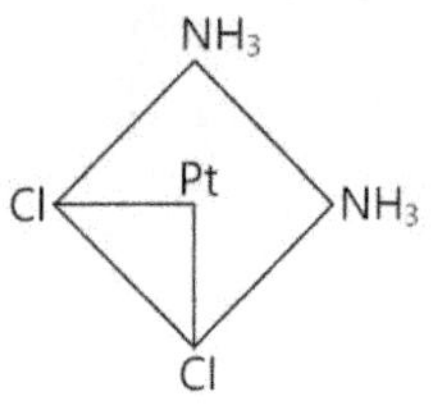

Therefore, Cl–Cl distance = $\sqrt{2}$ Pt – Cl distance is 3.28 Å

Hence Cl–Cl distance is 3.28 Å

Sol 9: Since , we know that size of cation is smaller than atomic size and size of anion is larger than atomic size. Therefore atomic radius of K > 1.34 Å and atomic radius of F < 1.34 Å

Sol 10: Noble gases are chemically inert and are therefore monoatomic so, it is not possible to calculate covalent radius of noble gases, instead, for noble gases, Vander waals radius is calculated which is always greater than the radii of precedent halogens (which is covalent radius)

Sol 11: While going from Aluminium to Gallium, 3d electrons are filled which poorly screen the outermost electrons from nuclear charge hence, even though the number of shells increases, an increase in effective nuclear charge will ensure that atomic size of Gallium is equal to the atomic size of Aluminium. Going further down from Ga to In- d-block will be there for all elements.

Hence, atomic radii will increase as usual.

Sol 12: We have $X_p = \frac{X_M}{2.8} = \frac{IE + EA}{5.6} = \frac{4 + 13}{5.6}$

$\Rightarrow X_p = 3.03$

Sol 13: In isoelectronic species, the increase in atomic radii means an increase in the effective nuclear charge on outermost shell electrons, as the number of electrons are the same. This increase in effective nuclear charge means that that the valence electrons are more firmly bound to the nucleus, hence it results in a decrease in ionic/atomic radii.

Sol 14: Ionisation energy if Cu > Ag is expected as ionization energy decreases down the group. But after Ag the 4f electrons start getting filled and since F orbital is the least penetrative it offers least screening and hence effective nuclear charge increases while going across lanthanoid series. This phenomona is called lanthanoid contraction and due to this $(IE)_{Ag} < (IE)_{Au}$. As Au has more effective nuclear charge than Ag.

Sol 15: Electronic configuration of C = [He] $2s^2 2p^2$

Electronic configuration of B = [He] $2s^2 2p^1$

(IE)1 of C > $(IE)_1$ of B as effective nuclear charge of B is greater than that for C

Electronic configuration of B^+ = [He] $2s^2$

Electronic configuration of C^+ = [He] $2s^2 2p^1$

$(IE)_2$ of B > $(IE)_2$ of C as B have a very stable fully filled orbital configuration which makes removal of electron extremely difficult.

Sol 16: Ionisation potential is the amount of energy required to remove the most loosely bound electron from an isolated gaseous atom $M(g) + IE \rightarrow M^+(g) + e^-$

Sol 17: Electronic configuration of K = [Ar] $4s^1$

Electronic configuration of Cu = [Ar] $3d^{10} 4s^1$

We see that on both K and Cu have the same outer shell of $4s^1$ but for Cu the effective nuclear charge will be more than K, as the 3d orbital do not provide very good screening from nuclear charge, therefore, its (IE) is high.

Electronic configuration of K^+ = [Ar]

Electronic configuration of Cu^+ = [Ar] $3d^{10}$

We note that K^+ has a very stable noble gas configuration compared to Cu whose configuration is not so stable. Hence, $(IE)_2$ of K is higher than $(IE)_2$ of Cu.

Sol 18: (i) For Ni, sum of four ionisation energies = 11.29 kJ/mol

For Pt, sum of four ionisation energies = 9.36 kJ/mol

Therefore, Ni can only exhibit +2 oxidation state.

Hence, most stable oxidation state for Ni: +2

Most stable oxidation state for Pt: +2, +4

(ii) Pt can easily form compounds in +4 oxidation state as $(IE)_1 + (IE)_2 + (IE)_3 + (IE)_4$ for Pt is not so high.

Sol 19: Electronic configuration of Mg: [Ne] $3s^2$

Electronic configuration of Na: [Ne] $3s^1$

Mg has a stable fully filled orbital configuration.

Hence (IE), of Mg is high

Electronic configuration of Mg^+: [Ne] $3s^1$

Electronic configuration of Na^+: [Ne]

Na^+ has an extremely stable noble gas configuration which makes removal of electron very difficult.

Hence, $(IE)_2$ of Na is high

Sol 20: Na^+ has higher value of IE because for Na due to removal of an electron, effective nuclear charge is much more compared to Ne, as there are more protons in Na^+ than Ne for same number of p electrons.

Exercise 2

Single Correct Choice Type

Sol 1 (A) X has a stable half-filled orbital configuration which makes removal of electron difficult and increases its IE.

Sol 2: (A) This is the true position of lanthanoids. They are kept below the periodic table to make it more compact.

Sol 3: (B) Similar ionic potentials is the main cause of diagonal relationship.

Sol 4: (C) Will from a diatomic molecule most readily because it has an outer shell configuration of $2s^2 2p^2$ and it needs just 1 electron to attain noble gas configuration, which it can achieve by sharing an electron with another atom to form a diatomic module.

Sol 5: (A) We know,

$$EN = \frac{EA + IP}{2}$$

$$\Rightarrow E_1 = \frac{EA + E_2}{2} \Rightarrow EA = 2E_1 - E_2$$

Sol 6: (A) This is ns^2np^4 configuration which is the configuration of ground state of S and Cl^+.

Sol 7: (C) O^- has no natural tendency to accept electrons because of coulombic repulsion between electrons , thus to add an electron to O^-, we have to give it energy from an external source.

Sol 8: (C) The process is

$$Na(s) \xrightarrow{SE} Na(g) \xrightarrow{IE} Na^+(g)$$

Therefore , total energy =SE +IE

Sol 9: (C) The addition of an extra electron to O^- requires addition of energy because of the coulombic repulsion offered by the extra electron to the incoming electron. For other processes, energy is evolved. As far, all of them, EA is positive.

Sol 10: (B) N_2 has no bond polarity and is a very neutral molecule compared to CN^-, which has a net negative charge and therefore will be much more reactive.

Previous Years' Questions

Sol 1: (B) Among isoelectronic species, greater the negative charge, greater the ionic size hence

$F^- < O^{2-} < N^{3-}$

Sol 2: (B) In a group, ionization energy decreases down the group:

Be > Mg > Ca

Sol 3: (B) Be^- is the least stable ion, Be($1s^22s^2$) has stable electronic configuration, addition of electron decreases stability.

Sol 4: (B, C, D) Incorrect: Electrons are not filled in sub-energy levels s, p, d and f in the same sequence.

(B) **Correct:** No. of valence shell electrons usually determine the stable valency state of an elements.

(C) **Correct:** Physical and chemical properties of elements are periodic function of atomic number which is the basis of modern, long form, of periodic table.

(D) **Correct:** Relative ionicity of the bond between any two elements is function of electronegativity difference of the bonded atoms which in turn has periodic trend in long form of Periodic table.

Sol 5: (A, B) Correct: For greater solubility, hydration energy must be greater than lattice energy.

(B) **Correct:** Greater lattice energy discourage dissolution of salt.

(C) **Incorrect:** When a salt dissolve, energy is required to break the lattice, which comes from hydration process.

(D) **Incorrect:** Explained in (A).

Sol 6: (True) In a group, size increases from top to bottom.

Sol 7: (False) Ionization potential decreases down the group but this is not the only criteria of reducing power.

Sol 8: (i) Mg^{2+}, O^{2-}, Na^{+} and F^{-} are all isoelectronic, has 10 electrons each. Among isoelectronic species, the order of size is cation < neutral< anion.

Also, between cations, higher the charge, smaller the size and between anions, greater the negative charge, larger the size. Therefore,

Ionic radii = $Mg^{2+} < Na^{+} < F^{-} < O^{2-}$

(ii) First ionization energy increases from left to right in a period. However, exception occurs between group 2 and 13 and group 15 and 16 where trend is reversed on the grounds of stability of completely filled and completely half-filled orbitals. Therefore : Ionization energy (1st) : Na < Al < Mg < Si

(iii) If the atoms are from same period, bond length is related directly to atomic radius. Therefore, Bond length : $N_2 < O_2 < F_2 < Cl_2$

Sol 9: Cl^{-}, S^{2-}, Ca^{2+}, Ar

Size : $Ca^{2+} < Ar < Cl^{-} < S^{2-}$

Explained in (i), question 1.

Sol 10: Zn : $3d^{10}\,4s^2$, Cu : $3d^{10}$, $4s^1$

The first ionization energy is greater for Zn but reverse is true for 2nd ionization energy.

Sol 11: $Li^{+} < Al^{3+} < Mg^{2+} < K^{+}$

Size decreases from left to right in a period and it increases from top to bottom in a group. Variation is more pronounced in group than in period.

Sol 12: (A) NO^{-} (16 electron system)

Bond order = 2.

$NO^{\oplus}$, CN^{-} and N_2 are isoelectronic with CO therefore all have same bond order (= 3)

Hence (A) is correct.

Sol 13: (C) The lower oxidation states for the group 14 elements are more stable for the heavier member of the group due to inert pair effect.

Sol 14: Na, F show only one non-zero oxidation state.

Sol 15: (B) The increasing order of atomic radii of the following Group 13 elements is Ga < Al < Tl < In

www.ingramcontent.com/pod-product-compliance
Lightning Source LLC
Chambersburg PA
CBHW082030200526
45171CB00015B/580
9781648285196